Neutron Transmutation Doping in Semiconductors

Neutron Transmutation Doping in Semiconductors

Edited by

Jon M. Meese

University of Missouri
Columbia, Missouri

Plenum Press • New York and London

Library of Congress Cataloging in Publication Data

International Conference of Neutron Transmutation Doping in Semiconductors, 2d,
    University of Missouri.

    Neutron transmutation doping in semiconductors.

    Includes index.
    1. Semiconductor doping—Congresses. 2. Neutron irradiation—Congresses.
3. Silicon—Defects—Congresses. I. Meese, Jon M. II. Title.
QC611.6.D6I57 1978   621.3815'2   79-395
ISBN-13: 978-1-4684-8251-5   e-ISBN-13: 978-1-4684-8249-2
DOI: 10.1007/978-1-4684-8249-2

Proceedings of the Second International Conference on
Transmutation Doping in Semiconductors held at the
University of Missouri, Columbia, Missouri, April 23—26, 1978

This work relates to Department of the Navy Research Grant N00014-78-G-0012
issued by the Office of Naval Research. The United States Government has a royalty-
free license throughout the world in all copyrightable material contained herein. As
long as this Proceedings is protected by copyright and is reasonably available to the
public for purchase, the Government shall not publish it for sale or authorize others
to do so.

© 1979 Plenum Press, New York
Softcover reprint of the hardcover 1st edition 1979
A Division of Plenum Publishing Corporation
227 West 17th Street, New York, N.Y. 10011

Preface

This volume contains the invited and contributed papers presented at the Second International Conference on Neutron Transmutation Doping in Semiconductors held April 23-26, 1978 at the University of Missouri-Columbia. The first "testing of the waters" symposium on this subject was organized by John Cleland and Dick Wood of the Solid-State Division of Oak Ridge National Laboratory in April of 1976, just one year after NTD-silicon appeared on the marketplace. Since this first meeting, NTD-silicon has become established as the starting material for the power device industry and reactor irradiations are now measured in tens of tons of material per annum making NTD processing the largest radiation effects technology in the semiconductor industry.

Since the first conference at Oak Ridge, new applications and irradiation techniques have developed. Interest in a second conference and in publishing the proceedings has been extremely high. The second conference at the University of Missouri was attended by 114 persons. Approximately 20% of the attendees came from countries outside the U.S.A. making the conference truly international in scope.

We at the University of Missouri enjoyed the opportunity to host the conference and meet with our warm guests. We believe that our efforts were successful. Perhaps the best indicator of the success of a conference is the willingness of the participants to attend and support further conferences on the same subject. At this meeting in Columbia, the next two conferences were tentatively arranged. In 1980, the NTD conference will be held in Copenhagen, hosted by TOPSIL, followed in two years by a fourth conference hosted by the U. S. National Bureau of Standards.

We would like to thank the following members of the Advisory Committee for their assistance and advice:

John W. Cleland, Oak Ridge National Laboratory
Gary M. Cook, University of Michigan
Ron R. Hart, Texas A&M University

Odgen J. Marsh, Hughes Research Laboratories
Herman J. Stein, Sandia Laboratories
George D. Watkins, Lehigh University
Richard F. Wood, Oak Ridge National Laboratory

We also wish to thank the Organizing and Planning Committee
for handling the many details which must be attended to for such
a meeting to take place:

Ronald R. Berliner, University of Missouri, Vice-Chairman
Steve L. Gunn, University of Missouri, Vice-Chairman
Sarah A. Hulett, University of Missouri, Conference Coordinator
Tom H. Blewitt, Argonne National Laboratory, Program Committee
Bobbie Stone, Monsanto Corporation, Program Committee
Mike D. McCarver, Monsanto Corporation, Program Committee
     and Publicity
Rose Mayfield, Monsanto Corporation, Displays and Publicity

We also gratefully acknowledge the financial support we
received for this conference from the following organizations:

Monsanto Corporation
Office of Naval Research
Air Force Office of Scientific Research
Rockwell International
University of Missouri Research Reactor
University of Missouri Extension Division

A special thanks must be given to Ola Montgomery who has been
invaluable throughout the organizing of the conference and who has
patiently prepared the final manuscripts for publishing. We also
wish to acknowledge the steadfast support of Bob Brugger and Don
Alger of the Research Reactor Facility and Ardie Emmons of the
Office of Research of the University of Missouri.

                                        Jon M. Meese
                                        Conference Chairman

Columbia, Missouri
December 1978

Contents

CHAPTER 5: BASIC PROCESSES--RADIATION DAMAGE AND DOPANT PRODUCTION

CHAPTER 6:   SUMMARY OF THE CONFERENCE

THE NTD PROCESS--A NEW REACTOR TECHNOLOGY

J. M. Meese

University of Missouri Research Reactor

Columbia, Missouri 65211

We wish to extend a warm welcome to all of you for attending and supporting this meeting. Without your interest and urging, this conference would not have taken place. In particular, we would like to thank John Cleland and Dick Wood of Oak Ridge National Laboratory for hosting the first NTD Conference and for suggesting that a second conference be held at the University of Missouri. Our sincere appreciation extends also to the Monsanto Corporation, the conference catalyst in the early planning stages, both financially and otherwise. We also thank the Air Force Office of Scientific Research and the Office of Naval Research for their encouragement and support and Rockwell International who supported the publishing of the abstracts.

The title I have chosen for this brief introduction, "The NTD Process--A New Reactor Technology," is perhaps a misnomer. The words "Reactor Technology" could easily be replaced by "Semiconductor Technology" or even "Radiation Technology." The title emphasizes, however, the impact this new technology is making on the reactor community. In a real sense, reactor facilities have had to make the greatest adjustments of all those involved in this new segment of the semiconductor industry.

The NTD process involves the cooperation of semiconductor materials specialists, device producers, radiation damage and defect specialists and reactor personnel. Of all the possible interactions among these groups, those with the reactor community have traditionally been the weakest. Reactor personnel have, therefore, had the greatest learning curves to overcome. It is to the credit of both the reactor community and the semiconductor industry that these difficulties have been overcome so readily in the few years since 1975 when NTD silicon first appeared on the market. An inspection of the conference abstracts indicates a number of joint contributions by semiconductor and reactor scientists suggesting that this new scientific association has been fruitful. The impact of NTD silicon on the semiconductor market since 1975 further supports this contention.

Because many of the papers in a topical conference such as this one tend to emphasis the unknown, it is appropriate for us to dwell briefly on those things we do understand about the NTD process in semiconductors. The transmutation doping process simply involves irradiation of an undoped semiconductor with a thermal neutron flux. The major advantages of the NTD process are illustrated schematically in Fig. 1. The homogeneity in NTD-Si is a result of a homogeneous distribution of silicon isotopes in the target material and the long range of neutrons in silicon. Doping accuracy is a result of careful neutron flux integration. The material improvements offered by the NTD process form the basis for semiconductor device improvement, the subject of several papers in this conference.

Research reactor facilities provide the best source of thermal neutrons for this purpose at the present time. These reactors are ideally suited for such projects because they have usually been constructed with sample irradiation as one of the prime design requirements. Several papers will describe, in considerable detail, such facilities. Although these reactor facilities provide a source of thermal (E $\sim$ 0.025 eV) neutrons, this thermal flux is always accompanied by a fast neutron component which is not useful in providing doping transmutations, but does produce radiation damage (displacements of atoms from their normal lattice sites) which must be repaired by annealing, the process of heating the irradiated material to temperatures sufficiently high that the irradiation produced defects become mobile and can be removed. A group of papers addresses these damage and annealing problems.

To understand the process further, we must be concerned with the interactions of neutrons, both thermal and fast, with the target material to be doped. Because neutrons are neutral particles, their range of penetration in most materials is usually very long. They interact only very weakly with atomic electrons through their magnetic movements. Being neutral, neutrons see no Coulombic barrier at the target nuclei and, therefore, even very slow neutrons may reach into the nucleus without difficulty. In fact, the slower the neutron velocity, the greater is the time of interaction between the neutron and the target nucleus. We, therefore, expect the probability of neutron capture by the target nuclei to be enhanced at low neutron energies.

This interaction is described in terms of a capture cross section, σ_c, where the number of captures per unit volume, N, is given by

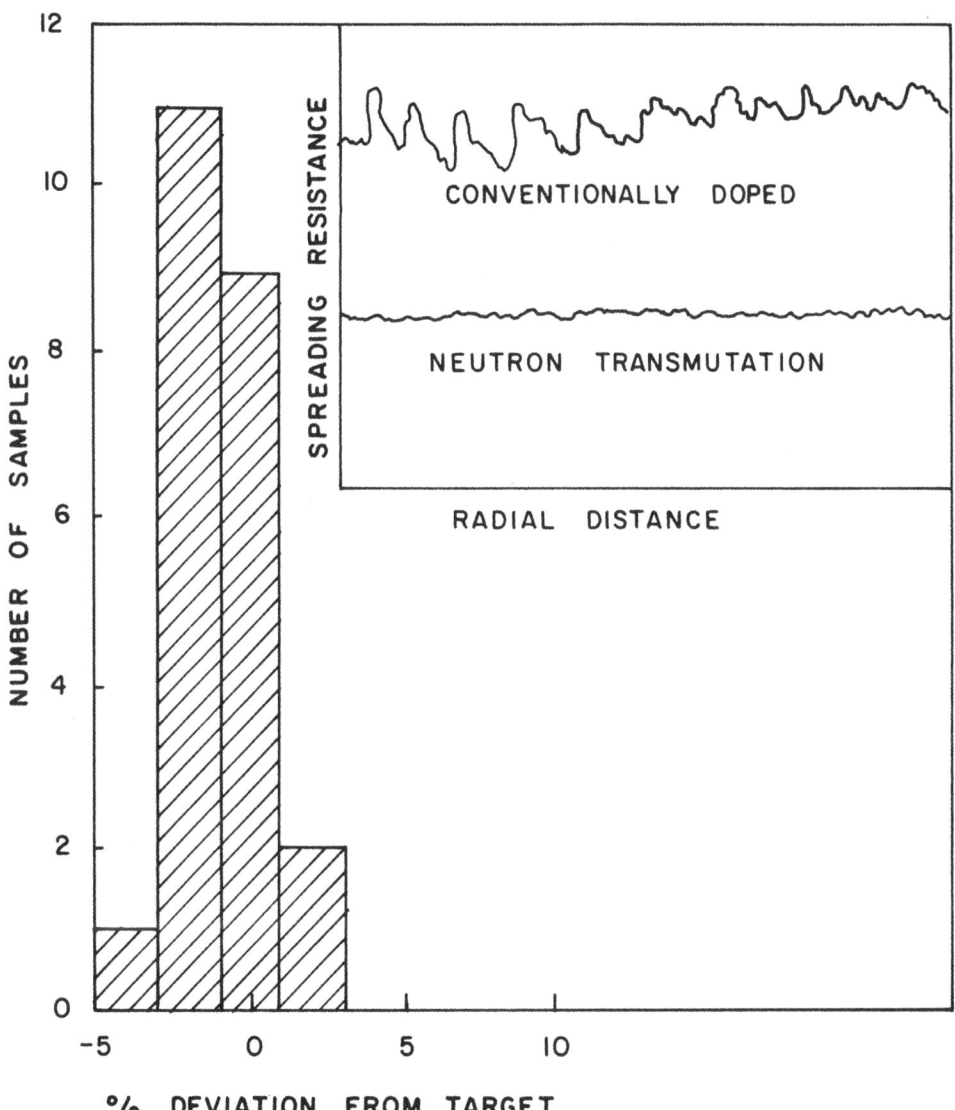

Fig. 1. Advantages of NTD process. Histogram of irradiation target accuracy obtained for commercial sample lot irradiated at MURR. Insert is a schematic representation of spreading resistance traces across a wafer diameter for conventionally doped and NTD Si.

$$N = N_T \sigma_c \Phi \tag{1}$$

where N_T is the number of target nuclei per unit volume, σ_c the capture cross section and $\Phi = \phi t$ is the fluence (flux times time) given in n/cm^2.

Figure 2 shows the capture cross section as a function of neutron energy for silicon as averaged over all three stable silicon isotopes.[1] Similar behavior is found individually for each silicon isotope. It can be seen in Fig. 2 that for low energies,

$$\sigma_c \sim E^{-\frac{1}{2}} \sim \frac{1}{v} . \tag{2}$$

For a given nuclear radius, 1/v is proportional to the interaction time. Therefore, the cross section represents a probability of interaction between the nucleus and neutron.

After neutron capture, the target nucleus differs from the initial nucleus by the addition of one nucleon and is a new isotope in an excited state which must relax by the emission of energy in some form. This emission is usually in the form of electromagnetic radiation (photons) of high energy called gammas. The time for decay of this excess energy by gamma emission can be very short (prompt gammas) or can take an appreciable time in which case a half-life, the time for the number of particles emitted per sec to decrease by a factor of two, can be measured. The gamma emission spectrum is characteristic of the nuclear energy levels of the transmuted target

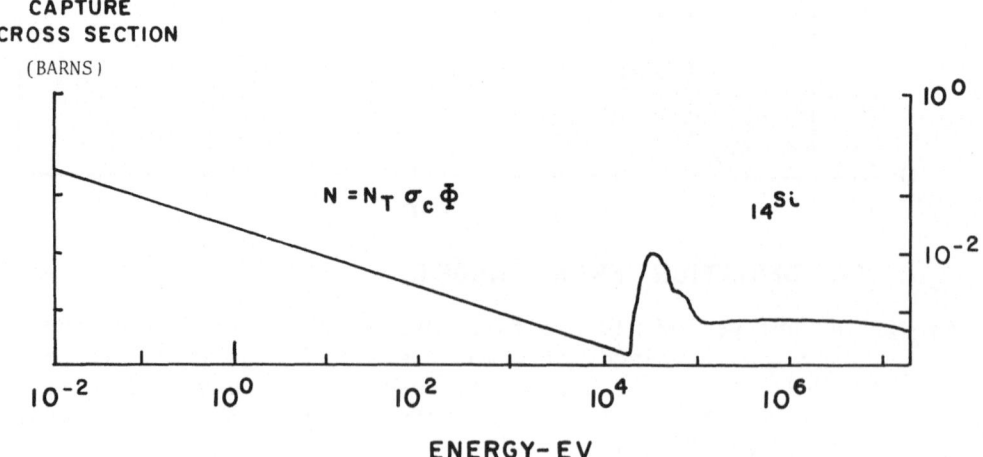

Fig. 2.   Neutron capture cross section as a function of neutron energy for natural silicon. (Data taken from ENDF/B Cross Sections, Brookhaven.)

nuclei and can be used as a powerful trace substance technique called neutron activation analysis (NAA), to detect quantatively impurity levels as low as 10^9 atoms/cm^3. A typical trace substance NAA gamma spectrum is shown in Fig. 3. Each emission line is characteristic of a particular nuclear transition of a particular isotope. A number of papers at this conference will make use of NAA as a powerful analytical tool.

The absorption of a neutron and the emission of gammas is represented by the notation

$$^A X \; (n, \gamma) \; ^{A+1} X \tag{3}$$

where (n, γ) represents (absorption, emission), A is the initial number of nucleons in the target element X before neutron absorption while A+1 is the number after absorption. It is possible for the product isotope ^{A+1}X to be naturally occurring and stable. In many cases, however, the product isotope is unstable. Unstable isotopes further decay by various modes involving the emission of electrons (β decay), protons, alpha particles, K-shell electron capture or internal conversion until a stable isotopic state is reached. These decays produce radioactivity and can be characterized by their half-lives,

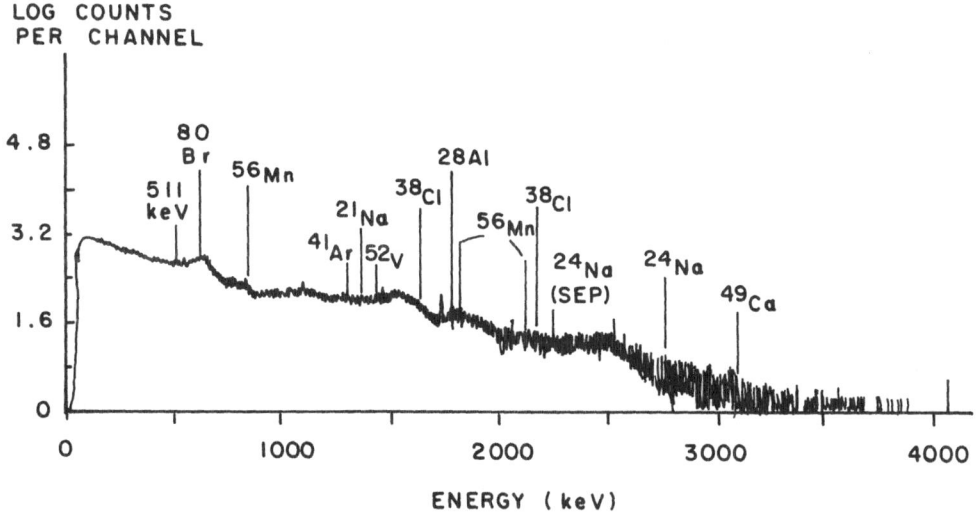

Fig. 3. Typical neutron activation analysis (NAA) gamma-ray spectrum to search for trace substances deposited on an air filter after 1 min. irradiation at MURR.

$T_{\frac{1}{2}}$. In the case of silicon, three stable target isotopes are transformed by (n,γ) reactions[2] as follows:

$$(92.3\%) \qquad {}^{28}Si(n,\gamma) \; {}^{29}Si, \; \sigma_c = 0.08 \text{ b}$$

$$(4.7\%) \qquad {}^{29}Si(n,\gamma) \; {}^{30}Si, \; \sigma_c = 0.28 \text{ b} \qquad\qquad (4)$$

$$(3.1\%) \qquad {}^{30}Si(n,\gamma) \; {}^{31}Si \rightarrow {}^{31}P + \beta^-, \; \sigma_c = 0.11 \text{ b}.$$

$$(T_{\frac{1}{2}} = 2.62 \text{ hr})$$

The relative abundance of each stable silicon isotope is shown in
parenthesis. The cross sections are expressed in barns ($1 \text{ b}=10^{-24}\text{cm}^2$).

The first two reactions produce no dopants and only redistribute
the relative abundances slightly. The third reaction produces ${}^{31}P$,
the desired donor dopant, at a rate of about 3.355 ppb per
$10^{18}n_{th}/cm^2$. This production rate is calculated using Eq. (1), the
${}^{30}Si$ capture cross section and the relative abundance of
${}^{30}Si$ ($N_T \simeq 5 \times 10^{22} \text{ Si/cm}^3 \times 0.031$).

In addition to the desired phosphorus production reaction and its
relatively short half-life for β^- decay, the reaction

$$^{31}P(n,\gamma) \; {}^{32}P \rightarrow {}^{32}S + \beta^-, \; \sigma_c = 0.19 \text{ b} \qquad\qquad (5)$$

$$(T_{\frac{1}{2}} = 14.3 \text{ d})$$

occurs as a secondary undesirable effect. The decay of ${}^{32}P$ is the
primary source of radioactivity in NTD float zone Si. Of course, any
undesirable trace impurities in the silicon starting material can
lead to abnormally long half-life activities which may require that
material be held out of production until exempt limits are reached.
These factors have stimulated several papers at this conference on
the subject of radiation protection.

Once the dopant phosphorus has been added to the silicon ingot
by transmutation of the ${}^{30}Si$ isotope, the problem remains to make
this radiation damaged and highly disordered material useful from an
electronic device point of view. Several radiation damage mechanisms
contribute to the displacement of the silicon atoms from their
normal lattice positions. These are:

1. Fast neutron knock-on displacements
2. Fission gamma induced damage
3. Gamma recoil damage
4. Beta recoil damage
5. Charged particle knock-ons from (n,p), (n,α), etc. reactions.

Estimates can be made of the rate at which Si atom displacements
are produced by these various mechanisms, once a detailed neutron

energy spectrum of the irradiation position is known, and these rates compared to the rate at which p⁺ ˅sphorus is produced.

The number of displaced ato...ɔ per unit volume per sec, N_D, is calculated from the equation

$$dN_D/dt = N_T \sigma \phi \nu \tag{6}$$

where N_T is the number of target atoms per unit volume, ϕ is the flux of damaging particles and ν is the number of displacements per incident damaging particle.

Of the damage mechanisms listed above, the gamma and charged particle mechanisms can be neglected compared with the other possible mechanisms. The cross section for gamma induced displacements in silicon is small while the cross sections for (n,p), (n,α), etc., are of the order of millibarns and have thresholds in the MeV range.

The fast neutron knock-on displacements can be calculated from the elastic neutron scattering cross section once the reactor neutron energy spectrum is known. Estimates of fission spectra and graphite moderated fission spectra can be found in the literature.[3]

Even if the fast neutron damage could be completely eliminated, the recoil damage mechanisms, which are caused by thermal neutron capture, still would produce massive numbers of displacements compared to the number of phosphorus atoms produced. In the case of gamma recoil, a gamma of energy $\hbar\omega$ carries a momentum $\hbar\omega/c$ which must equal the Si isotope recoil momentum MV. The recoil energy

$$E_R = \tfrac{1}{2}MV^2 = \tfrac{1}{2}\frac{(\hbar\omega)^2}{MC^2}$$

is, therefore, departed to the silicon atom of mass M for each gamma emitted. An average over all possible silicon isotope gamma emissions and cross sections yields an average recoil energy of 780 eV[4] which is significantly higher than the Si displacement energy.

A similar effect is encountered for ^{31}Si β⁻ decay. The β⁻ carries a momentum

$$p = \sqrt{E_\beta^2 - (m_o c^2)^2}/c \equiv MV.$$

Therefore,

$$E_R = \tfrac{1}{2}MV^2 = \tfrac{1}{2}[E_\beta^2 - (m_o c)^2]/Mc^2.$$

For a β⁻ emitted with an energy of 1.5 MeV, E_R = 33.2 eV or roughly twice the displacement threshold.

From the above considerations, a very rough estimate of the numbers of displacements per phosphorus atom produced can be made. The results are shown in Fig. 4. While the absolute numbers of displacements should not be taken literally, the relative magnitudes of the amounts of damage produced by these various mechanisms are probably order of magnitude correct. An inspection of Fig. 4 indicates that the gamma recoil mechanism is significant relative to the quantity of phosphorus produced even in highly moderated reactors. We are led to the inescapable conclusion that transmutation doping will always produce significant amounts of radiation damage which must be repaired in some way. These defects introduce defect levels into the band gap which cause free carrier removal and a reduction in carrier mobility and minority carrier lifetime. Several papers will address these effects.

The defects produced by neutron irradiation are removed by thermal annealing as discussed previously. It is at this point in the process where disagreement as to the best procedure is likely to be the greatest. The spectrum of possible defect structures and their energetics is impressively large and incompletely understood. Therefore, annealing procedures will be based on art rather than exact science. They will also tend to become proprietary for this reason.

POSITION / DAMAGE PARTICLE	IN CORE	IN POOL
FAST NEUTRON	4.06×10^6	1.38×10^4
FISSION GAMMA	3.64×10^3	36.4
GAMMA RECOIL	1.29×10^3	1.29×10^3
BETA RECOIL	2.76	2.76
TOTAL DISP / (P)	4.06×10^6	1.51×10^4

Fig. 4. Number of displaced silicon atoms per phosphorus produced for various damage mechanisms shown for an in-core fission spectrum and a graphite moderated spectrum.

This is unfortunate since it is precisely in this area that fundamental knowledge is needed to produce the best possible product.

Although carrier concentration and mobility recovery are easily obtainable by various annealing procedures, minority carrier lifetime recovery is very elusive at present. Progress on this front is being made, however, and will be discussed in several papers in this conference. Many of us concerned with this problem feel that very soon the lifetime problems will again return to the device manufacturers' diffusion furnaces where they belong.

In closing, we should mention that transmutation doping can be applied to many semiconductor systems. An example is found in a paper on GaAs in this conference. As device designers continue to push material parameters, the need for transmutation doping in materials other than silicon is likely to continue to grow.

To summarize then, neutron transmutation offers both advantages and disadvantages over conventionally doped silicon as shown in Fig. 5. The steady growth of the NTD-silicon market suggests that the advantages are outweighing the disadvantages and we hope that this series of conferences will continue and contribute to this growth.

ADVANTAGES

1. PRECISION TARGET DOPING (= 1 % OR BETTER) .

2. BETTER AXIAL AND RADIAL UNIFORMITY.

3. NO MICRORESISTIVITY STRUCTURE.

DISADVANTAGES

1. IRRADIATION COSTS.

2. REDUCTION IN MINORITY CARRIER LIFETIME.

3. RADIOACTIVE SAFEGUARDS CONSIDERATIONS.

Fig. 5. Advantages and disadvantages of the NTD process.

REFERENCES

1) D. E. Cullen and P. J. Hlavac, ENDF/B Cross Sections, Brookhaven
    National Laboratory, Upton, N. Y.,(1972).
2) M. Tanenbaum and A. D. Mills, J. Electrochem. Soc. 108, 171
    (1961).
3) D. S. Billington and J. H. Crawford, Jr., Radiation Damage in
    Solids, Princeton University Press, Princeton, N. J., (1961),
    Chapter 2.
4) M. V. Chukichev and V. S. Vavilov, Sov. Phys. Solid State 3,
    1103 (1961).

DETECTION AND IDENTIFICATION OF POTENTIAL IMPURITIES ACTIVATED BY NEUTRON IRRADIATION OF CZOCHRALSKI SILICON

Bobbie D. Stone

Monsanto Commercial Products Company, St. Peters, MO 63376

Donald B. Hines

Monsanto Industrial Chemicals Company, St. Louis, MO 63166

Steve L. Gunn and David McKown

University of Missouri Research Reactor, Columbia, MO 65211

ABSTRACT

A worst-case experiment was carried out wherein a matrix of Czochralski-grown silicon crystals irradiated to the five ohm-cm level was analyzed for radioactive isotopes that might result from activation of impurities initially present in the crystals. A computer study was used to predict the nuclides most likely to occur. Results from the two methods of analysis--radioactivity decay rate and gamma spectroscopy--were consistent with each other. Methods of measurements, calibrations and calculations to assure that radio-activity levels do not exceed the Nuclear Regulatory Commission exempt concentrations standards are presented.

The neutron transmutation doping process for semiconductor grade silicon is complicated by the fact that ^{31}P formed by the primary transmutation of ^{30}Si also captures a neutron to give rise to a beta emitter, ^{32}P. These familiar consecutive reactions are summarized in Fig. 1. While the beta decay of ^{31}Si is very short-lived and presents no problem with residual radioactivity, that of ^{32}P is moderately long lived and does result in measurable amounts of radioactivity under certain conditions. The amount of ^{32}P present depends primarily, of course, on the amounts of ^{31}P produced and to some degree on neutron flux and the amount of phosphorus in the silicon originally. Since

$$^{30}Si + n(thermal) \longrightarrow {}^{31}Si$$

$$^{31}Si \xrightarrow{\quad t\frac{1}{2} = 2.6 \text{ hr.} \quad} {}^{31}P + \beta^-$$

$$^{31}P + n(thermal) \longrightarrow {}^{32}P$$

$$^{32}P \xrightarrow{\quad t\frac{1}{2} = 14.3d \quad} {}^{32}S + \beta^-$$

Fig. 1. Consecutive thermal neutron reactions during neutron transmutation doping of Si.

the resistivity of silicon is inversely proportional to total phosphorus content, the amount of radioactivity in the final product will
vary inversely with final resistivity.

The Nuclear Regulatory Commission defines levels of radioactivity
for each artifically produced nuclide that can be released by Reactor
facilities to persons or firms who do not hold NRC licenses to possess
radioactive by-product materials. This exempt concentration of ^{32}P
is 2×10^{-4} microcuries per gram which corresponds to about six times
the background count obtained with standard laboratory type Geiger
or flow proportional counters. If the resistivity desired for the
NTD silicon is such that the exempt concentration for ^{32}P is exceeded,
the material must be allowed to decay to this level before it can be
transferred from the Reactor facility. Figure 2 shows decay time
required as a function of final resistivity calculated from published
values for cross sections and decay characteristics.

Even though levels of radioactivity up to exempt concentrations
may be released by the Reactor facility to individuals, NRC regulations also require that those who process and distribute substances containing radioactive by-products at any level must be
licensed under the exempt concentration section of the Code of
Federal Regulations. It is incumbent on the holder of such a license
to make measurements and maintain records to show that the material
he handles and distributes is at or below the exempt concentration
of the nuclides involved. This in turn means that the nuclides must
be identified and the efficiency of counting with the monitoring
equipment used must be determined.

This work was undertaken to detect and identify any impurities
in silicon that might be activated to significant levels. Since
the Czochralski technique for pulling silicon crystals is inherently
less pure than the floating zone method, Czochralski crystals were
chosen as a "worst-case" for this experiment.

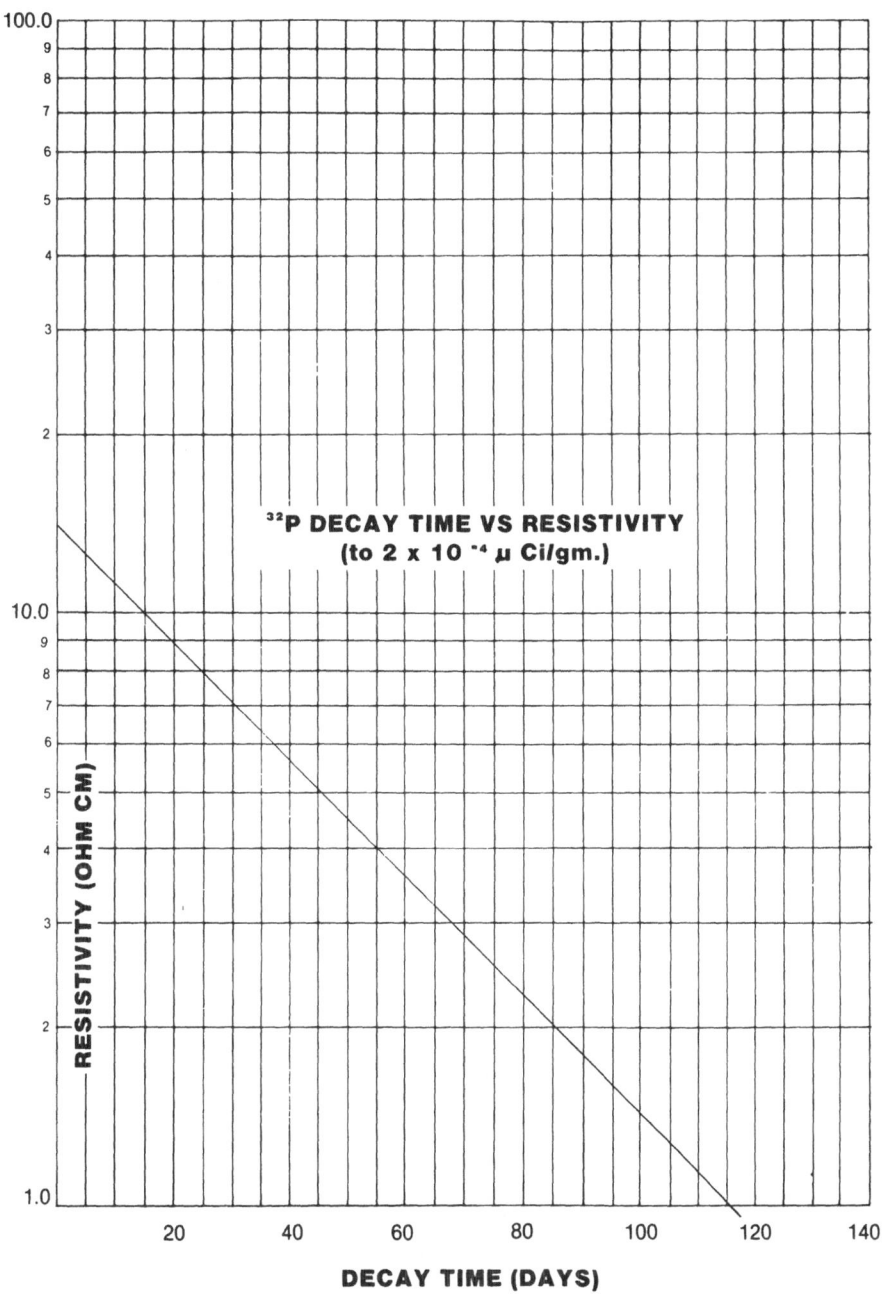

Fig. 2.   Decay time to an exempt limit of 2 x 10^{-4} µCi/gm vs. target resistivity.

Table 1. Computer-calculated levels of radioactivity due to matrix impurities in silicon at maximum levels reported.

Element	Max. Conc. (ppma)	Radioactive Nuclide	u Ci/gm. At discharge	u Ci/gm After 14 days	NRC Exempt Level
Boron	.010 (a)	^{12}B	5.8×10^{-4}	0	--
Carbon	0.45 (a)	^{14}C	1.97×10^{-8}	1.97×10^{-8}	3×10^{-3}
Oxygen	20.00 (a)	^{19}O	7.14×10^{-4}	0	--
Fluorine	0.24 (b)	^{20}F	3.90×10^{-2}	0	$(^{18}F) 8 \times 10^{-3}$
Sodium	0.07 (b)	^{24}Na	4.84×10^{-1}	8.7×10^{-8}	2×10^{-3}
Sulfur	0.045 (a)	^{33}p	1.78×10^{-3}	1.2×10^{-3}	(2×10^{-4})
Chlorine	0.07 (a)	^{35}S	4.61×10^{-3}	4.1×10^{-3}	6×10^{-3}
Potassium	0.04 (a)	^{42}K	4.02×10^{-2}	2.8×10^{-10}	3×10^{-3}
Calcium	0.05 (b)	^{49}Ca ^{47}Ca	1.45×10^{-3}	0	5×10^{-4}
Iron	0.30 (b)	^{55}Fe ^{59}Fe	1.01×10^{-3}	9.9×10^{-4}	8×10^{-3}
Copper	0.36 (b)	^{64}Cu	1.5×10^{-1}	1.9×10^{-7}	3×10^{-3}
Silver	0.30 (b)	^{110}Ag ^{111}Ag	2.9×10^{-2}	7.3×10^{-3}	3×10^{-4}
Magnesium	0.004 (c)	^{27}Mg	4.9	0	0
Aluminum	0.017 (c)	^{28}Al ^{29}Al	2.0×10^{-2}	0	0
Titanium	0.004 (c)	^{51}Ti	4.3×10^{-4}	0	0
Chromium	0.02 (a)	^{51}Cr	1.46×10^{-3}	1.0×10^{-3}	2×10^{-2}
Manganese	0.003 (c)	^{54}Mn ^{56}Mn	7.1×10^{-1}	7.6×10^{-6}	1×10^{-3}
Nickel	0.07 (c)	^{63}Ni ^{65}Ni	1.65×10^{-2}	2.17×10^{-5}	1×10^{-3}
Zinc	0.002 (c)	^{65}Zn ^{69}Zn	5.3×10^{-3}	3.5×10^{-5}	1×10^{-3}
Gallium	0.002 (c)	^{70}Ga	4.1×10^{-2}	0	4×10^{-4}
Germanium	0.004 (c)	^{70}Ge	0	0	2×10^{-2}
Molybdenum	0.011 (c)	^{93}Mo ^{99}Mo	5×10^{-3}	1.5×10^{-4}	2×10^{-3}
Tin	0.005 (c)	$^{117 m}$Sn ^{121}Sn	4.7×10^{3}	2.3×10^{-4}	2×10^{-4}
Antimony	0.005 (c)	^{124}Sb	6.1×10^{-3}	5.2×10^{-3}	2×10^{-4}
Barium	0.004 (c)				0
Tungsten	0.012 (c)	^{181}W ^{185}W ^{187}W	2.0	3.1×10^{-4}	7×10^{-4}
Lead	0.007 (c)				
Bismuth	0.004 (c)				
Colbalt	0.07 (c)	^{60}Co	2.7×10^{-2}	2.7×10^{-2}	5×10^{-4}
Phosphorus	.001 (a)	^{32}P	4.5×10^{-6}	2.2×10^{-6}	2×10^{-4}

(a) Maximum actual values determined at Monsanto by Spark Source Mass Spectrometry
(b) Maximum values reported by others
(c) Not detected in numerous Monsanto analyses by SSMS. Detection limit given.

Although semiconductor grade silicon is one of the purest substances known in commerce--indeed its utility is directly based on
its extreme purity--several elements have been detected at the parts
per million level by a variety of analytical techniques. The most
common of these techniques have been spark source mass spectroscopy
and neutron activation analysis. A list of possible contaminant
elements and assumed concentrations based on actual analyses or
detection limits was compiled and fed into the University of Missouri
computer program used to predict radioactivities that can rise from
sample impurities. A flux density of 2.5 x 10^{13} neutrons/cm^2/sec
and a final resistivity of 5 ohm-cm were assumed. The computer
program produced a list of nuclides and the concentrations of each
that could be expected from the assumed input concentrations. In
Table 1 the assumed contaminants and concentrations are listed along
with the calculated radioactivity level at discharge and after 14
days decay time. The NRC exempt concentration level is also listed
for comparison purposes. The isotopes underlined are those calculated to yield radioactivities $\geq$ the exempt level after 14 days decay.

The experimental matrix to search for these potential impurities
(shown schematically in the last figure) was chosen to give the
maximum feasible range of materials and still explore what was considered significant variables in the production process. First, five
separate batches of polycrystal were chosen from different production
units. Both resistance heated and radio frequency heated crystal
pullers were available, so one 2-inch diameter crystal from each
batch of poly was pulled in each type of crystal puller. Finally,
since nearly all impurities in silicon have segregation coefficients
less than unity, and would be expected to be most concentrated in the
tang end of the crystal, samples were taken from each end of the
crystal. After coding by grinding combinations of flats and notches,
each piece was sliced into 21-mil thick slices. Additionally, the
efficiency of four different pre-irradiation cleaning methods was
tested. The slices were wrapped in commercial aluminum foil in coin-
roll fashion, sealed in welded aluminum cans, and irradiated in a
flux of about 3.6 x 10^{13} neutrons/cm^2/sec to give a total fluence of
5 x 10^{18} neutrons/cm^2 corresponding to about 4 ohm-cm. Immediately
after removal from the reactor, the four groups of slices were
measured as batches by gamma spectroscopy and the results are shown
in Table 2. It can be seen that there was considerable difference
in the levels of surface radioactivity between the different pre-
irradiation cleaning methods. The HF/HNO$_3$ etch was the best pre-
irradiation cleaning method by far. The post irradiation cleaning
method was effective in removing all the nuclides except antimony
and possibly chromium and iron. Individual wafers were then counted
with a Geiger type beta probe and the decay curve was found to follow
that of ^{32}P. However, repetitive etching experiments proved that
over 80% of the radioactivity could be removed by etching with a
mixture of hydrofluoric, nitric and acetic acids to remove a total
thickness of about 0.1 mil. It appears then that phosphorus was a

Table 2.   Total radioactivity on freshly irradiated slices as
measured by gamma spectroscopy.

(μ Ci/gram x 10⁵)

Isotope	Batch A		Batch B		Batch C		Batch D	
	Before	After	Before	After	Before	After	Before	After
¹⁹⁸Au	93	6	1500	7	24	2		
¹⁹⁹Au			24	3	5	2		
¹⁹⁶Au			4	2	2			
¹¹⁰Ag (m)	5	1	1	-				
¹²²Sb	9	2	21	12	11	6	21	13
¹²⁴Sb	5	1	4	4	2	2	10	7
⁸²Br			2	15				
⁵⁹Fe	33	4	2	2	2	1		
²⁴Na			4	-				
⁶⁰Co			18	-				
⁵¹Cr	24	0	5	2				
⁶⁵Zn	28	3						
⁵⁴Mn	5	1						

"Before" and "After" refer to cleaning of wafers with NH_4OH-H_2O_2 solution followed by HC1 - H_2O_2 solution.

Pre-irradiation cleaning methods for four batches were:

    (A) Mechanical scrubbing only
    (B) Mechanical scrubbing followed by NH_4OH-H_2O_2 and HC1-H_2O_2
    (C) Mechanical scrubbing followed by conc. H_2SO_4-HNO_3 solution @ 70° C.
    (D) Bright etching with HF-HNO_3 solution followed by HC1-H_2O_2 solution.

major contaminant on the surface of the wafers. All of these experiments showed that it is necessary to remove surface contamination before attempting to analyze the matrix and that etching
about 0.05 mil from each side with the hydrofluoric-nitric acid
mixture is an effective removal method.

After establishing the methods of analysis and the procedure
for reliably removing surface radioactivity, the experiment was
repeated twice more with replicate wafers from the same rods. In
one of these repetitions, the wafers were again irradiated to about
5 ohm-cm, the surface radioactivity removed by the etching techniques
and the wafers were counted at 1-2 week intervals under carefully
controlled counting conditions using a Nuclear Measurement Corporation flow proportional counter consisting of a Model DS-1AP scaler
and a PCC-11A detector. The detector was equipped with a 2-inch
diameter sample chamber covered by an aluminumized Mylar window.
Background counts were made under conditions identical to the sample
count and the instrument was calibrated each time by counting a
standard source. Counting was continued until the net values equaled
about one-half background.

When the logs of the net counts were plotted vs. time, four of
the slices followed the ³²P decay curve exactly shown by the bottom

line in Fig. 3. Another four wafers followed the ^{32}P curve through
about six half lives and then began to diverge slightly, but the
differences were too small to quantify. The other twelve wafers
gave curves resembling to some degree the plot of points shown on
the top curve in Fig. 3, which represents the one of the twelve with
the most radioactivity. In all twelve, departure from the ^{32}P curve
was detectable. These were subjected to computer regression analysis
and the results are shown in Table 3. This table give the y intercept and the slopes of the component curves along with the confidence
levels for these numbers, and the range of half lives corresponding
to the slopes. The short-lived nuclide is ^{32}P, but the precision
of the method is not sufficient to identify the second nuclide involved with certainty.

The irradiation experiment was repeated a second time irradiating to a fluence of 1.2×10^{19} neutrons/cm^2 corresponding to a
resistivity of about 2 ohm-cm. After four days, the wafers were
analyzed by gamma spectroscopy before cleaning and after each of
two etches with a hydrofluoric-nitric mixture with results as
shown in Table 4.

It can be seen that the sodium, bromine, chromium, and zinc
isotopes were all removed with one etch, but the silver and gold
isotopes were slightly more persistent in that they survived the
first etch to a large degree. Only ^{124}Sb could be identified in
the matrix after the second etching. The nuclear data for all the
isotopes detected are shown in Table 5. It is interesting to note
also that antimony was a significant surface contaminant as evidenced
by the decrease in radioactivity after the first etch being much
larger than could be accounted for by decay.

It appears then that ^{124}Sb is the nuclide responsible for the
"tailing" observed in the decay curves, since its half life of 60
days is consistent with the result of the analysis of the decay
curves. No evidence was uncovered indicating the presence of any
other impurity.

With the isotope accounting for the "tailing" of the decay curve
identified as ^{124}Sb, the next step was to consider its concentration
relative to the NRC exempt concentration to determine if additional
decay time over that required for ^{32}P would be necessary. To determine the counting efficiency for a particular isotope, one must
count a known amount of that radioisotope under standard conditions.
If a suitable standard of that radioisotope is unavailable, then a
substitute radioisotope of similar energy may be used. It is also
necessary to determine the absorption characteristics of silicon
relative to the particular energies involved. Experimentally, this
is done by counting the radioactive sources with various thicknesses
of silicon between it and the detector. No long-lived isotope with
a beta decay energy similar to that of ^{32}P (which is 1.70 MeV) exists,

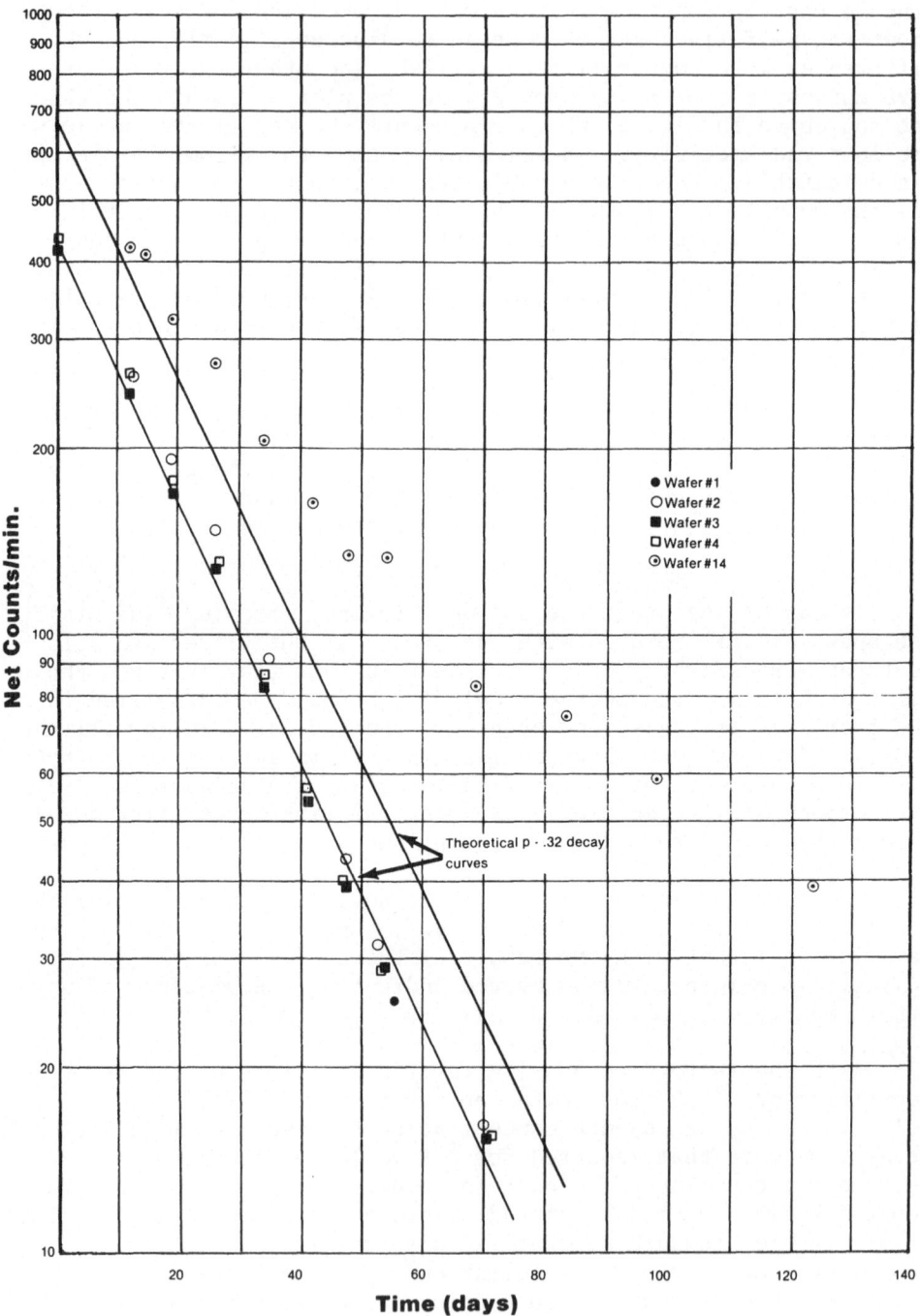

Fig. 3.   Decay of radioactivity from selected slices compared to calculated rate for ^{32}P.

Table 3. Results of regression analysis of radioactive decay of irradiated Czochralski wafers.

$$\text{Counts} = A_1 e^{-B_1 t} + A_2 e^{-B_2 t}, \quad \text{where } t = \text{time in days}$$

Wafer No.	A_1	B_1	$t_{1/2}$ (days) min.	mid.	max.	A_2	B_2	$t_{1/2}$ (days) min.	mid.	max.	$A_1 + A_2$
5	311 ± 58	.068 ± .018	8	10	14	217 ± 44	.016 ± .0016	40	48	518	
7	419 ± 244	.055 ± .015	20	13	17	88 ± 270	.0103 ± .0270	15	34	--	507
8	459 ± 182	.051 ± .024	9	14	26	38 ± 210	.0161 ± .0488	--	43	--	497
10	393 ± 88	.062 ± .018	9	11	16	139 ± 90	.0191 ± .0058	28	36	52	432
11	397 ± 78	.054 ± .012	11	13	16	139 ± 86	.0177 ± .0050	30	39	55	436
12	504 ± 42	.045 ± .006	14	15	18	108 ± 36	.0081 ± .0024	65	85	121	610
13	468 ± 38	.049 ± .006	13	14	16	61 ± 40	.0100 ± .0052	46	69	144	529
14	458 ± 48	.054 ± .010	11	13	16	213 ± 44	.0139 ± .0016	45	50	56	671
15	481 ± 76	.043 ± .012	13	16	22	38 ± 82	.0076 ± .0162	--	91	--	519
16	456 ± 34	.051 ± .006	12	14	15	118 ± 30	.0120 ÷ .0018	50	58	68	574
*19	306 ± 368	.059 ± .034	7	12	27	162 ± 372	.0288 ± .0210	--	24	--	468
*20	476 ± 38	.050 ± .006	12	14	16	10 ± 40	.0035 ± .0213	--	198	--	486
Avg.	427	.053 ± .004	12.2	13	14.2	--	.014 ± .004	39	48	69	--

*Fit obtained using square.root of counts.

Table 4. Results of gamma spectroscopic analysis of wafers irradiated to ∿ 2 ohm cm before and after acid etchings (all results in disintegration/second per wafer).

Sample	^{122}Sb			^{124}Sb			^{198}Au			^{110m}Ag			^{24}Na			^{82}Br			^{51}Cr			^{65}Zn		
	a	b	c	a	b	c	a	b	c	a	b	c	a	b	c	a	b	c	a	b	c	a	b	c
1	250	9	--	--	--	<1	214	16	--	<20	<5	--	3798	--	--	--	--	--	--	--	--	68	--	--
2	448	8	--	--	--	<1	462	61	--	<20	<5	--	3397	--	--	50	--	--	238	--	--	116	--	--
3	1186	13	--	--	--	<1	527	17	--	<22	--	--	3516	--	--	55	--	--	--	--	--	--	--	--
4	381	12	--	--	--	<1	272	11	--	<29	12	--	3882	--	--	40	--	--	991	--	--	119	--	--
5	825	<5	--	--	--	<1	1446	83	--	111	8	--	4374	--	--	154	--	--	1031	--	--	228	--	--
6	631	<5	--	--	--	<1	1339	47	--	35	--	--	3504	--	--	78	--	--	160	--	--	133	--	--
7	371	15	--	--	--	<1	279	13	--	23	--	--	3945	--	--	--	--	--	--	--	--	57	--	--
8	536	37	--	--	--	2.6	432	36	--	--	--	--	4701	--	--	125	--	--	208	--	--	59	--	--
9	749	25	--	--	--	<1	756	35	--	2137	52	--	4615	--	--	354	--	--	282	--	--	262	--	--
10	1104	143	--	--	10	11	685	29	--	4972	5	--	3767	--	--	472	--	--	--	--	--	142	--	--
11	756	78	--	--	--	4.6	574	23	--	41	8	--	4127	--	--	67	--	--	145	--	--	162	--	--
12	675	264	--	--	25	18	195	7	--	<17	--	--	3559	--	--	49	--	--	109	--	--	--	--	--
13	--	--	--	--	--	--	--	--	--	--	--	--	--	--	--	--	--	--	--	--	--	--	--	--
14	1385	705	--	--	61	52	242	22	--	29	--	--	4053	--	--	32	--	--	135	--	--	--	--	--
15	765	201	--	--	24	14	312	20	--	23	--	--	3447	--	--	53	--	--	230	--	--	--	--	--
16	626	281	--	--	32	21	259	26	--	--	--	--	4294	--	--	75	--	--	126	--	--	101	--	--
17	659	10	--	--	--	<1	1554	55	--	32	--	--	3121	--	--	71	--	--	177	--	--	145	--	--
18	886	--	--	--	--	<1	695	88	--	665	9	--	3726	--	--	66	--	--	236	--	--	88	--	--
19	509	--	--	--	--	<1	356	24	--	35	5	--	3725	--	--	--	--	--	109	--	--	--	--	--
20	313	--	--	--	--	<1	570	62	--	41	--	--	3263	--	--	50	--	--	106	--	--	--	--	--

Under each nuclide:

1st column - After 4 days and no etching
2nd column - After 4 days and 1 etch
3rd column - After 20 days and 2 etches

Table 5. Nuclear data for isotopes observed on slice surfaces.

Isotope	γ energy (kev)	Half-life	β-energies
^{124}Sb	603	60.1 days	2.3 mev (21%, 0.6 mev 65%)
^{122}Sb	564	2.8 days	2.0 mev (30%, 1.40 mev. 63%)
^{198}Au	412	2.7 days	.096 mev (99 + %)
^{110m}Ag	658	253 days	0.53 mev (36%, 0.09 mev, 64%)
^{82}Br	776	35 hours	0.44 mev (100%)
^{51}Cr	320	28 days	Electron capture, v x-rays
^{65}Zn	1115	245 days	2 mev β + (50%) Electron capture (50%)

so that bracketing of the energies with two different sources is necessary. In Fig. 4 the counting efficiencies, i.e., the ratio of the actual counts per unit time to the disintegrations per unit time calculated for the certified source, are plotted against thickness of silicon for ^{210}Bi and for ^{234}Pa with peak energies of 1.17 and 2.34 MeV respectively. The point at which the efficiency becomes zero is referred to as the infinite thickness for the particular energy involved. Integration of these curves between zero and any thickness will give the average counting efficiency for that particular thickness. In Fig. 5, average counting efficiency and infinite thickness are plotted as functions of peak energies to allow extrapolation to the ^{32}P case. These values are then substituted into the equation shown in Fig. 6 to determine the net counting rate that corresponds to the exempt concentration. Thus, for a 2 inch diameter slice at least 72 mils thick, 193 net counts per minute corresponds to the NRC exempt limit for ^{32}P using the counting equipment specified. For the 21 mils thick slice used here, the corresponding value is about 150 counts/minute.

Now the peak energy for ^{124}Sb corresponds almost exactly to that of our ^{234}Pa source and one can take the values of counting efficiency directly off its curve. When a similar calculation is applied to ^{124}Sb, the corresponding figure is 217 counts/minute for the exempt concentration for a thick sample or about 150 counts/minute for a 21 mil slice. In the worst case examined here, the ^{124}Sb was at about 1.2 times this level about 70 days after discharge from the reactor when the ^{32}P level was at the exempt level. Note incidentally that this 70 days required for ^{32}P to reach the exempt level is about 20 days longer than predicted by the computer study (Fig. 2). Thus, in the worst case found, the presence of antimony would cause extension of the required decay time of about 12 days. The gamma spectroscopy method, however, showed only about two-thirds this much ^{124}Sb in the worst case after calculating to the 5 ohm-cm irradiation level.

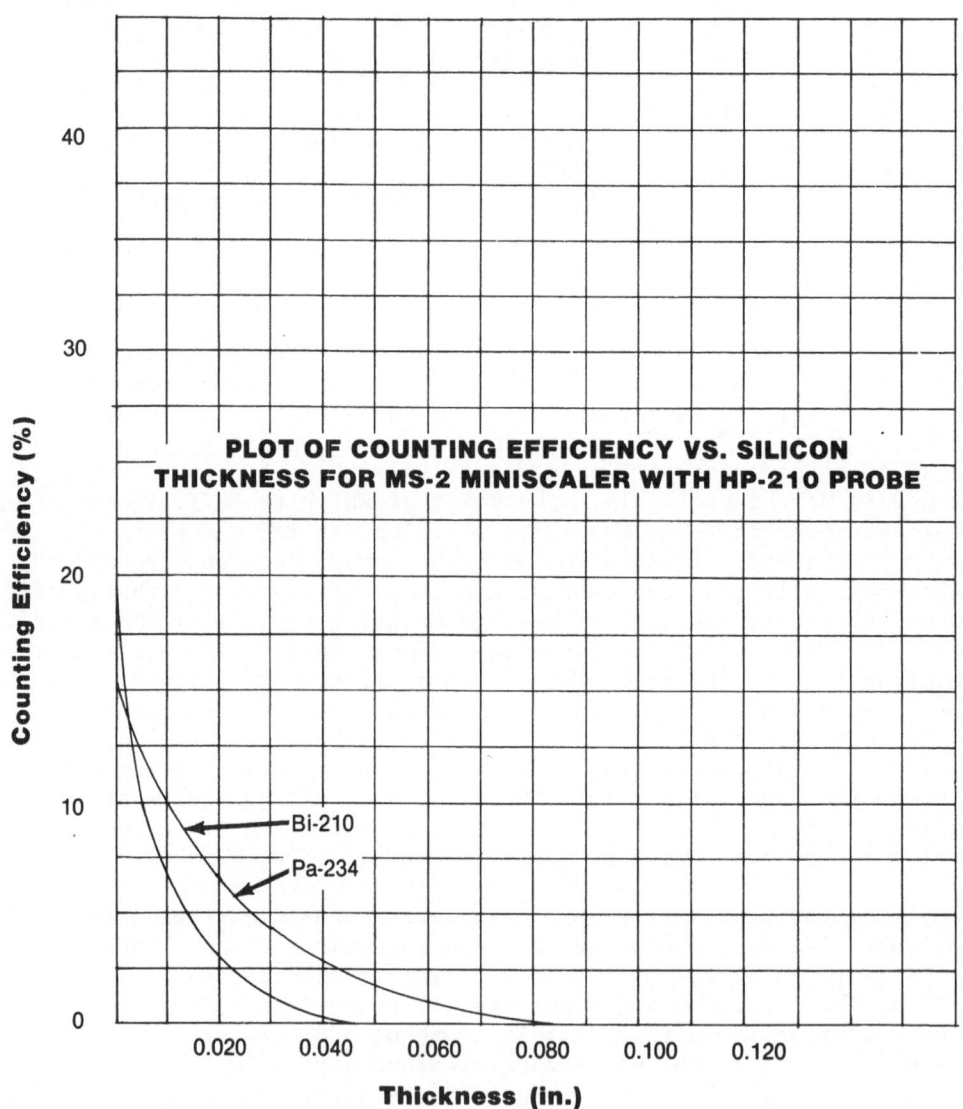

Fig. 4. Counting efficiency vs. silicon thickness using the 1.17 MeV decay of ^{210}Bi and the 2.34 MeV decay of ^{234}Pa.

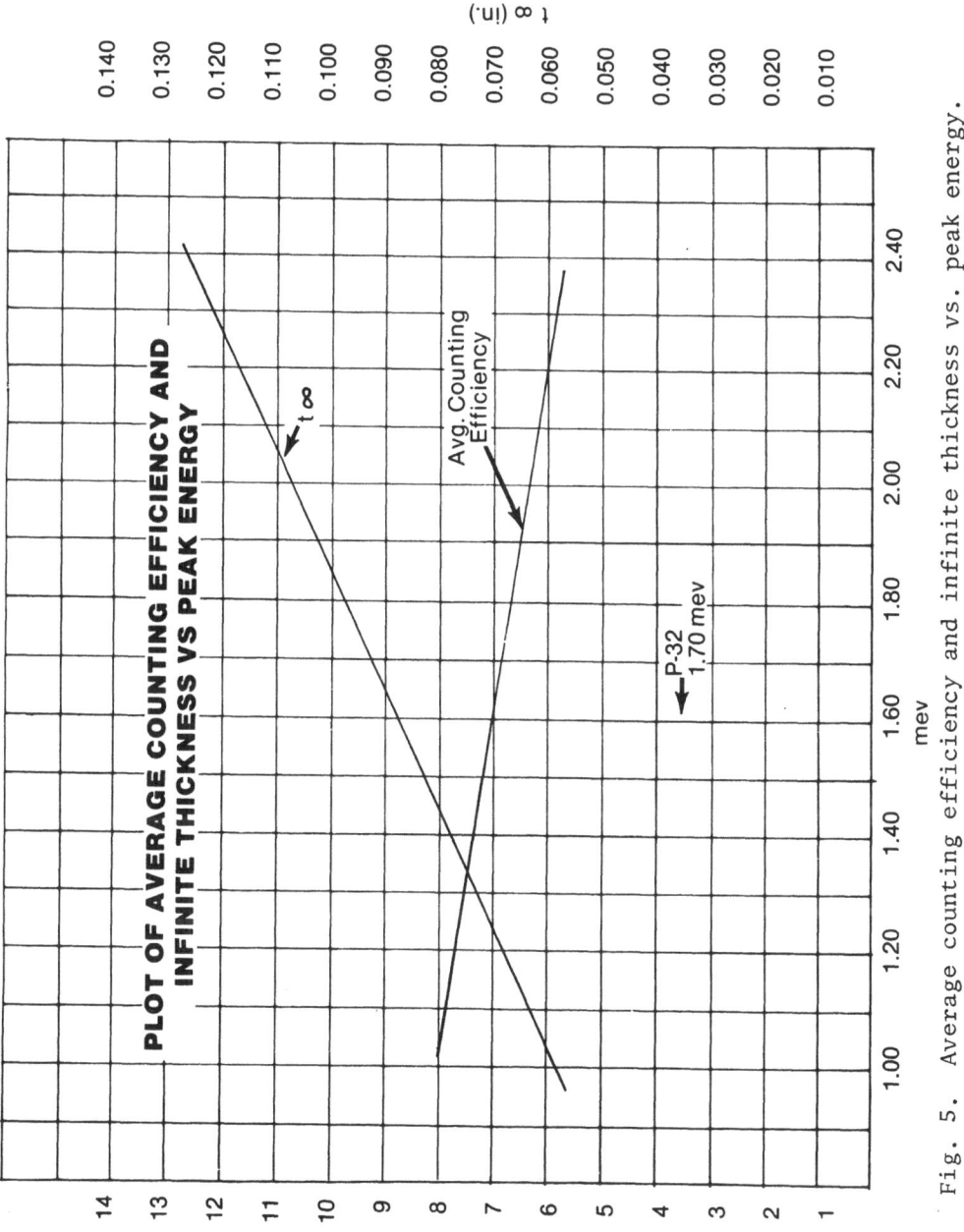

Fig. 5.   Average counting efficiency and infinite thickness vs. peak energy.

Ex. Conc. $(\mu Ci/g.) = \dfrac{Net\ CPM}{(Ec)(Kci)(D\ell\ Silicon)\ t\ \infty}$

where:

Net CPM = Sample counts/minute — Background counts/minute

Ec = Average counting efficiency

Kci = Counts/μCi = 2.22 x 10^6/minute

Dℓ = Density of Silicon at diameter (g/in.)

t∞ = Infinite thickness (in.)

Assuming that Diameter of Silicon ≤ Diameter of Probe Face

For Phosphorus -32 using Eberline MS-2/HP-210 System,

$$2 \times 10^{-4} = \dfrac{Net\ CPM}{(0.048)(2.22 \times 10^6)(216\ g/in.)(.072\ in.)}$$

Net CPM = 193

Fig. 6.   Relationship between counting rate and exempt concentrations.

In Fig. 7, the results of the radioactivity analyses are related to the experimental matrix. It is apparent that resistivity measurements before and after irradiation are not sensitive enough to detect the antimony which is a donor in silicon. Before irradiation, the antimony is obscured by boron contributed by the Czochralski crucible and all the crystals were p-type. If antimony could be detected by this method, the crystals with the highest level of radioactivity should have the highest initial resistivity, but this is not the case. Likewise, after irradiation, those highest in antimony should be the lowest in resistivity, but again the measurement method is not sensitive enough to detect it.

Finally, it is clear that all the crystals pulled in the resistance heated puller were essentially free of antimony. Crystal number 5 was anomalous in this respect in that no antimony was detected by gamma spectroscopy, but a departure from the ^{32}P decay curve did occur. In this experiment two different RF pullers were used (denoted in Fig. 7 as number 42 and 43) and both contributed some antimony although number 43 was much worse. Thus, the impurity was almost certainly introduced in the crystal pulling step, which is not surprising in view of the number of crystals heavily doped with antimony that are pulled on occasion. This study shows how necessary it is to avoid chance contamination in crystals to be irradiated.

In summary, we searched for radioactive nuclides in Czochralski silicon irradiated to about 5 ohm-cm that were predicted based on levels of several impurities that have been found in silicon by various methods of instrumental analysis. The rate of beta decay and gamma spectroscopy were used to analyze for the isotopes in

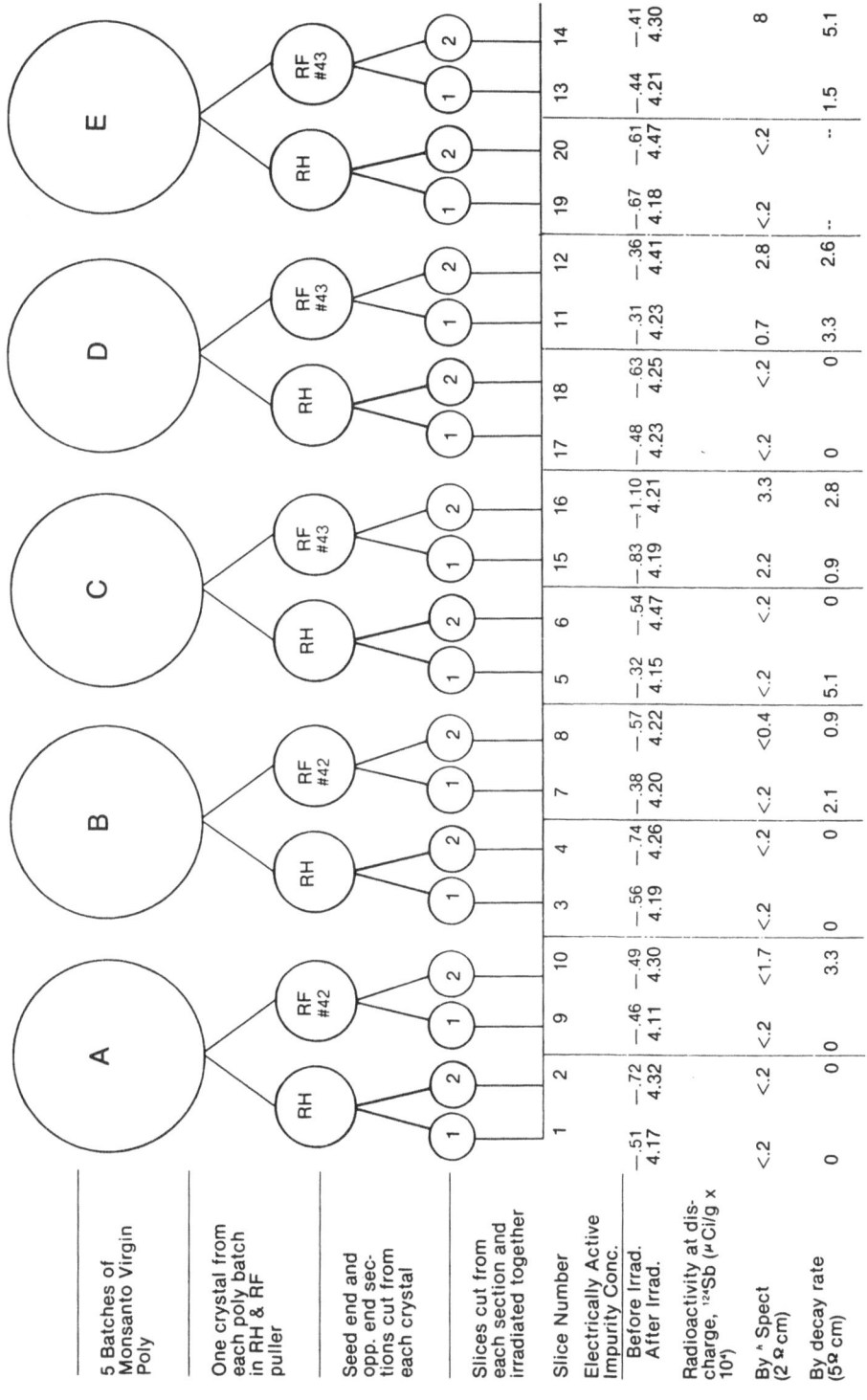

Fig. 7. Results of radioactivity and resistivity measurements related to the sample matrix.

question. It was shown that it is absolutely necessary to chemically
remove some of the silicon surface to separate surface from matrix
impurities. Only ^{124}Sb could be identified in the matrix at worst
case levels about equal to that of ^{32}P when the latter has decayed
to the NRC exempt concentration level. The data indicate that the
antimony was introduced during the crystal growing step. This
could be traced to the pullers in question having been used to pull
antimony-doped crystals in the past. The results show that antimony
must be excluded from incorporation into crystals to be irradiated
to an extent far below that that would affect resistivity.

NUCLEAR TRANSMUTATION DOPING FROM THE VIEWPOINT OF RADIOACTIVITY FORMATION

E. W. Haas and J. A. Martin

Kraftwerk Union AG

Erlangen, Germany

ABSTRACT

The doping of semiconductor materials by nuclear transmutation is associated with the formation of radioactivity. In the case of semiconducting silicon, Si-31 activity, surface contamination and P-32 activity stemming from the silicon are of significance. While the Si-31 activity will have decayed to zero after about 4 days, the surface contamination can be removed by an etching treatment applied to the silicon. The P-32 activity produced by further activation in the silicon can be ignored up to a concentration of 2 μCi/kg. At higher concentrations, observance of the 10 μCi exempt limit must be checked. The formation of P-32 activity can be controlled to a certain extent by a suitable choice of neutron flux density. It is shown that high neutron flux densities are preferable to low ones from economic considerations.

Despite theoretically good possibilities for the use of nuclear transmutation for doping other semiconductor material as well, such as germanium, arsenic or III-V compounds, this application appears to founder on the considerably greater radiation protection problems involved.

1. INTRODUCTION

I had the opportunity of lecturing at the AIME Conference in Princeton in 1975 on the development of the neutron transmutation doping method applied to semiconducting silicon at Siemens AG and Kraftwerk Union AG.[1] We have pressed on systematically with the development of this doping method and are in a position to state

that in the doping range from 5 to 500 Ω-cm there are no longer
problems of any sort today as far as the precise attainment of objectives, reproducibility and homogeneity of doping are concerned.
The complete rectification of radiation damage has even yet not been
fully mastered in the case of irradiation carried out in a neutron
field with a high proportion of fast neutrons.

The quantity of silicon doped by ourselves with the nuclear
transformation process is shown in Fig. 1. The relatively small
volume is accounted for by the fact that only the material needed
for the development departments is shown.

In this lecture we shall deal mainly with the radioactivity
occurring when silicon is doped. The radioactivity problems arising
when other semiconducting materials are doped will be commented on
in brief.

The consistent properties of components made from silicon doped
by neutron transmutation and the resulting reduction in the failure

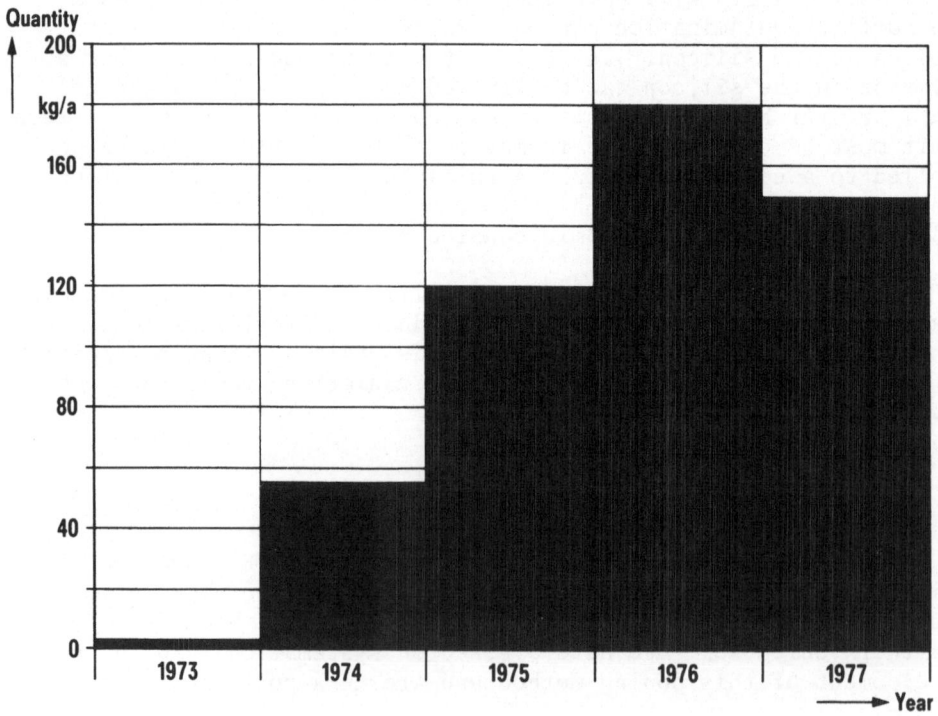

Fig. 1.   Production of NBH silicon for device development within
           Siemens AG.

rate led to considerations concerning the manufacture of yet more
simple devices from material doped in this way.

The large quantities of silicon which would then have to be
doped and the higher dopant concentrations required for many components give rise to problems regarding the residual activity of the
doped material. There are essentially two causes operating here:

A. Activation products on the silicon surface stemming from foreign
matter already there and from foreign matter deposited in the reactor.

B. The P-32 activity brought about by further activation of the
P-31 produced by nuclear transmutation.

2. RADIOACTIVITY ON THE SILICON SURFACE

After the 3 to 4 day decay period needed for adequate decay
of the Si-31, the doped Si rods still exhibit contamination by
longer-lived radionuclides on the surface. In the case of irradiation in swimming-pool reactors, these types of activity considerably
exceed the permissible amounts or concentrations in individual cases.
Fig. 2 shows an example of this. The high W-187 activities stemmed
here from the sawing of the silicon into rods; the cause of protac-
tinium activities could not be fully clarified. Contamination reduced by about 1 order of magnitude is observed in those rods that
are doped in dry irradiation channels.

Removal of these surface activities is generally possible by
means of the removal of a small quantity of material by etching.
This can take place before the transfer of the doped material to
the manufacturing stage either at the reactor station itself or in
a suitably equipped laboratory in the processing plant.

3. P-32 ACTIVITY IN SILICON

According to the radiation protection recommendations of the
IAEA, "radioactive material" shall mean any material having a
specific activity greater than 2×10^{-3} µCi/g.[2] Weakly radioactive
material with an activity below this concentration limit can be
handled in any desired quantity. If the limit is exceeded, the
value 10 µCi applies for P-32 as the maximum permissible activity
for exemption from notification, registration or licensing.[3] With
doping of over 5 Ω-cm, this necessitates considerable decay times
in the handling of larger quantities of silicon, as shown in Fig. 3.
The reduction of this P-32 activity is possible to a limited extent
when optimum neutron flux densities are used for the irradiation.

Surface area: $\sim 200\ cm^2$

Radionuclide →	Ce-141 μCi	Ru-103 μCi	Zr-95 μCi	Nb-95 μCi	Co-60 μCi	Cr-51 μCi	W-187 μCi	Au-198 μCi	Pa-233 (+U-233) μCi
Exempt Limit μCi →	10	10	10	10	10	100	0,1	10	10
Sample: 1	0,01	0,001	0,007	0,003	0,003	0,008	3,2	0,06	0,6
2	0,01	0,002	0,005	0,004	0,012	0,017	5,0	0,02	0,1
3	0,01	0,002	0,006	0,005	0,005	0,016	4,7	0,02	72,2
4	0,01	–	0,004	0,003	0,005	0,015	6,5	0,02	34,2
5	0,03	0,003	0,008	0,007	0,005	0,05	3,0	0,15	0,4
6	0,02	0,003	0,008	0,006	0,009	0,02	3,1	0,09	19
7	0,008	0,002	0,005	0,003	–	0,006	–	0,003	1,8
8	1,5	0,8	2,0	4,7	0,4	–	–	–	17

■ = Exempt limit exceeded

Fig. 2.   Radionuclides detected at the surface of silicon rods.

	Non radioactive material	Radioactive material handling without notification, registration or licensing
Activity	$< 2\mu$ Ci/kg	10μ Ci
Limitation of quantity	Non	Activity of the quantity of silicon must be less than 10μ Ci
Example	Handling of unlimited quantities of NBH silicon up to 5 Ω cm	Handling of a maximum of (Φ: 1x10^{13}) 0.2 kg 1 Ω cm Si or 10.0 kg 1 Ω cm Si after 81 days decay time

Fig. 3.   Handling of NBH silicon with ^{32}P volume activity.

The formation of P-32 was calculated for the determination of these optimum irradiation conditions from the simplification of the Bateman equation published by Maenhaut and de Beeck.[4] The time element t appearing in the equation was replaced here by the expression shown in Eq. (1) where t is the time for P-31 generation, Φ is the neutron flux and C_{ph} is the concentration of P-31.

$$t = \frac{C_{ph}}{2.06 \times 10^{-4} \; \Phi} \tag{1}$$

This mathematical expression represents the known computational basis for the neutron transmutation doping of silicon.[1] Using Eq. (2) it is possible to calculate the P-32 activity per unit mass of silicon as a function of the neutron flux density.

$$S_4 = \left(\frac{W_1 H_1 \sigma_1 \sigma_3 L \Phi^2}{A_1} \right) \left(\frac{\lambda_4}{\lambda_2 (\lambda_4 - \lambda_2)} \right)$$
$$\left\{ \left[\left(e^{-\lambda_2 C_{ph}/K\Phi} - 1 \right) + \frac{\lambda_2 C_{ph}}{K\Phi} - \frac{(\lambda_2 C_{ph}/K\Phi)^2}{2} \right] \right.$$
$$\left. - \left(\frac{\lambda_2}{\lambda_4} \right)^2 \left[\left(e^{-\lambda_4 C_{ph}/K\Phi} - 1 \right) + \frac{\lambda_4 C_{ph}}{K\Phi} - \frac{(\lambda_4 C_{ph}/K\Phi)^2}{2} \right] \right\} \tag{2}$$

Meaning of symbols

W_1 = Mass of the irradiated silicon   in g
H_1 = Isotopic abundance of ^{30}Si in %/100
σ_1 = n,γ cross section of ^{30}Si in cm^2
σ_3 = n,γ cross section of ^{31}P in cm^2
L   = Avogardo's number   in mol^{-1}
A_1 = Atomic weight of ^{30}Si in $g \; mol^{-1}$
Φ = Thermal neutron flux density   in $s^{-1} cm^{-2}$
C_{ph}= ^{31}P concentration   in cm^{-3}
λ_2 = Decay constant of ^{31}Si in s^{-1}
λ_4 = Decay constant of ^{32}P in s^{-1}
S_4 = Specific ^{32}P activity   in $s^{-1}g^{-1}$
K   = 2.06×10^{-4}

The resulting relationship between neutron flux density and P-32 activity per kg of silicon is represented graphically for various resistivity values in Fig. 4. The following statements can be deduced from this.

(1) For doping up to 5 Ω-cm, neutron transmutation doped silicon can be regarded as non-radioactive material, irrespective of the neutron flux densities at which irradiation was carried out. In the case of doping higher than 5 Ω-cm, the choice of neutron flux

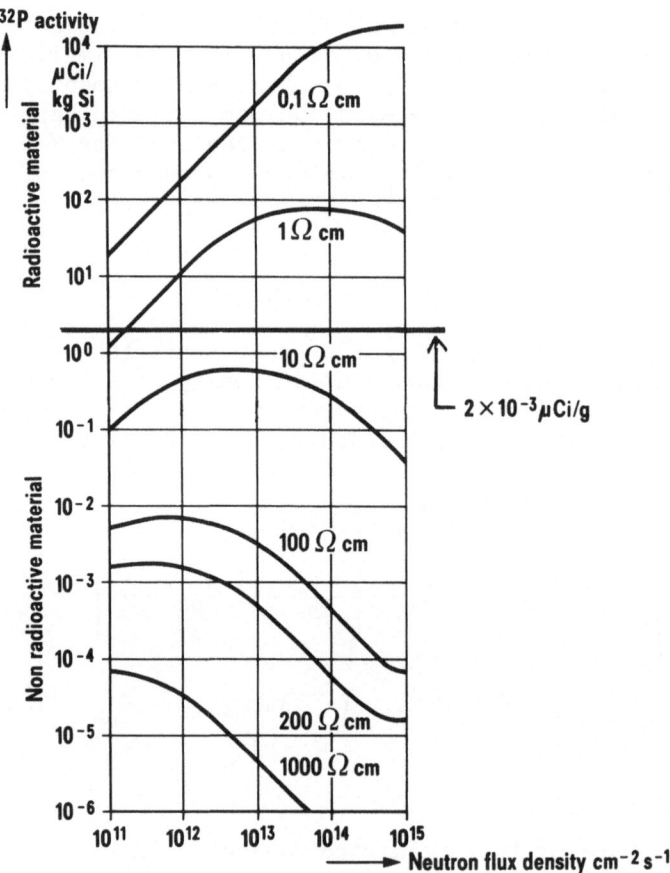

Fig. 4.   Relationship between neutron flux density and ^{32}P activity
per kg of silicon for various resistivities.

density determines to a limited extent whether or not the doped
silicon must be classified as radioactive material and whether or
not storage periods are required, in any given case, for the reduction of the P-32 activity to less than the permissible limit of
10 μCi.

(2) In the resistivity range above 10 Ω-cm, the P-32 activity could
be considerably reduced by the use of lower neutron flux densities.
The reason for this lies in the fact that the production of P-32
varies with the square of the neutron flux. However, the reduction
of activity by the use of lower neutron flux densities has to be
paid for by a marked increase in irradiation time, so that this
manner of application becomes unattractive from the view point of
cost.

(3) The formation of P-32 activity can also be reduced when high neutron flux densities are used. In the range of neutron flux densities considered here, from 10^{11} to 10^{15}cm^{-2}s^{-1}, the P-32 activity in the case of silicon with resistivities below 0.1 Ω-cm can be reduced by 1-2 orders of magnitude, with respect to the most unfavorable case. The cause of this lies in the short irradiation times. Owing to the yet incomplete decay of the Si-31 activity within the irradiation time, there are less P-31 atoms available for further activation to P-32.

Figure 5 shows examples of application for both cases. If one considers the total time required which will allow the P-32 activity formed in each case to fall below the 10 µCi exempt limit, then it emerges in general that the use of higher neutron flux densities has advantages. From the viewpoint of the obligation to keep the amount of radioactive material to a minimum[5], one should bear in mind the reduced formation of activity at high neutron flux densities in the case of material of higher resistivity also.

The extremely short irradiation times which are sometimes involved in the use of high neutron flux densities would require the provision, at the reactor stations, of irradiation equipment with short insertion and removal times.

Objective of doping	Neutron flux density [cm^{-2}s^{-1}]	Irradiation time [d]	Specific ^{32}P activity [µCi/kg]	Decay time [d]	Total time expended [d]
10 Ω cm	10^{12}	27	0.46	0	27
	10^{13}	2.7	0.62	0	2.7
	10^{14}	0.27	0.27	0	0.27
	10^{15}	0.027	0.04	0	0.027
1 Ω cm	10^{12}	300	11	2	302
	10^{13}	30	56	36	66
	10^{14}	3	78	42	45
	10^{15}	0.3	37	27	27.3
0.1 Ω cm	10^{12}	3000	175	59	3059
	10^{13}	300	1750	107	407
	10^{14}	30	12000	146	176
	10^{15}	3	19000	156	159

Fig. 5. Effect of neutron flux density on the storage time of NBH silicon.

Neutron flux density $[\text{cm}^{-2}\text{s}^{-1}]$	Irradiation time [min]	P conc. calculated from irradiation conditions $[\text{cm}^{-3}]$	^{32}P-activity calculated from irradiation conditions $[\mu\text{Ci} \times \text{kg}^{-1}]$	Measured value of ^{32}P activity $[\mu\text{Ci} \times \text{kg}^{-1}]$
1.8×10^{11}	15410	3.4×10^{14}	2.9×10^{-3}	$\leq 3 \times 10^{-3}$
4.6×10^{13}	212	1.2×10^{14}	1.1×10^{-2}	$\leq 3 \times 10^{-2}$
4.2×10^{13}	196	1×10^{14}	7.2×10^{-3}	$\leq 2 \times 10^{-2}$
4.6×10^{13}	788	4.5×10^{14}	3.5×10^{-1}	2.7×10^{-1}
2.9×10^{13}	5820	2.4×10^{15}	$1.5 \times 10^{+1}$	$1.5 \times 10^{+1}$
3×10^{13}	2340	9.8×10^{14}	2.4	1.8
7.9×10^{13}	590	5.7×10^{14}	5×10^{-1}	7.3×10^{-1}

Fig. 6.   Comparison of calculated and measured ^{32}P activities in silicon.

Semiconductor	Nuclear reaction	Type of doping	Resulting dopant concentration atoms/cm^3	
Si	^{30}Si (n,γ) ^{31}Si (β) ^{31}P	n		$2.06 \times 10^{-4}\ \Phi t$
Ge	^{70}Ge (n,γ) ^{71}Ge (ϵ) ^{71}Ga	p	$3.0 \times 10^{-2}\ \Phi t$	$1.8 \times 10^{-2}\ \Phi t$
	^{74}Ge (n,γ) ^{75}Ge (β) ^{75}As	n	$1.1 \times 10^{-2}\ \Phi t$	
	^{76}Ge (n,γ) ^{77}Ge (β) ^{77}As (β) ^{77}Se	n	$1.1 \times 10^{-3}\ \Phi t$	
Se	^{74}Se (n,γ) ^{75}Se (ϵ) ^{75}As	p		$9 \times 10^{-3}\ \Phi t$
Ga As	^{69}Ga (n,γ) ^{70}Ga (β) ^{70}Ge	n	$2.5 \times 10^{-2}\ \Phi t$	$1.7 \times 10^{-1}\ \Phi t$
	^{71}Ga (n,γ) ^{72}Ga (β) ^{72}Ge	n	$4.4 \times 10^{-2}\ \Phi t$	
	^{75}As (n,γ) ^{76}As (β) ^{76}Se	n	$9.9 \times 10^{-2}\ \Phi t$	
Ga P	^{69}Ga (n,γ) ^{70}Ga (β) ^{70}Ge	n	$2.8 \times 10^{-2}\ \Phi t$	$8.2 \times 10^{-2}\ \Phi t$
	^{71}Ga (n,γ) ^{72}Ga (β) ^{72}Ge	n	$4.9 \times 10^{-2}\ \Phi t$	
	^{31}P (n,γ) ^{32}P (β) ^{32}S	n	$4.7 \times 10^{-3}\ \Phi t$	

Fig. 7.   Possibilities for semiconductor doping by irradiation with thermal neutrons.

The accuracy of the computational determination of P-32 activity could be checked against measured data, but only for a small range of values of Φ and t, of course. It can be seen from Fig. 6 that in the doping range investigated, from 2 to 150 Ω-cm, the P-32 activities calculated from the formula and those derived from measurement are in adequate agreement.

3. THE FORMATION OF ACTIVITIES IN OTHER SEMICONDUCTING MATERIALS

In semiconducting substances other than Si, thermal neutrons also cause reactions to take place which can be taken advantage of to dope these materials. Figure 7 shows a summary for the materials Ge, Se, GaAs, and GaP. The attainable dopant concentrations are

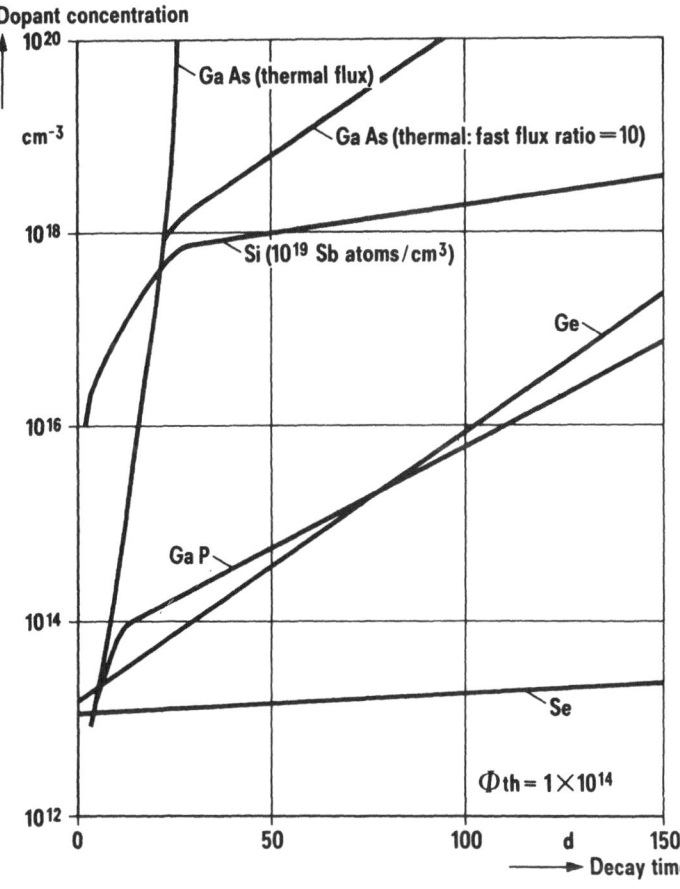

Fig. 8.   Decay time for 1 g of neutron irradiated material.

almost 100 times higher in Ge, for example, than in silicon, and almost 1000 times higher in GaAs than in silicon. Unfortunately neutron irradiation leads in most substances to radionuclides with prolonged half-lives. This means that at an appropriate dopant level, in the gramme range alone decay times of weeks or even months have to be put up with (Fig. 8). As the GaAs example shows, here too more favorable decay conditions can be created in individual cases by a suitable choice of irradiation conditions.

The difficulties pointed out in the case of semiconducting sub-stances other than Si make it easy to understand why the extensive commercial application of the neutron transformation method has to date been restricted to silicon.

REFERENCES

1)   Schnoller Haas, Conference on Preparation and Properties of Electronic Material, Princeton, Aug. 1975.
2)   Safety Series No. 6, (IAEA, Vienna, 1973), p. 8.
3)   Safety Series No. 9, (IAEA, Vienna, 1967), p. 32.
4)   W. Maenhaut, J. P. de Beeck, J.Radioanal. Chem. 5, 115 (1970).
5)   Safety Series No. 38, (IAEA, Vienna, 1973), p. 139.

APPLICATION OF NTD SILICON FOR POWER DEVICES

Hans Mork Janus

Topsil A/S

Frederikssund, Denmark

ABSTRACT

Industrial application and manufacturing of neutron doped silicon was initiated in large scale in 1975. Since then the range of applications has been widened drastically in concordance with the usual price pattern of semiconductor products.

The yield impact in the device industry has completely changed the projected float zone silicon consumption for at least the next five years.

A survey of these phenomena is presented.

1. INTRODUCTION AND HISTORIC LANDMARKS

In the last two decades, semiconductor silicon has passed through a number of step functions in quality. Starting from the introduction of the float zone crystal growing process and passing through the dislocation free pulling and the elimination of swirls, we arrive at the method of homogenous phosphorus doping of silicon by neutron transmutation.

This last process was in reality as old as the float zone pulling process, having been originally pointed out by Lark-Horovitz[1] in 1951.

The process was exploited for applications by Tanenbaum & Mills[2] in 1961 and during the sixties by several others.[3-4]

37

Due to material problems other than resistivity inhomogeneity, it
did not really find its applications until some German high power
device manufacturers in 1973 demonstrated its superiority for the
manufacturing of 4 kV high current thyristors and diodes.[5]

In 1974 the first commercial NTD (neutron transmutation doped)
silicon was introduced[6] and already in 1975 several power device
manufacturers were each using on the order of one hundred kilograms
per year.

From 1976 the process as a commercial method for homogenous
phosphorus doping became widely accepted and is now in use to
some extent by nearly all float zone semiconductor silicon manufacturers. In 1976 the first NTD silicon conference was held as
a one day seminar in Oak Ridge, Tennessee.

The fast acceptance of the method and the homogenous material
may be best demonstrated by the fact that in 1977 approximately
20% of the world's float zone silicon production for discrete
power devices was manufactured by neutron transmutation doping
(more than twenty tons out of approximately 100 tons per year).
The importance of the process is further manifested by the present
three day conference solely on this subject. In addition to power
device applications, we see today areas like solar cells, particle
detectors, VLSI devices and exploitations of the same process
used for doping III-V compounds.

Several power manufacturers have presently an annual NTD
silicon comsumption in excess of one million dollars.

2. NTD PROCESS EVOLUTION

As the NTD method was introduced in 1973-74, it was merely
serving a highly specialized production of semiconductor silicon
for the large area, single wafer high power devices. As such, no
general production advantages were taken into account and the
early NTD silicon was specially manufactured by high purity poly
silicon deposition followed by (vacuum) zone passes and float zone
crystal pulling of the required quality for the final neutron
irradiation and annealing.

In Fig. 1 is shown the process flows for the conventionally
doped float zone and NTD silicon, respectively, for today's
industrial production. The increased flexibility for the NTD
production is evident. It is also noted that the "raw" material
in the sense of non-customer oriented material is pushed from the
silicon chlorides to undoped float zone silicon crystals.

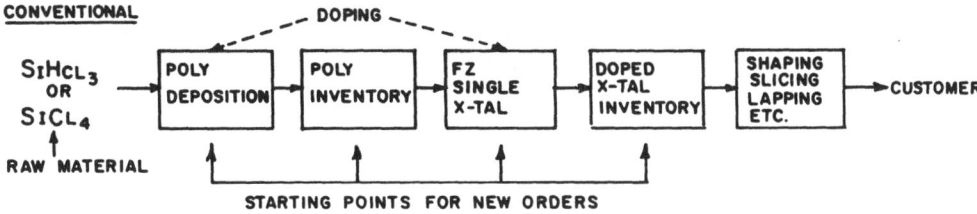

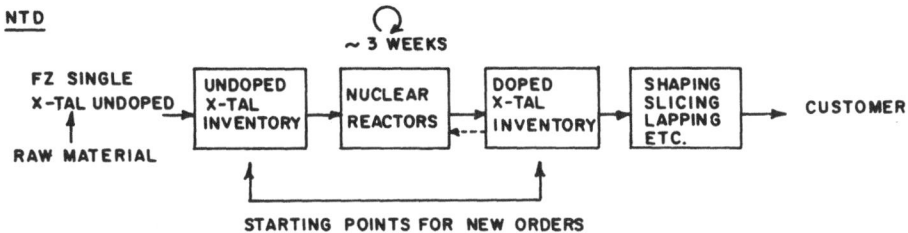

Fig. 1.   Process flows for manufacturing of float zone silicon
chemically (conventionally) doped and neutron transmutation
doped (NTD) respectively.

In cases of commonly used resistivity and diameter ranges,
doped silicon may be kept in a non-customer oriented inventory
for both production methods. For NTD crystals, however, the ingot
homogeneity yields a more precisely defined inventory and very
little resistivity determined fall-outs will occur for any process
steps.

The simplified crystal pulling without chemical doping and
the improved inventory control allows to a certain degree the
manufacturer of silicon to compensate for the cost of the neutron
irradiation itself.

3.   NTD SILICON PRICE EVOLUTION

For new products, the price will be given by 1) comparable
existing products, 2) cost/benefit for the user, and 3) strategic
reasoning. Hence, initial pricing may not necessarily reflect
cost whereas the product, as it becomes of commodity nature, usually
will be priced in relation to the costs involved.

Figure 2 shows an example of relative pricing for NTD silicon
used for the same device type from 1974 to 1978. The low initial

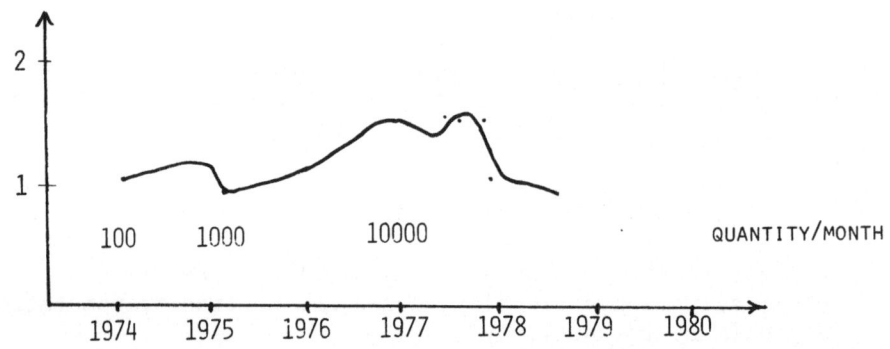

NTD SILICON PRICE EVOLUTION

1. FOR NEW PRODUCT PRICE IS SET BY:

    A) COMPARABLE EXISTING PRODUCT
    B) COST - BENEFIT FOR USER
    C) STRATEGIC REASONS

2. FOR COMMODITY PRODUCTS PRICE IS SET BY COSTS OF
   LEADING SUPPLIER.

Fig. 2. NTD silicon price evolution. Application for the same
device type. Tightening of specification occurs in 1976.

price was strategic in nature and equal to that of conventional
float zone silicon. The price increase from 1975-77 was caused
by tightening of specifications. After 1977 the price is seen
to follow the general learning curve. In a later section we shall
comment on the expected future cost trends.

4. NTD SILICON APPLICATION PATTERN IN THE POWER INDUSTRY

    As previously mentioned, the devices triggering the application
of the NTD method were the high power devices for high voltage
direct current transmission lines. The benefits of NTD silicon were
the possibility of making devices with larger power dissipation
per unit area and breakdown voltages up to 4-5 kV with reasonably
good yields.[5]   The homogenous resistivity caused a uniform
avalanche breakdown across the device and likewise the forward
voltage drop could be minimized by reducing the base width, as the
punch through condition also occurred evenly over the complete area.

    This development took place in 1973-74 and possibly earlier,
first in Europe and later in the United States and Japan.

It was evident that the NTD method was mandatory for the above mentioned devices. It was, however, surprising that the volume production of NTD silicon was caused by production control factors for manufacturing of medium power diodes and thyristors rather than by design considerations. Device houses in Europe and the United States in 1975 simultaneously discovered that the spectrum of devices in the same family but with different test parameters could be targeted with considerably higher precision using NTD silicon. In the case of multidice wafers, all devices from one wafer were produced with very similar specifications.

These facts allowed better inventory control and a production output profile resembling market demand. In a following section we shall further investigate these considerations.

One year later, in 1976-77, the manufacturers of triple diffusion power transistors for television in particular began to implement NTD silicon. Here, the reasoning again was a tight control of the final device parameters as a function of the input material. Further, it was realized that a redesign of the transistors, taking into account the resistivity homogeneity, could result in considerably more devices per unit area, the same conclusions made by the first high power device designers.

A number of other power devices previously considered as material-insensitive have in the last year been converted to NTD silicon. We may here mention stacked diodes (fewer junctions for the same voltage) for television applications and very lately, automotive rectifiers and ignition transistors for reasons of reliability and centering of final product parameters because fall-out specifications are of little commercial interest.

5. SILICON SPECIFICATION ENGINEERING WITH REGARDS TO FINAL DEVICE PARAMETERS

In the following we shall investigate more deeply the reasons for the narrow targeting possibilities in the case of NTD silicon usage. In Fig. 3 we demonstrate some typical material specifications as given by power device producers or designers. A strong tendency for more controlled resistivity distribution is clearly seen. The necessity of the inclusion of radial variation measured by four-point probe[(δ edge - δ center)/ δ center] and striation measured by spreading resistance probe [(R max - R min)/(R max + R min)] is shown in Fig. 4.

In Fig. 4(a) we notice the specification of, for example, 100 ohm-cm center resistivity (four-point probe) with plus/minus ten percent tolerance may result in areas of local resistivity up to 175 ohm-cm unless specification of striations and radial gradient

NTD-SILICON APPLICATION PATTERN

SPECIFICATION ENGINEERING (POWER, N-TYPE)

		CENTER RES.	RAD.VAR.	STRIATION
BEFORE 1970		25%	UNSPEC.	UNSPEC.
1970 - 74		15-25%	10-25%	UNSPEC.
1974 -	CONV.	10-25%	10-15%	25-50%
	NTD	3-25%	1-5 %	5-10%

NTD ALLOWS SPECIFICATION OF RESISTIVITY FOR ALL DEVICES IN A BATCH - MULTIDICE OR SINGLE WAFER - TO BE DEFINED AS NEEDED.

Fig. 3.   Evolution of typical resistivity specifications as given by power device manufacturers

SPECIFICATION ENGINEERING

RESISTIVITY MAXIMA FOR DIFFERENT FZ PRODUCTS (CENTER ± 10%)

100 OHM CM	INCL. CENTER TOL.	INCL. RAD. GRAD.	INCL. STRIATION
CONV. FZ	110	125	175
DOPING FACTOR 5	110	120	140
DOPING FACTOR 20	110	113	117

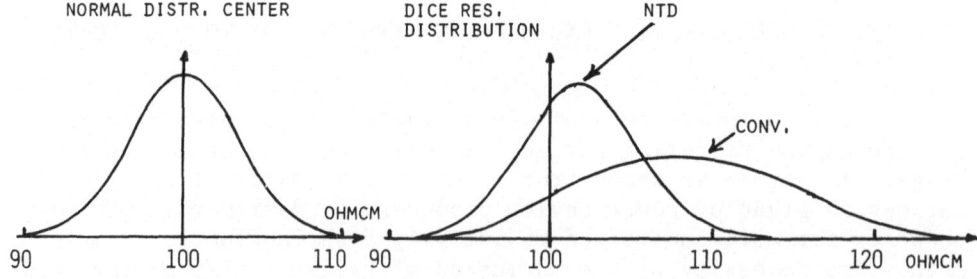

Fig. 4.   Specification engineering: (a) Resistivity maxima for typical float zone silicon with minimum resistivity in center. Radial gradient is measured by 4 point probe and striations by spreading resistance probe. (b) Distribution of die resistivity for normally distributed center resistivities and parabolic radial gradient of 15% (conventional float zone) and 3% (NTD) respectively.

is given. Before 1970 several major power device manufacturers requested center resistivity ranges only.

Figure 4(b) shows the individual die resistivity (four-point probe) distribution when the wafer center resistivities are normally distributed and the radial variation in the conventional float zone silicon is assumed parabolic with 15% variation. Obviously the individual die resistivity distribution is broadened by a factor of two because we find more dice per resistivity unit as we go out from the wafer center towards the edges. Furthermore, the average die resistivity is observed to shift approximately 8% up. The NTD silicon die resistivity distribution resembles the center resistivity distribution to a much greater extent.

The above reasoning explains the strong centering of final device parameters, e.g., the breakdown voltage, the saturation voltage and others, in the case of NTD silicon application.

It should be noted that a switch to NTD silicon can often change the average resistivity.

6.   YIELD IMPACT IN POWER DEVICE MANUFACTURING

The industrial concept of yield is often misleading in the sense that yield is defined as saleable product output compared to theoretical maximum (total yield). In the following we shall define yield as the number of products with given specifications compared to the above maximum (target yield).

In this sense the drastic yield impact of NTD silicon is understood even for traditionally high total yield (80-90%) devices like medium to low power thyristors and power transistors. Figure 5 shows typical distributions of breakdown voltages for the differently doped float zone materials. The increase of target yield, as shown, from 60% to 85% has been commonly observed and was the reason described in Section 4 for the early success of NTD silicon in the medium power device industry.

7.   NTD IMPACT ON THE FLOAT ZONE SILICON MARKET

Due to the yield impact in the device manufacturing, the silicon usage per device is decreasing. It is predicted that in terms of wafer area, the float zone silicon consumption for discrete power devices will be flat for the next five years in spite of an estimated growth of 15% per year for power devices.*

Twenty percent of this silicon market was converted to NTD in 1977 and it is expected that more than 50% will be converted by 1980.

YIELD IMPACT IN DEVICE MANUFACTURING

1)
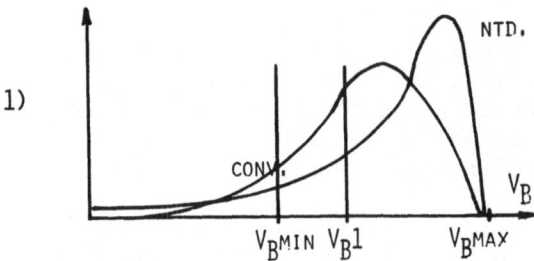

$$\text{TOTAL YIELD } V_B > V_B^{MIN} \sim 90\%$$

$$\text{YIELD CONV. } V_B > V_B^1 \sim 60\%$$

$$\text{YIELD NTD } V_B > V_B^1 \sim 85\%$$

2) OPTIMUM DESIGN   SMALLER AREA PER DEVICE
                        MORE DICE PER WAFER E.G.
                        POWER TRANSISTORS (2KV) + 20% DICE

3) SORTING OF WAFERS   TAILORING OF OUTPUT MIX.

Fig. 5.   Total yield and target yield. Breakdown voltage distrib-
          ution for conventional float zone and NTD silicon
          respectively.

Another market trend for power silicon is the transition to
larger diameters. This development has been pushed by the NTD
method because large diameter wafers of conventionally doped
material resulted in dice or pellets with minimum resistivities at
the edge rather than centered.

The self-absorption of neutrons in silicon results in less
than 5% inhomogeneity across a 100 mm wafer.

Finally, NTD silicon has minimized to a great extent inven-
tories of both virgin wafers and finished devices because of the
improved target yield and parameter centering. This development
is, of course, essential for the market and business stability.

                8.   COST TRENDS FOR NTD SILICON

The cost involved in the manufacturing of NTD silicon of
large diameter is more or less equally split between:

     a.   poly-silicon (capital and power intensive)

     b.   neutron irradiation (capital, nuclear fuel and skilled
           labor intensive)

     c.   crystal pulling

All these cost factors decrease with increasing diameter; marginally, however, for diameters greater than three inch. Unless a major change in technology is introduced, poly-silicon and neutron irradiation will have traditional inflationary trends. Reduction in crystal pulling will, to some extent, occur with the introduction of automated equipment.

In conclusion, NTD silicon costs and prices will be cut as a result of new wafering technologies only. The reduction in NTD silicon cost for the power device industry is to be expected mainly as a result of total and target yield increases.

REFERENCES

*   The total float zone silicon production may be growing con-
     siderably depending on applications in the IC industry. VLSI
     is a potential market for NTD silicon.
1)   K. Lark-Horovitz, "Nuclear-bombarded Semiconductors" in Semi-
     conductor Materials, Conf. Univ. Reading. London, Butter-
     worths, (1951), p. 47.
2)   M. Tanenbaum and A. D. Mills, J. Electr. Soc. 108, 171 (1961).
3)   J. Messier, Y. Le Coroller, and J. M. Flores, IEEE Trans.
     Nucl. Sci., 11, 276 (1964).
4)   C. M. Klahr and M. S. Cohen, Nucleonics 22, 62 (1964).
5)   M. S. Schnoller, IEEE. Trans. Electr. Devices ED-21, 313
     (1974).
6)   H. M. Janus and O. M. Malmros, IEEE Trans. Electr. Dev. ED-23,
     797 (1976).

THE ADVANTAGES OF NTD SILICON FOR HIGH POWER SEMICONDUCTOR DEVICES

Roger W. Phillips

National Electronics Division, Varian Associates, Inc.

Geneva, IL 60134

ABSTRACT

The float zone silicon used in manufacturing high power semi-conductor devices is known to have resistivity variations. These variations are large enough to create an unwanted distribution around given design values such as blocking voltage, turn-on characteristics, on-state voltage drop, reverse recovery, and turn-off time. Both domestic and foreign vendors now supply NTD (neutron transmutation doped) silicon. The significant advantage of this process of manufacturing n-type silicon is it provides a narrow resistivity range and low radial resistivity gradient. A more desirable distribution of electrical characteristics and a higher blocking voltage distribution results in devices made using NTD silicon. The advantages of NTD silicon on these device parameters will be discussed in detail.

It is important to understand that to manufacture a large area power semiconductor, the whole slice is used to make only one device. A typical power device structure is shown in Fig. 1. This is in contrast to integrated circuits in which a multitude of devices are made on one slice. A 50mm SCR or rectifier is made using one whole slice 50mm in diameter. This silicon slice must be virtually defect free and with uniform doping densities.

A difference in the doping level in only a single spot determines whether a device will attain the electrical parameters for which it was designed. The manufacturer of power semiconductors has control of the p- and n-type diffusions needed to make a device.

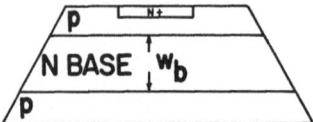

Fig. 1. Typical power device structure.

However, they rely upon the silicon manufacturer to provide them
with defect free, uniformly doped silicon.

Float zone n-type silicon has been the primary material used
for manufacturing power semiconductors. The float zone silicon
used in manufacturing high power devices is known to have resistivity
variations. These variations occur from slice to slice as well as
within an individual slice. These variations are large enough to
create an unwanted distribution around given design values such as
blocking voltage, turn-on characteristics, on-state voltage drop,
reverse recovery, and turn-off time of SCR's (Semiconductor Controlled Rectifiers). Both domestic and foreign vendors now supply
NTD (neutron transmutation doped) silicon. The significant advantage of this process of manufacturing n-type silicon is it provides
a narrow resistivity range and low radial resistivity gradient. A
more desirable distribution of electrical characteristics and a
higher blocking voltage distribution results for devices made using
NTD silicon.

I am sure the actual manufacturing process used for fabricating
neutron doped silicon crystal will be explained in the following
papers at this conference, and therefore, will not be dealt with
here. This paper will deal more with the actual manufacturing
advantages in using NTD silicon for producing power devices.

The uniform doping density of NTD silicon is contrasted to that
of float zone silicon as indicated in Fig. 2. The float zone process
typically yields silicon slices with four to ten times higher radial

FZ	NTD
≤ 20%	≤ 5% 4-POINT PROBE
≤ 50%	≤ 5% SR PROBE

Fig. 2. Doping uniformity obtainable with conventional float zone
         (FZ) and neutron transmutation doped (NTD) Si.

resistivity gradient than the NTD process. The silicon suppliers
will be glad to tell you that even lower radial resistivity gradients are available.

Many users of SCR's are now demanding higher voltage, lower on-
state voltage drop (V_{tm}), smaller reverse recovery charge (Q_{RR}),
faster turn-off time (tq), and, to facilitate paralleling devices,
uniform gate characteristics. NTD silicon is beneficial to all of
these parameters.

The blocking voltage is basically a function of the N-base
resistivity and the N-base width of the PNP structure of the SCR
(Fig. 3). The higher the ohm-cm the larger base width that is
required to support the voltage. In Fig. 4 a target of 4000 Volts

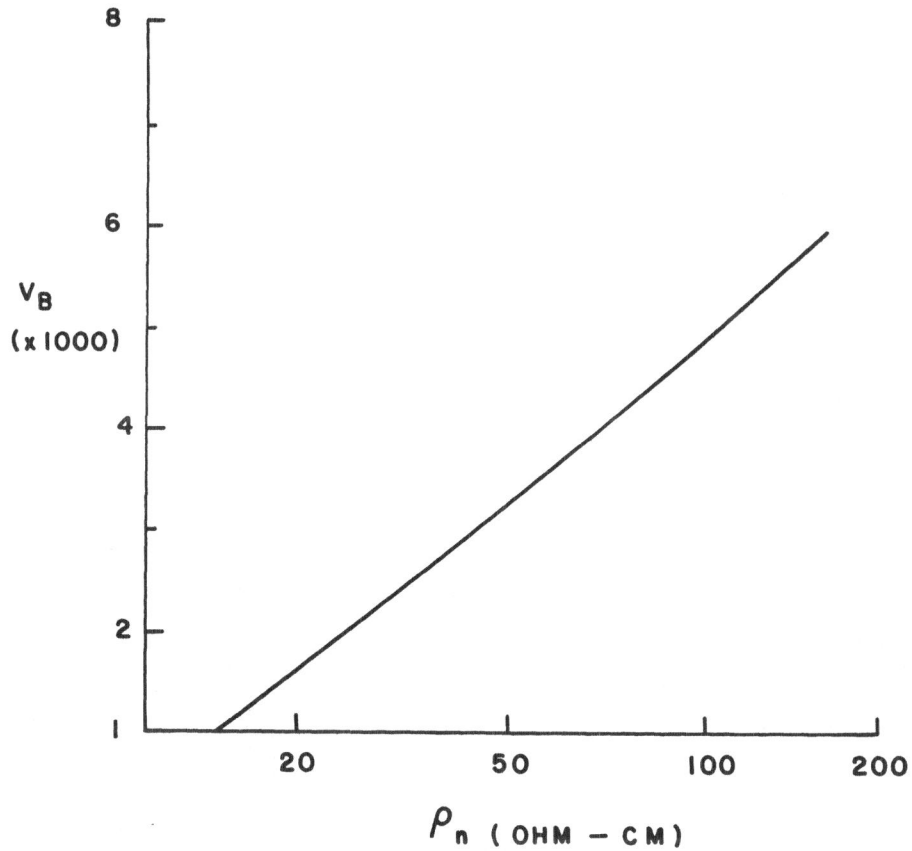

Fig. 3. Blocking voltage V_B of an SCR as a function of N-base
         resistivity.

±20% RESISTIVITY VARIATION

-20%	136 Ω CM	3400V
TARGET	170 Ω CM	4000V
+20%	204 Ω CM	4600V

±5% RESISTIVITY VARIATION

-5%	162 Ω CM	3900V
TARGET	170 Ω CM	4000V
+5%	178 Ω CM	4100V

Fig. 4.   Effect of resistivity variation on blocking voltage
          variation.

requires 170 ohm-cm silicon. If the resistivity variation in a
silicon slice is ±20%, as is possible with float zone material, a
voltage degradation to 3400 Volts is predicted to occur. The use
of ±5% resistivity variation silicon as provided by the neutron
transmutation method maintains the voltage obtainable to within 100
Volts. This increase of 300 Volts has been attained experimentally.

Since the N-region base width is proportional to the resistivity
of the silicon, the increase of 200 to 300 Volts that NTD silicon
provides over float zone can be used to design a given voltage
device using slightly lower ohm-cm and thus slightly thinner slices.
The uniform resistivity of NTD silicon allows the N-region base
width to be decreased from 50μm to 24μm. The thinner device that
results will have a lower on-state voltage drop while still maintaining the design blocking voltage. Fig. 5 shows the relationship
of V_{tm} versus W_b for a high voltage 50mm SCR at 1000 amps and 2000
amps. Data was taken using 170 ohm-cm material.

Devices used in series require the use of snubber circuitry
to dynamically balance the reverse recovery (Q_{RR}) of the SCR's.
Smaller components for the snubber circuitry can be used if the
reverse recovery of the SCR's is held within a narrow band. Within
a diffusion run the band of Q_{RR} has been held within ±10% using
NTD silicon.

Uniform doping density of the starting material reduces the
probability of hot spots occurring in a device. Turn-off time

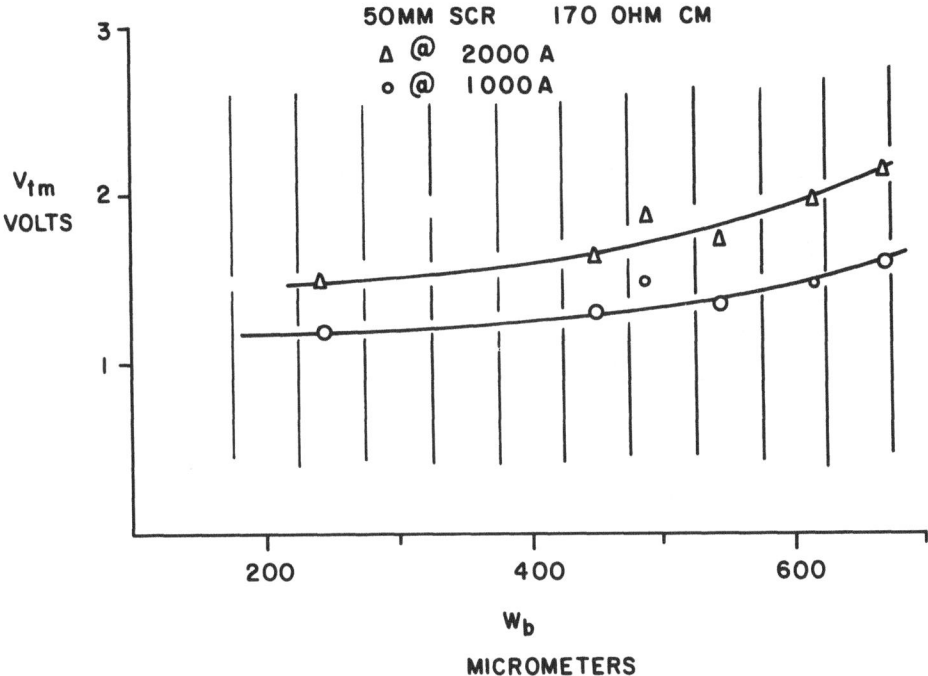

Fig. 5. On-state voltage (V_{tm}) vs. device base width (W_b) at
various currents.

benefits from this decreased probability. Again within a diffusion
run a narrower band of turn-off times are achieved with NTD silicon.

Turn-on characteristics are also maintained within a narrower
band by using NTD silicon. Users will find this useful in series
and parallel device applications.

The fact that NTD silicon has a more uniform doping density
throughout a wafer and from wafer to wafer leads to narrower yield
distributions around the design values desired for high power semiconductor devices, be they SCR's or rectifiers.

The manufacturing economics of using NTD silicon to replace
float zone type is very favorable. Since the most important factor
in the production of power devices is yield to design levels, the
small added cost for NTD silicon is more than offset by increased
yields.

REFERENCES

1)   K. Platzoder and K. Lock, IEEE Trans. Electron Devices, ED-23,
      No. 8, August 1976.
2)   M. J. Hill, P. M. Van Iseghem, and W. Zimmerman, IEEE Trans.
      Electron Devices, ED-23, No. 8, August 1976.
3)   A. Senes and G. Sifre, "Stabilization of Transmutation Doped
      Silicon," 1977 ECS Spring Meeting, Abstract No. 171.
4)   H. Herzer, "Neutron Transmutation Doping," 1977 ECS Spring
      Meeting, Abstract No. 168.
5)   S. K. Ghandhi, Semiconductor Power Devices (Wiley, New York,
      1977).
6)   R. Van Overstraeten and H. De Man, Solid State Electronics,
      Vol. 13, "Measurement of the Ionization Rates in Diffused
      Silicon p-n Junctions," pp. 583-608 (Pergamon Press, New York,
      1970).

NTD SILICON ON HIGH POWER DEVICES

C. K. Chu and J. E. Johnson

Westinghouse Electric Corporation

Youngwood, PA

ABSTRACT

High power rectifiers and thyristors are generally fabricated
with float zone silicon material. The resistivity variations of
this type of material produce non-uniform electrical characteristics
from device to device, especially in blocking voltage, forward voltage drop, and switching characteristics. Large area NTD (neutron
transmutation doped) silicon[1] has become available from different
suppliers since 1975. This type of silicon has a narrow range of
resistivity across the whole area. During the last few years, we
have evaluated and manufactured 50mm to 67mm diameter high resistivity NTD silicon for high power devices. Devices fabricated from
this type of material resulted in a higher voltage distribution and
a narrower distribution of electrical characteristics.

1. INTRODUCTION

High power silicon devices, especially those of large area and
high blocking voltage, need uniform high-resistivity, high lifetime
and defect-free silicon material. Conventional float zone silicon
material usually has high lifetime and low defects, but the resistivity variation can go as high as ± 25%. NTD silicon generally
meets the requirements of power semiconductor devices. The resistivity variation of NTD material can go as low as ± 4% which is
very desirable for power device manufacturers, however, the cost is
high and the availability of this type of material is low. NTD
material with a resistivity variation ± 10% and a minimum lifetime
of 200 μsec is readily available and meets the design and cost
requirements.

Since 1975, we have investigated NTD silicon for fabricating power devices. We started with 3000 Volt design, 50mm diameter phase control, high power thyristors. During 1976, we worked on 2000 Volt and 1400 Volt, 50mm diameter, fast speed devices. Lately, we have been working on 67mm high power thyristors and rectifiers with NTD material. Today, all of the above mentioned devices are manufactured in the production line with good results.[2]

2. EXPERIMENTAL PROCEDURES

During 1975, we experienced difficulties in manufacturing 3000 Volt, 50mm diameter, high power thyristors with reasonable yield. Therefore, we started to investigate NTD silicon for this product. We also provided a control group by using regular float zone silicon material. The major specifications of these silicon materials are shown in Table 1. These materials were processed in identical fashion. The entire fabricating procedure can be divided into six major process steps as outlined in Fig. 1. Typical values of the electrical characteristics of the finished devices are shown in Table 2.

The fast switch devices we were working toward were a 50mm diameter, 1200 Volt and a 2000 Volt with 40 and 60 μsec maximum turn-off time. The turn-off time is controlled by an electron irradiation and annealing process.[3,4] Typical values of electrical characteristics of these fast switch devices are shown in Table 3.

Table 1. Specifications of silicon materials.

	N-TYPE RESISTIVITY	RADIAL RESISTIVITY GRADIENT	MINIMUM LIFETIME (μSECS)
FLOAT-ZONED SILICON (CONTROL GROUP)	185 ± 25%	± 7.5%	300
SUPPLIER A (NTD SILICON)	150 ± 3%	± 3%	100
SUPPLIER B (NTD SILICON)	150 ± 3%	± 3%	300

FAST SPEED	PHASE CONTROL	Section Views
P—Type Diffusion	Same	
N—Type Diffusion	Same	
Anode and Cathode Metallization	Same	
Angle Bevel	Same *	
Electron Radiation and Annealing	No	
Junction Etching and Coating	Same	
Evaluation and Testing	Same	

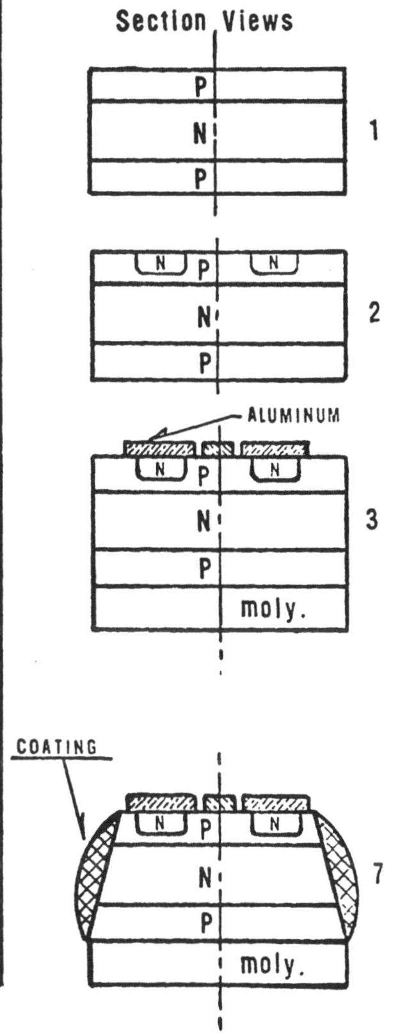

* same as View 7 without coating.

Fig. 1. Fabrication procedure for phase control and fast speed thyristors with NTD silicon.

Table 2. Electrical characteristics of 50mm diameter high power
          thyristors (typical values).

PARAMETERS	CONTROL GROUP	SUPPLIER A	SUPPLIER B
V_{FB} (V) @ 25°C < 10 MA	2870	2900	3025
V_{FB} (V) @ 125°C @ 50 MA	2285	2900	2925
V_{RB} (V) @ 25°C < 10 MA	3060	3450	3466
V_{RB} (V) @ 125°C @ 50 MA	2130	2850	3060
V_{TM} (V) @ 1500A @ 25°C	1.86	2.15	1.91
Q_{RR} (μC)	540	550	390

During the development of 67mm diameter high power thyristors
and rectifiers, only NTD material was used. The design voltages for
the thyristors and rectifiers are 1600 Volts and 2200 Volts. The
fabricating steps for the thyristors were similar to those used for
the 50mm diameter, 3000 Volt devices. Typical values of electrical
characteristics are shown in Table 4. The fabricating steps for
the rectifiers are shown in Fig. 2. The electrical characteristics
are shown in Table 5.

BLOCKING VOLTAGE

3.   DISCUSSION OF RESULTS

3.1 Blocking Voltage

NTD silicon with tight control of resistivity did give a higher
voltage distribution as shown in Table 2. Due to the uniformity of
resistivity, it is possible to use slightly lower resistivity starting material and still maintain a satisfactory high yield at the
design voltage.[5] The distribution of voltages about the design
center is also narrower.

Table 3. Electrical characteristics of 50mm diameter NTD high
power fast speed thyristor (typical value).

PARAMETERS	1200 VOLT DESIGN	2000 VOLT DESIGN
V_{FB} (V) @ 25 °C $\leq$ 10 MA	1300	2100
V_{FB} (V) @ 125° C $\leq$ 60 MA	1200	2000
V_{RB} (V) @ 25° C $\leq$ 10 MA	1300	2100
V_{RB} (V) @ 125° C $\leq$ 60 MA	1200	2000
V_{TM} (V) @ 1500A @ 25 °C	1.7	2.2
Q_{RR} (μC)	150 μC	190 μC
T_Q (μS)	35	45

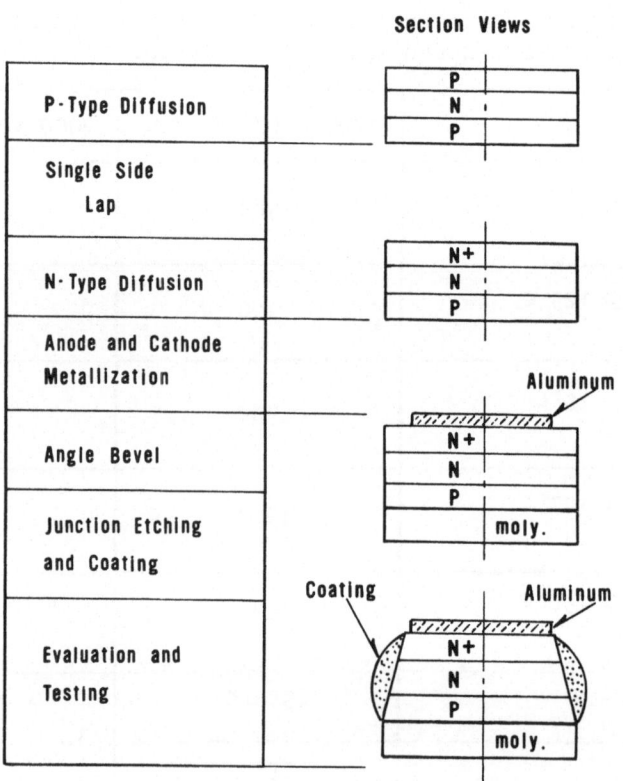

Fig. 2.   Fabrication procedure for power rectifiers.

Table 4. Electrical characteristics of 67mm diameter NTD high
          power thyristors (typical values).

PARAMETERS	1600 VOLT DESIGN	2200 VOLT DESIGN
V_{FB} (V) @ 25°C < 10 MA	1700	2400
V_{FB} (V) @ 125°C < 75 MA	1700	2000
V_{RB} (V) @ 25°C	1800	2500
V_{RB} (V) @ 125°C < 75 MA	1700	2000
V_{TM} (V) @ 3000A (25°C)	1.40	1.56

3.2 Forward Voltage Drop

The forward voltage drop for high power thyristors is a function
of N-base width (W_n). By using NTD silicon for fast switch thyristors or phase control rectifiers and thyristors, the N-base
width could be reduced to 50 µm. The blocking voltage will be
maintained but the forward voltage drop will be lowered. The relationship of forward voltage drop and N-base width for either phase
control or high speed 50mm diameter thyristors is shown in Fig. 3.

3.3 Reverse Recovery Charge

The reverse recovery charge is proportional to the turn-off
time and inversely proportional to the forward voltage drop as shown
in Fig. 4. Table 2 shows that devices made with NTD material have
lower Q_{rr} values than comparable float zone material, although the
forward voltage drop is about the same. This could be caused by
the lattice damage which is not completely annealed out during the
high temperature diffusion processing. We also found that the Q_{rr}
value is proportional to W_n as shown in Fig. 4.

Table 5.   Electrical characteristics of 67mm diameter high power
           rectifiers (typical values).

PARAMETERS	1600 VOLT DESIGN	2200 VOLT DESIGN
V_{RB} (V) @ 25° C < 10 MA	2200	2800
V_{RB} (V) @ 175° C < 75 MA	2200	2800
V_{TM} (V) @ 25° C 3000 A	1.0	1.15

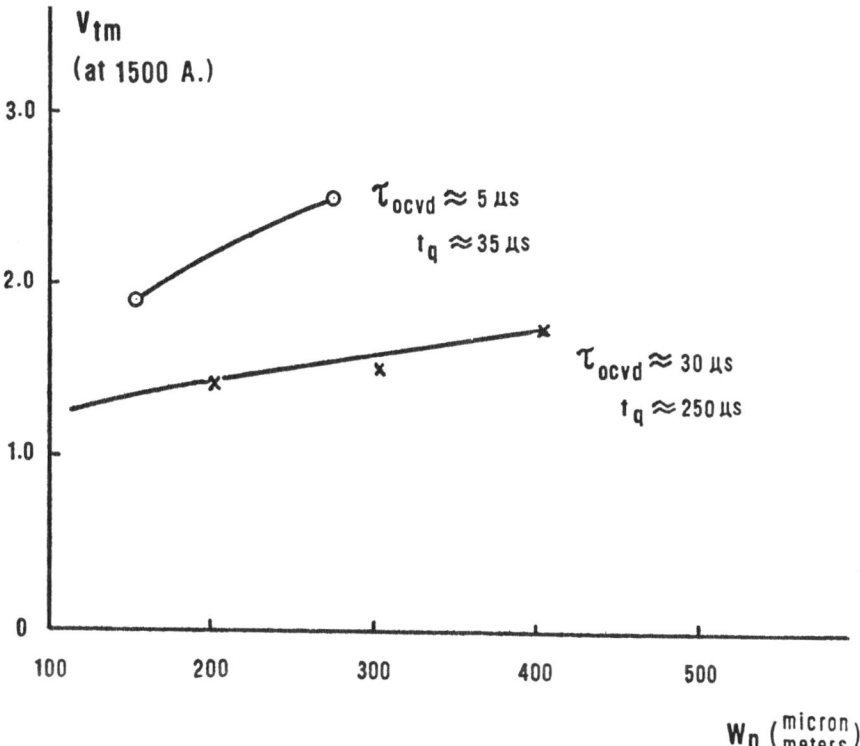

Fig. 3. V_{tm} vs. W_n for 50mm high power thyristors.

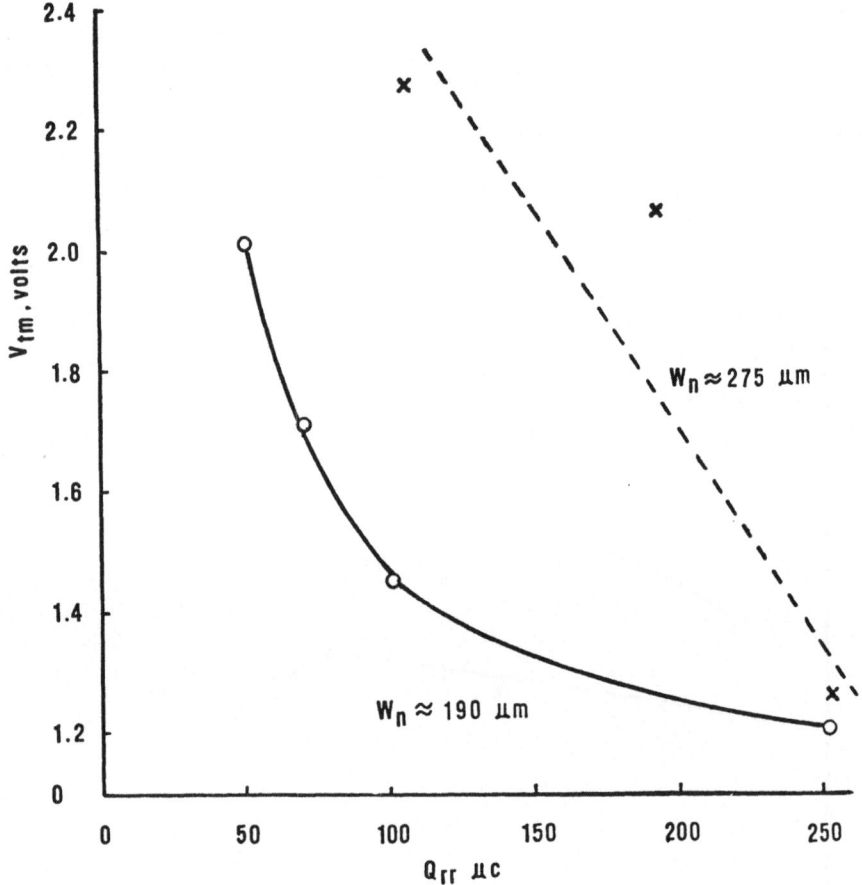

Fig. 4. V_{tm} vs. Q_{rr} for fast speed high power thyristor.

3.4 Turn-off Time

Table 2 shows that the average value of turn-off time of phase control devices made with NTD silicon and float zone material was identical. For fast switch devices made with NTD silicon, the turn-off time distributions were very narrow. In general, they were 35 µsec with 80% of the products in this region.

4. CONCLUSION

Devices fabricated with NTD silicon display a narrow distribution about a given design value. Higher power large area devices can be fabricated with NTD material with reasonable yield.

REFERENCES

1)   M. Tanenbum and A. D. Mills, J. Electrochem Soc., <u>108</u>, 171, (1961).

2)   C. K. Chu, J. E. Johnson, and W. H. Karstaedt, "The Impact of NTD Silicon on High Power Thyristors and Applications," IAS Paper, 27-D, 1977.

3)   C. K. Chu, J. Bartko, and P. E. Felice, "Electron Radiated Fast Switch Power Thyristor," 1975 Annual IAS Meeting Record, 75CH0999-31A, p 180.

4)   C. K. Chu and J. E. Donlon, "Annealing Effects on Electron Irradiated and Gold Diffused Thyristors for Fast Switch Application," 1976 IEEE-IAS Meeting Record, 76CH1122-1-1A, p 51.

5)   K. Platzoder and K. Loch, IEEE Trans. Electron Devices, <u>ED-23</u>, No. 8, August 1976.

ROLE OF NEUTRON TRANSMUTATION IN THE DEVELOPMENT OF HIGH SENSITIVITY EXTRINSIC SILICON IR DETECTOR MATERIAL*

H. M. Hobgood, T. T. Braggins, J. C. Swartz, and
R. N. Thomas

Westinghouse Research and Development Center

Pittsburgh, PA 15235

ABSTRACT

The monolithic CCD type sensors, which are currently under development for high performance infrared detection and imaging systems, utilize extrinsic indium or gallium doped silicon substrates. Considerations of detector operating temperature and sensitivity require that residual shallow acceptor impurities, such as boron, in these highly doped silicon substrates be compensated as closely as possible by donor impurities. Highly uniform and accurately known phosphorus concentrations can be introduced into the silicon lattice by neutron transmutation, making the technique attractive for producing precisely compensated, p-extrinsic silicon detector material of high infrared sensitivity.

We report here the successful use of neutron transmutation for compensating residual acceptors in highly indium doped silicon grown by both crucible pulling and float zoning techniques. As grown Czochralski crystals were found by low temperature Hall measurements to contain relatively high boron concentrations in the mid-10^{13} to $10^{14} \mathrm{cm}^{-3}$ range, whereas significantly lower concentrations (low $10^{12} \mathrm{cm}^{-3}$) could be maintained by float zone doping. The improved purity of these float zone Si:In crystals enables very low net compensation densities to be achieved by irradiation, since uncertainties associated with usual reactor transmutation practices become relatively unimportant. Transmuted phosphorus concentrations ranging from 10^{12} to 10^{14} cm^{-3} were investigated and compensation densities, $N_D - N_A$, as low as 2×10^{12} cm^{-3} were achieved in irradiated highly doped Si:In crystals after suitable damage annealing.

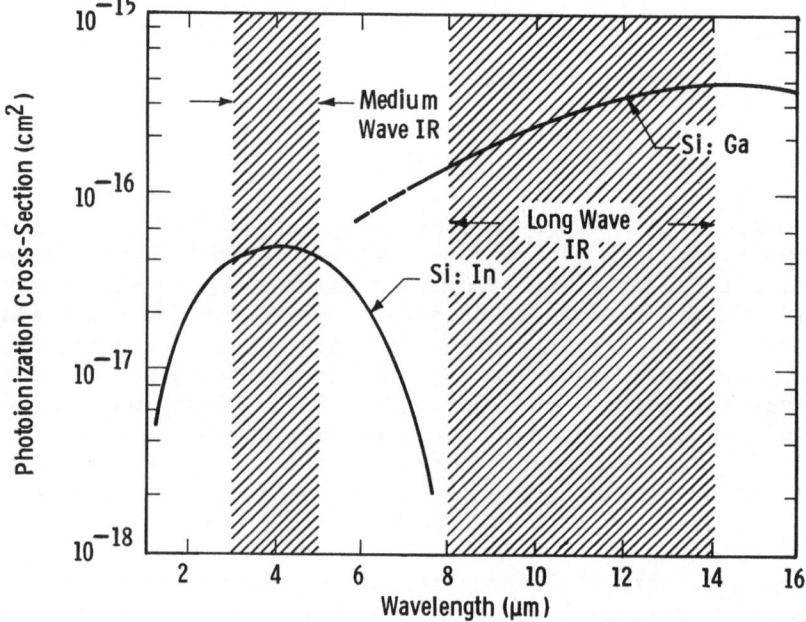

Fig. 1.   Photoionization cross section vs. wavelength for Si:In
and Si:Ga infrared detector material. Shaded areas
represent approximate atmospheric transmission bands.

Activity due to In transmutation products approaches negligibly small levels ($\sim 10^{-3}$ μC/gmSi) in these samples.

Significant improvements in infrared detector performance have been demonstrated with neutron compensated indium doped silicon. Peak responsivities (at 5.9 μm) up to 100 A/W at 50K and 10^3 V/cm detector bias have been measured, corresponding to high mobility-lifetime products in the $> 2 \times 10^{-3}$ cm^2/V range. These measured responsivities are considerably higher than those previously reported for extrinsic silicon detectors.

1. MATERIALS CONSIDERATIONS

Since 1970 neutron transmutation of silicon has been increasingly utilized for production of silicon with uniform, tight resistivity specifications for power device fabrication. Many studies have demonstrated the improved power device performance, particularly improved breakdown characteristics, which has resulted from the use of NTD silicon substrates.

Neutron transmutation of silicon has recently been applied to the preparation of infrared sensitive detectors of extrinsic silicon with improved sensitivity.[1-2] The importance of these extrinsic silicon devices lies in the fact that the high density detector arrays fabricated from the extrinsic substrates can be fully integrated with CCD-type signal processors formed on the surface into a single monolithic silicon chip.

The monolithic silicon CCD-type sensors which are currently being developed for high performance infrared detection and imaging systems, utilize indium doped silicon substrates for the medium wave IR band corresponding to the (3-5 μm) atmospheric transmission window, as shown in Fig. 1. For the long wavelength (8-14 μm) atmospheric transmission window, gallium-doped silicon substrates are used as the IR sensor.

In these applications, extrinsic substrate material of extremely high quality is required. Fig. 2 illustrates these materials requirements. For maximum optical carrier generation in the substrate and for efficient low-noise injection of the photocharge into the potential wells of the CCD registers, high responsivity is essential. High responsivity requires that the product of the quantum efficiency η, carrier mobility μ, and lifetime τ be maximized. In terms of material parameters in indium doped silicon, for example, this calls for material of the highest indium content consistent with maintaining high crystal perfection (to assure that mobility is not adversely affected).

The carrier lifetime dependence of extrinsic silicon detector material is most easily discussed by referring to the schematic

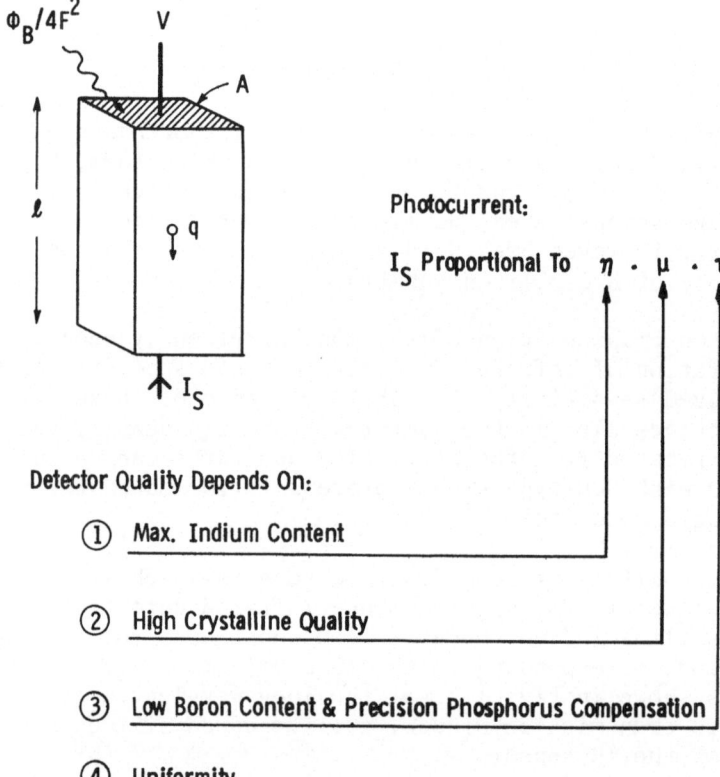

Fig. 2.   Materials requirements for high quality extrinsic Si:In
          detectors.

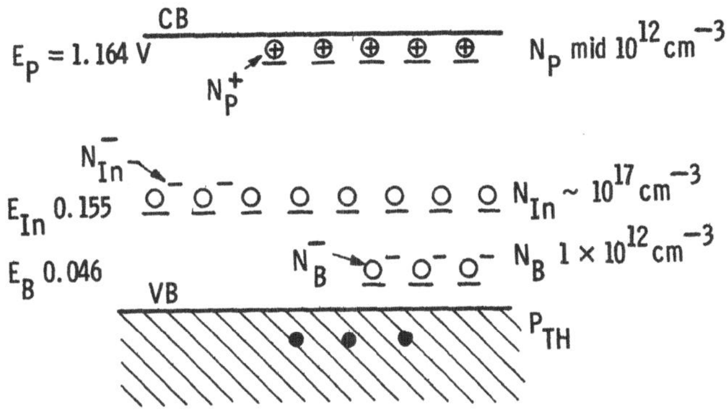

$E_P = 1.164$ V — CB — $\oplus$ $\oplus$ $\oplus$ $\oplus$ $\oplus$ N_P^+ N_P mid 10^{12} cm^{-3}

N_{In}^-

E_{In} 0.155 $\underline{O}^-$ $\underline{O}^-$ $\underline{O}$ $\underline{O}$ $\underline{O}$ $\underline{O}$ $\underline{O}$ $\underline{O}$ $N_{In} \sim 10^{17}$ cm^{-3}

E_B 0.046 $N_B^- \rightarrow \underline{O}^-$ $\underline{O}^-$ $\underline{O}^-$ N_B 1×10^{12} cm^{-3}

VB P_{TH}

Photoconductive Lifetime τ prop^n to 1/(No. of recombination centers)

For Si: Ga, $T \lesssim 30$ K

$$\tau \propto 1/\left(N_{Ga}^- + N_B^-\right) = 1/N_P$$

For Si: In, $T \gtrsim 40$ K

$$\tau \propto 1/N_{In}^- = 1/(N_P - N_B)$$

Fig. 3. Schematic energy diagram of Si:In containing boron and compensating phosphorus impurities.

energy band diagram of Fig. 3. It is widely recognized that the
shallow acceptors, such as boron, which are unavoidably present
as residual impurities in silicon, must be compensated as closely
as possible by shallow donor impurities, such as phosphorus, in
order to achieve detector material with high majority carrier lifetime. Undercompensation imposes serious penalties, however, since
thermal generation from shallow acceptors will result in excess
noise, requiring significantly lower focal plane temperatures for
background limited detection. On the other hand, excessive over-
compensation is equally deleterious, since the carrier lifetime
is degraded and results in poor detector responsivity. Precision
compensation of these residual acceptor impurities is, therefore,
a key factor in the production of high lifetime extrinsic silicon
detector material. In this respect, the neutron transmutation of
silicon offers an attractive technique for achieving the precise
compensation densities required in p-extrinsic silicon of high
responsivity.

2. MATERIALS GROWTH

The precision with which extrinsic silicon infrared detector
material can be compensated is dependent upon the residual
acceptor impurity concentration (e.g. boron and aluminum). The
residual acceptor concentration is primarily the result of the
growth method employed. Two types of growth techniques are generally used for the production of extrinsic silicon detector
material: 1) the Czochralski method, and 2) the float zone method.

The Czochralski growth of Si:In material is a well established
method in which large diameter <100> crystals can be grown with
high indium concentrations. The variable temperature Hall-effect
curve of Fig. 4 is typical for as-grown Czochralski Si:In material
pulled from a high purity quartz crucible. The high temperature end
of the curve is dominated by the freezeout of the indium whereas
the low temperature end displays the freeze-out of the residual
boron levels. The principal disadvantage and limiting factor of
the Czochralski method in the growth of extrinsic silicon detector
material is the introduction of boron by dissolution of the
crucible during growth. Boron concentrations in the 1×10^{14} cm^{-3}
range are typical for growths carried out in standard quartz
crucibles. As illustrated by the Hall-effect curves in Fig. 5 for
unintentionally doped Czochralski silicon, residual boron concentrations less than 10^{14} cm^{-3} (shaded area in Fig. 5) can be
obtained by utilizing improved purity synthetic quartz crucibles
and high grade as-deposited starting polysilicon charges; however,
boron concentrations less than 1×10^{13} cm^{-3} are difficult to
achieve and represent the current limits of purity for Czochralski
growth of silicon. In addition, with conventional in situ compensation normally used in Czochralski growth of extrinsic silicon

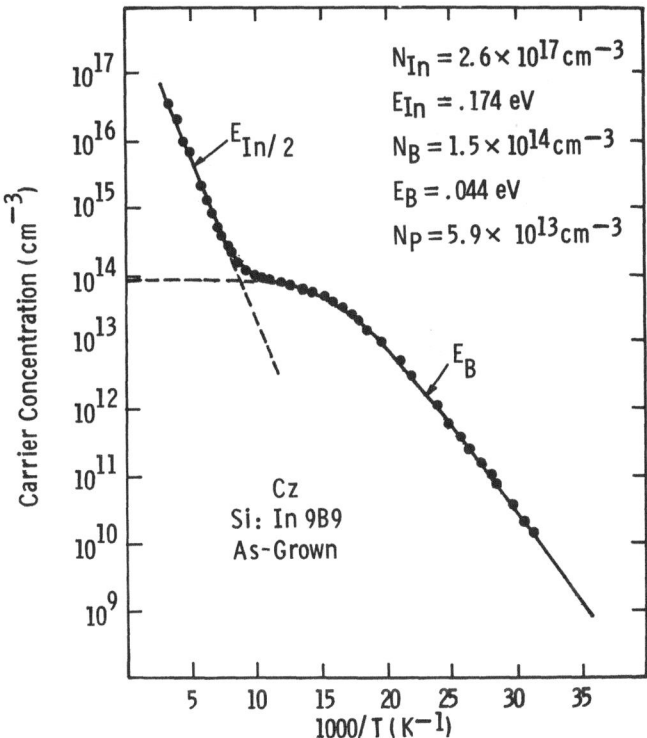

Fig. 4. Carrier concentration versus reciprocal temperature for uncompensated as grown Si:In 9B9.

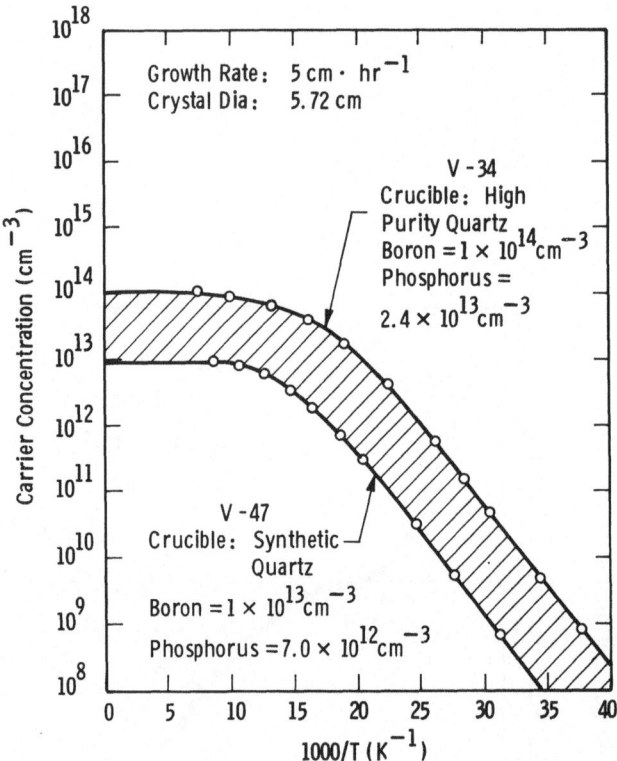

Fig. 5. Variable temperature Hall effect data giving residual
         impurity concentrations for undoped growth from quartz
         and synthetic quartz crucibles.

detector material, close compensation (i.e. mid 10^{12} cm^{-3}) is
difficult to attain, and often the compensation is close only over
a relatively small length of crystal, owing to the axial variation
of the residual impurities. The net compensation actually realized
varies from growth to growth and is dependent upon the purity of
the crucible and the amount of phosphorus lost to evaporation
during melt-down and growth.

    In contrast to crucible grown silicon, extremely high purities
can be routinely achieved in undoped silicon crystals grown by the
float zone technique. Silicon with resistivities up to 20,000 Ω-cm
p-type can be obtained corresponding to net acceptor concentrations
of less than 1.5 x 10^{12} cm^{-3}. However, the application of this
technology to the preparation of extrinsic indium doped silicon by

float zone entails some special problems. In particular, high
indium concentrations must be incorporated uniformly throughout
the crystal, while at the same time preserving the low 10^{12} cm^{-3}
background boron concentrations characteristic of this growth
method.

The growth of indium doped float zone silicon is modeled after
the "starting charge only" method, first developed by Pfann,[3] and
involves a single pass zoning operation in which an elemental
indium charge is uniformly distributed throughout a high purity,
multi-vacuum passed, silicon rod. The low distribution coefficient
of indium in silicon (1 x 10^{-3}) would normally insure an axially
uniform indium distribution. However, indium has a relatively
high vapor pressure (7 Torr) at the melting point of silicon which
results in a signficant loss of indium from the molten zone during
crystal growth. This evaporation loss can lead to variations in
the axial indium distribution of as much as a factor of 10 over
the crystal length. To minimize vaporization, the Si:In growths
are carried out in an argon atmosphere at a pressure of 10^3 Torr.
The axial indium distribution can be made uniform by maximizing the
growth speed in order to further reduce the evaporation losses by
minimizing the total evaporation time. Figure 6 shows the axial
indium distributions corresponding to three growth rates demonstrating that the highest zoning speed enables indium concentrations up
to 2 x 10^{17} cm^{-3} with an axial uniformity of ± 15% over crystal
lengths of 8 cm to be attained.

In spite of the very high purities potentially achievable in
multi-passed float zone silicon, our earlier float zone growths of
indium doped crystals showed that significant contamination was
introduced into the material during the indium doping pass. The
effects of such impurity contamination are illustrated in Fig. 7
where Hall effect analyses of two samples cut several centimeters
apart from one such crystal are shown. Both boron and aluminum
impurities were found to be present at concentrations in the mid
to high 10^{13} cm^{-3}. Thus, the residual shallow acceptor impurity
content of these early float zone crystals was not much better
than that obtained in crucible grown material where dissolution of
the quartz crucible in molten silicon is known to contribute
significant amounts of boron and other impurities.

Contamination introduced by improper handling and cleaning
procedures was suspected as being the main source of the residual
shallow acceptor impurities in our early indium doped float zone
crystal growths. In view of this problem, considerable efforts
were made to develop ultraclean handling, seeding, and doping
techniques. The axial indium uniformity and residual impurity
content of a more recent crystal is shown in Fig. 8. A series of
"dummy doping" experiments were performed using scrupulously

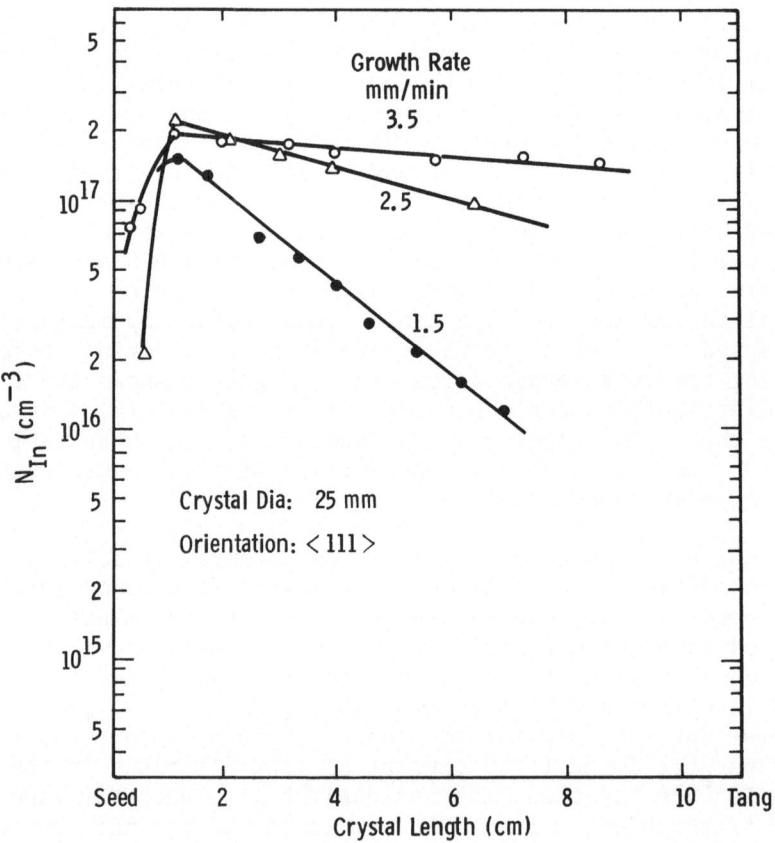

Fig. 6.   Effect of growth rate on axial uniformity of indium concentration in float zone silicon crystals. (Pressure
1 atm argon.)

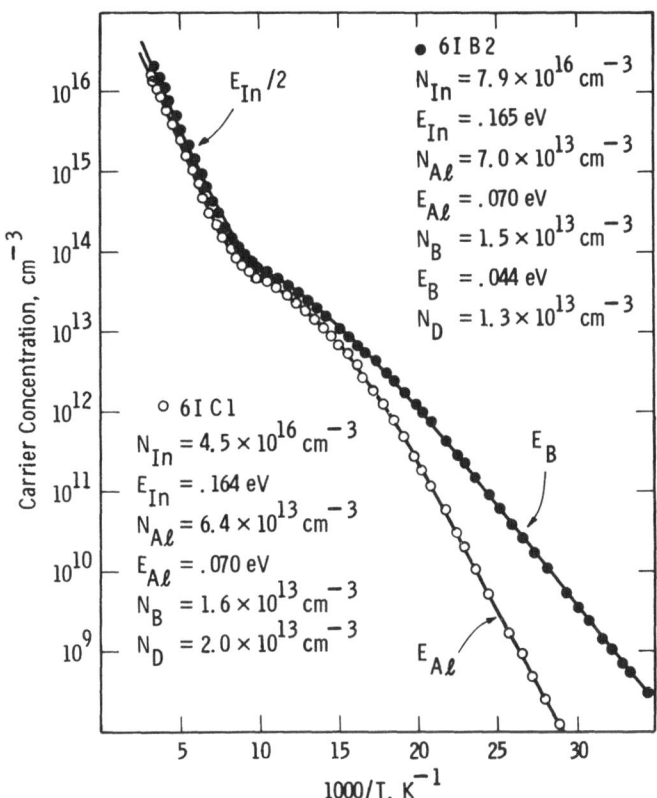

Fig. 7. Carrier concentration versus reciprocal temperature for as-grown float zone samples 6InB2 and 6 InCl. Solid lines are computed from theoretical expressions and fit to experimental data shown by coded points.

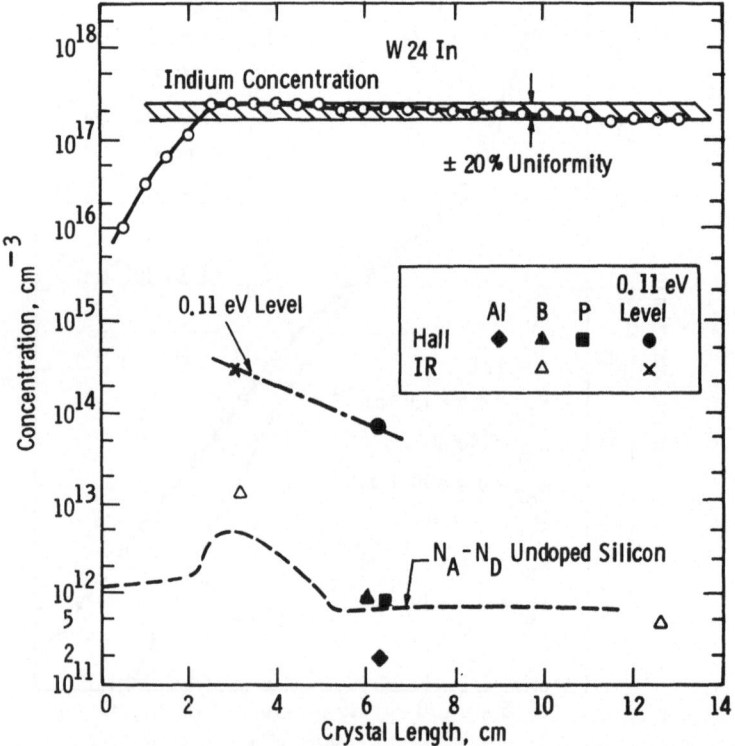

Fig. 8.   Distribution of indium and background impurities in a
          float zone indium doped crystal as determined from
          electrical resistivity, IR spectroscopy and Hall effect
          data.

clean procedures in which the crystal growth and handling procedures of the high purity vacuum passed starting material were
identical with indium doped growth runs except that the insertion
of the indium charge was omitted. It was observed that some
introduction of impurities does indeed occur at the point where
the high purity rod is sectioned prior to insertion of the indium
(see dashed line in Fig. 8); however, these impurities are ex-
hausted within about 2 cm of growth and the background impurity
level characteristic of the high purity vacuum passed starting
rod (low 10^{12} cm^{-3}) is maintained with further growth. These
results confirm the need for absolute cleanliness for high purity
float zone doping.

A large number of <111>, 25 mm diameter crystals up to 25 cm
in length, with indium concentrations in the (1 to 3) x 10^{17} cm^{-3}
range and with shallow acceptor concentrations below 5 x 10^{12} cm^{-3}
have been grown to date. The exceptional purity of these highly
doped Si:In crystals represents at least an order of magnitude
improvement over present day crucible grown extrinsic silicon
detector material and makes them ideally suited for precise compensation by neutron transmutation.

3. NEUTRON TRANSMUTATION

The neutron irradiations were carried out at the University of
Michigan Research Reactor in cooperation with Dow-Corning Corporation. Details concerning the reactor and irradiation
techniques have been described previously.[1-2] Of considerable
significance for precision compensation of extrinsic silicon
detector material are the 1) radial and axial uniformity of the
transmuted phosphorus concentrations, and 2) the accuracy with
which a desired target concentration can be introduced into a
silicon crystal. Our earlier studies[1] indicate that transmuted
phosphorus uniformities of ± 5% axially (over 6 to 12 inch crystal
lengths) and ± 2% radially (over 2 inch diameter crystals) can be
attained with the present reactor facility. The measured resistivity uniformity of neutron transmuted silicon will depend on the
ratio of the transmuted phosphorus to the net impurity concentration present in the sample initially. Four-point probe measurements
were performed to assess the radial slice uniformity and are shown
in Fig. 9. An improvement in uniformity in going from the high
resistivity to the lower resistivity material is evident. For
precise compensation where $N_P/N_B \rightarrow 1$, radial uniformities of ± 5
to 10% have been observed.

The transmutation process is capable of introducing quite
accurately known phosphorus concentrations. In our experiments the
relationship between the targeted phosphorus concentration vs. the
measured phosphorus concentration introduced by the transmutation

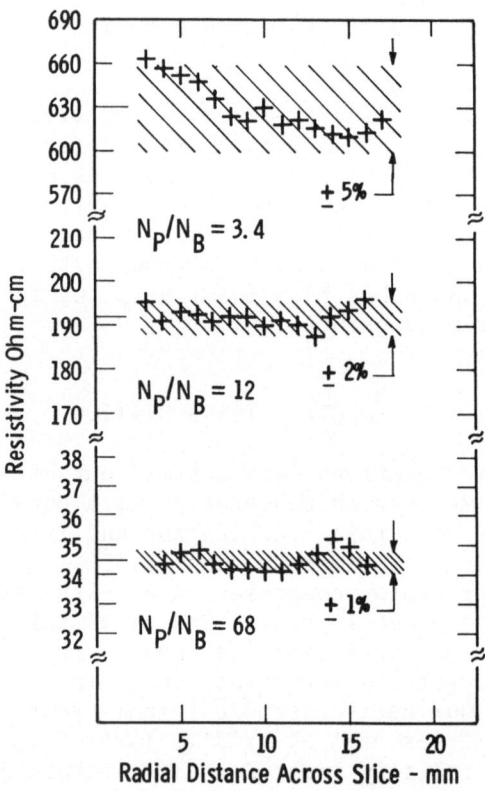

Fig. 9. Four point probe resistivity profile across float zone silicon slices after nuclear transmutation doping to different phosphorus levels. Data taken at 1 mm intervals.

was determined from resistivity measurements on high-purity control samples irradiated simultaneously with the Si:In crystals. The resistivity measurements were performed after the samples had been annealed under conditions which yielded stable resistivity values. As illustrated in Fig. 10, our studies indicate that for high purity float zone silicon starting material, relatively constant resistivities corresponding to stabilized net donor concentrations are obtained following isothermal anneals of up to 4 hrs. or longer at 850°C. Figure 11 shows that the measured concentrations of transmuted phosphorus are within ± 20% of the targeted value. This accuracy has enabled very low net donor densities (low^{12} cm^{-3} range) to be achieved in indium doped silicon of very high purity in our present studies. Further improvements may be possible, however, through careful calibration and monitoring of the neutron fluence during irradiation.

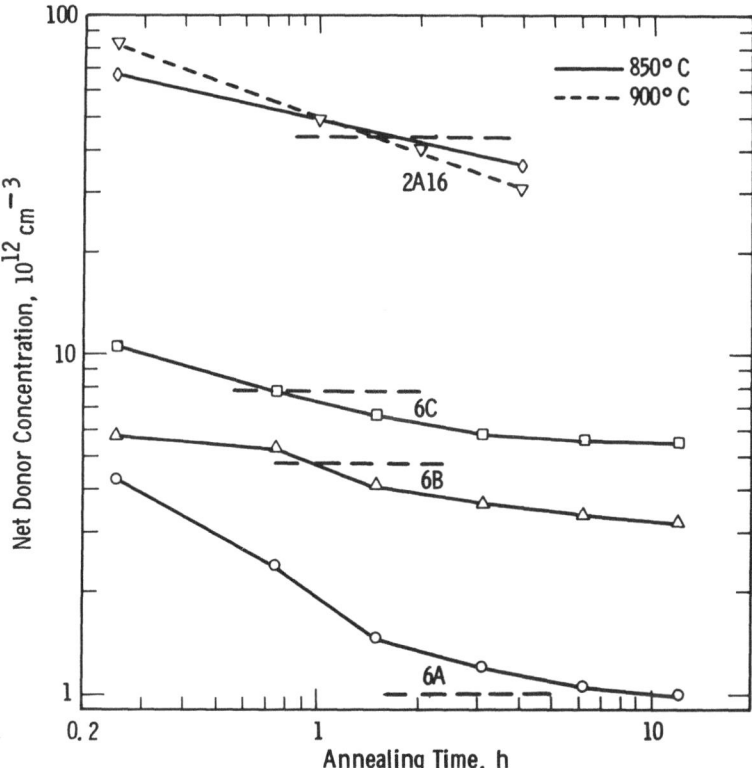

Fig. 10. Isothermal annealing of neutron irradiated high resis-
          tivity float zone silicon. The horizontal dashed line
          for each sample gives the expected post-irradiation
          donor concentration.

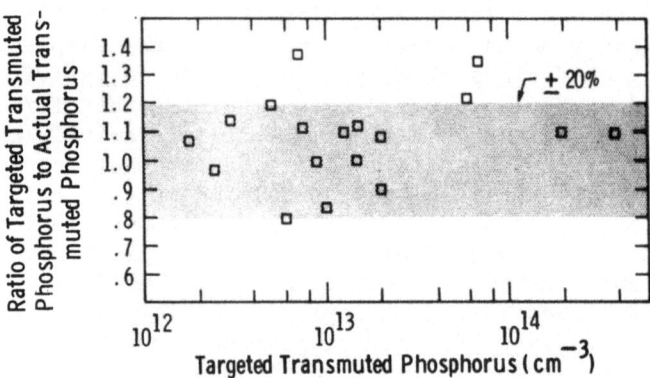

Fig. 11.   Accuracy with which a targeted phosphorus concentration
           can be introduced by neutron transmutation doping.

        An additional requirement of neutron transmuted Si:In
detector material having high indium concentrations is that it is
highly desirable to have very low radioactivity levels associated
with any transmutation products of the major dopant indium.
Examination of the probable thermal neutron reactions with indium
indicates the following reaction to be of primary interest owing
to the long half-life (50 days) of the intermediate ^{114m}In
isotope:

$$_{49}In^{113} + n \rightarrow {}_{49}In^{114m} \rightarrow {}_{49}In^{114} + \gamma \rightarrow {}_{50}Sn^{114} + \beta^-$$

   (8 barns)   (50 days)   (72 sec.)

        The residual activity measured in our Si:In crystals as a
function of various concentrations of transmuted phosphorus is
shown in Fig. 12.   For irradiations typically required to achieve
close compensation of float zone Si:In detector material, the
residual activity is in the low 10^{-3} µC/gm Si range and would
enable many kilograms of material to be handled safely, without
resorting to special handling or safety precautions beyond
normal safe laboratory practices.

        It is also to be noted that all of the transmutation reactions
involving the major dopant indium yield the neutral impurity, tin,
as the final transmutation product, therefore, giving rise to no
electrically active impurities.   However, a consequence of the
long half-life ^{114m}In isotope is that ionizing radiation continues
to be emitted which, through absorption in the silicon, results in
the production of electron-hole pairs which can give rise to

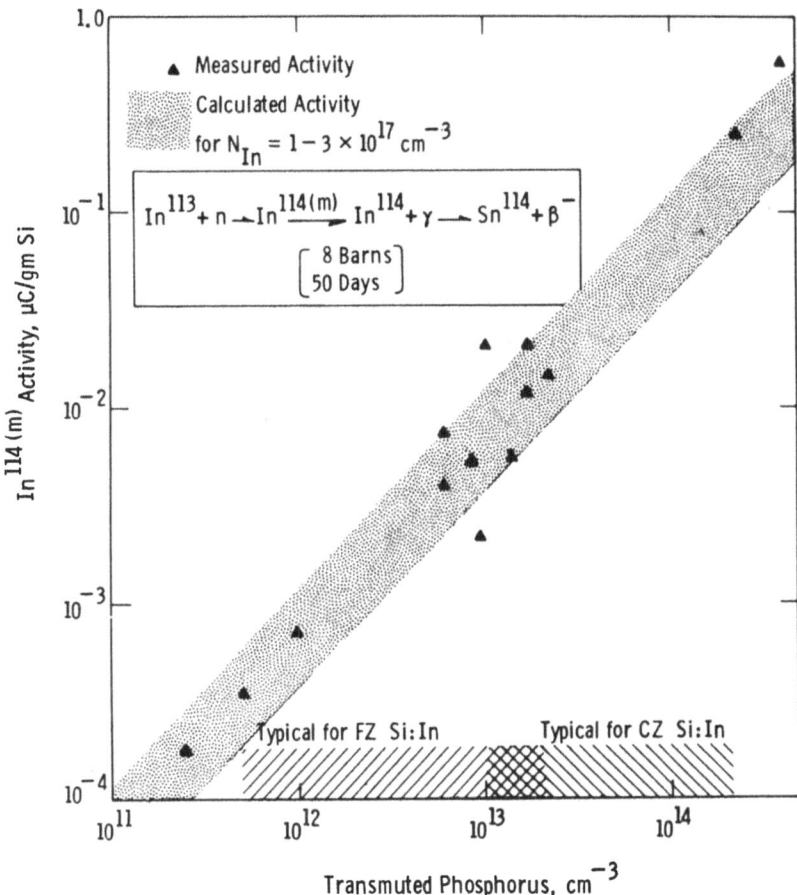

Fig. 12. Measured and calculated radioactivity from ^{114m}In as a function of phosphorus created by neutron transmutation of Si:In.

undesirable effects in infrared detector arrays. A worst case estimate for this problem shows that the radioactivity associated with an irradiation of Si:In material containing 2×10^{17} cm^{-3} indium such that a transmuted phosphorus concentration of 5×10^{12} cm^{-3} is achieved, gives rise through ionizing radiation (γ and β^{-}) to an electron-hole pair generation rate of less than 1.7×10^{8} cm^{-3} sec^{-1}. For a typical CCD imager, the equivalent generation rate would be produced by a photon flux of 1.7×10^{7} photons/cm^{2} sec. Thus, we conclude from this worst case estimate, that the

influence of the residual ^{114m}In activity is negligible relative
to even the lowest photon background fluxes of practical interest
for CCD sensors.

4.   EVALUATION OF TRANSMUTATION COMPENSATED Si:In DETECTORS

Simple photoconductive test structures of the type where the
electric field is applied transversely to the optical-absorption
direction have been fabricated from neutron compensated samples of
both Czochralski and float zone Si:In detector material.   It has
been shown previously[2] that the typical relative spectral response
measured for neutron compensated Czochralski Si:In detectors is
similar to that observed with conventionally compensated material
and is in good agreement with theoretical Lucovsky model for deep
impurities [4-5] (see Fig. 13).   Peak responsivities in the 1 to 5
A/W range at 60K have been observed in both neutron compensated
and conventionally compensated Czochralski Si:In detectors.   From
the measured responsivity R and noise current i_n as a function of
temperature, the detectivity D* can be calculated from

$$D^* = R \; A^{\frac{1}{2}}/i_n \tag{1}$$

where A is the optical area of the detector.   The experimental
variation of D* with detector temperature for neutron-compensated
Czochralski Si:In material is shown in Fig. 14.   A close fit between experiment and theory (solid line in Fig. 14) is observed
and illustrates that background limited detector operation is
achieved at temperatures only sightly below that predicted
theoretically.

The evaluation of neutron compensated Si:In has recently been
extended to detectors fabricated from closely compensated float
zone detector material.   Figure 15 shows a typical Hall-effect
curve for high purity float zone Si:In material in which a net
compensation of 2.1×10^{12} cm^{-3} was achieved by neutron transmutation doping.   The curve with the open circles corresponds to
the as-grown material and exhibits a low temperature tail characteristic of the uncompensated shallow acceptor impurities (boron
and aluminum) which are present in this sample at a concentration
of 1×10^{12} cm^{-3}.   The closed circles show data taken after neutron
irradiation demonstrating that the shallow acceptors present in the
as-grown Si:In have been closely compensated to a measured net
compensation density of 2×10^{12} cm^{-3}.   The low net donor density
achieved in this material is a direct consequence of the high
purity (low residual acceptor concentration) Si:In starting material
prepared by the float zone technique.

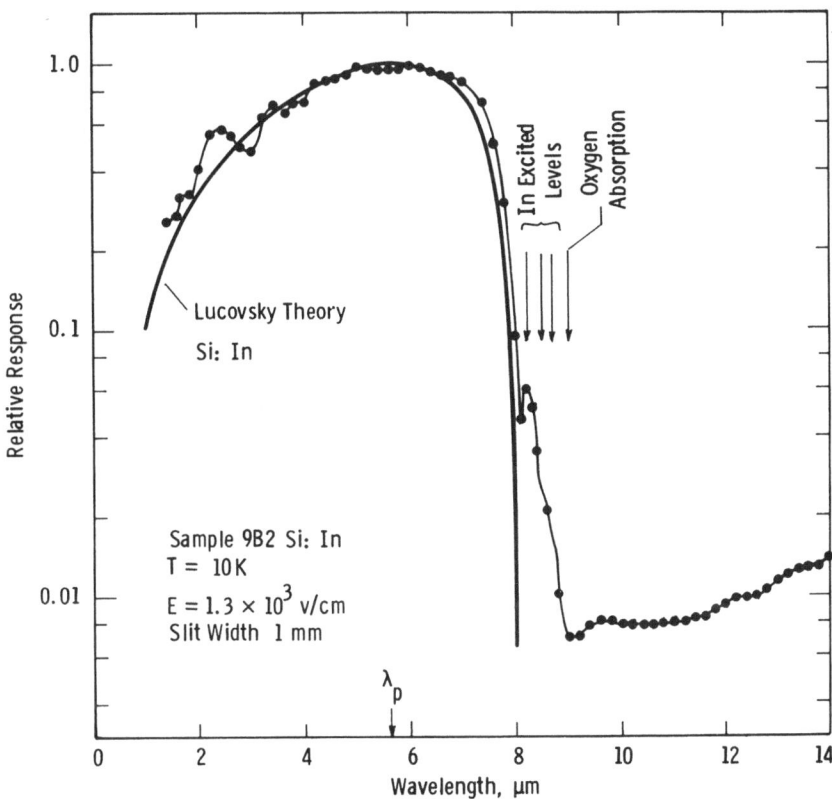

Fig. 13. Experimental and theoretical relative spectral response for neutron compensated Czochralski Si:In sample with $N_{In} = 2.5 \times 10^{17}$ cm^{-3}, $N_P - N_B = 1.6 \times 10^{14}$ cm^{-3}, and $N_B = 1.3 \times 10^{14}$ cm^{-3}.

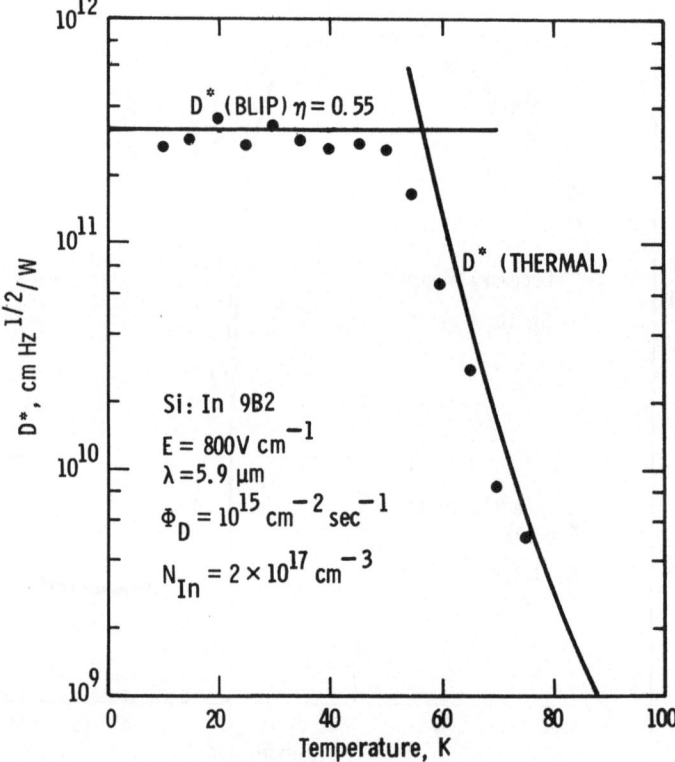

Fig. 14. Measured and calculated detectivity D* as function of
          temperature for neutron compensated Si:In.

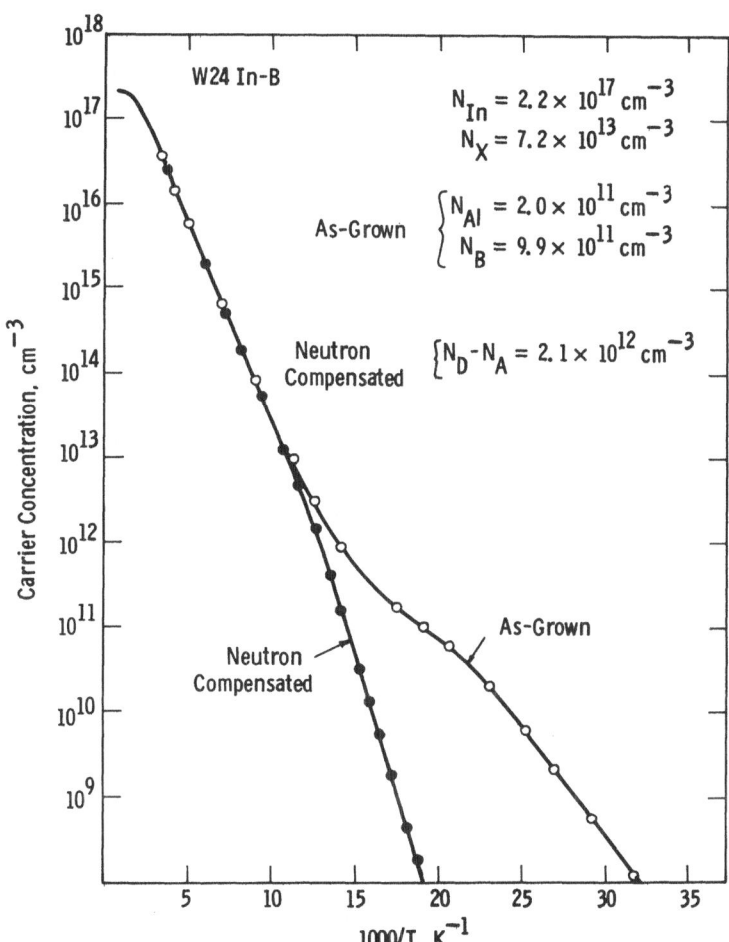

Fig. 15. Carrier concentration vs. reciprocal temperature for
sample W24 In-B. Shallow acceptors present in as-grown
Si:In have been compensated by neutron transmutation.

The temperature dependence of the peak responsivity for three float zone Si:In samples neutron irradiated to net compensation densities of 1.2×10^{13} cm^{-3}, 5.5×10^{12} cm^{-3}, and 2.1×10^{12} cm^{-3} is shown in Fig. 16. For sample 13 D ($N_D - N_A = 5.5 \times 10^{12}$ cm^{-3}), responsivities of 39 A/W and 6.5 A/W have been measured at a bias field of 980 V/cm at temperatures of 50K and 20K, respectively. Also for the same temperatures, responsivities of 106 A/W and 9.1 A/W have been measured for sample 24 B ($N_D - N_A = 2.1 \times 10^{12}$ cm^{-3}, at a bias field of 870 V/cm. These exceptionally high responsivities can be directly attributed to the long photocarrier lifetimes in the extrinsic silicon material--a consequence of the very low net compensation densities which were achieved by transmutation.

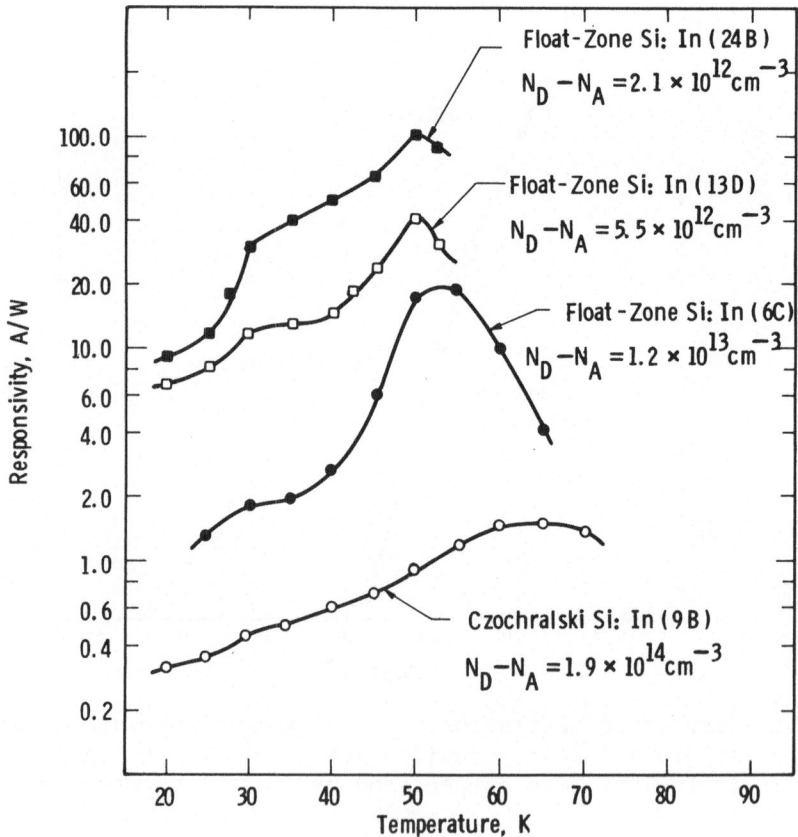

Fig. 16.   Responsivity vs. temperature for four samples with
           different net compensation densities. Note increase in
           responsivity for decreasing compensation.

In contrast, the donor densities which have been achieved in crucible grown silicon are at least an order of magnitude higher and the measured responsivities are typically below 5 A/W.

The peak responsivity is given by

$$R = \frac{q\lambda}{hc} \; G_o \eta \tag{2}$$

with

$$G_o = \frac{\mu\tau E}{d} \tag{3}$$

where λ is the wave length, μ is the mobility, τ is the lifetime, E is the electric field, η is the quantum efficiency, and d is the detector sample thickness. The photocarrier lifetime can be estimated from the photoconductive gain G_o as determined from the responsivity measurements. For sample 24 B [$N_{In} = 2.2 \times 10^{17}$ cm^{-3}, $\eta = 0.64$, $d = 0.058$ cm, $\mu = 8.8 \times 10^3$ cm^2/V.sec, $E = 870$ V/cm] τ is calculated to be 263 nsec at 50K. At this temperature, the photoconductive lifetime is determined principally by recombination at the ionized indium levels (see Fig. 3), since photo-excited holes captured on the shallow boron levels are immediately re-emitted into the valance band by thermal excitation. The carrier lifetime may therefore be written as

$$\tau = \frac{1}{BN^-_{In}} = 1/B(N_D - N_A) \tag{4}$$

where N^-_{In} is the total ionized indium concentration which is equal to the net compensation density, $N_D - N_A$, and B is the recombination coefficient. Taking $B \simeq 2 \times 10^{-6}$ cm^3/sec at 50K and $N^-_{In} = N_D - N_A = 2 \times 10^{12}$ cm^{-3} for sample 24 B, the lifetime is calculated to be 240 nsec, which is in reasonable agreement with the value determined from the measured responsivity. The inverse dependence of carrier lifetime on the net donor density in indium doped silicon detectors at 50K and $E \simeq 1000$ V/cm is demonstrated in Fig. 17. All of the measured lifetimes (determined from responsivity measurements) agree well with the calculated lifetimes (as determined from Eq. (4)) represented by the shaded area. These results demonstrate conclusively that high lifetime extrinsic silicon can be obtained by proper control of the net compensation density. In addition, these results confirm the absence of other additional recombination or trapping levels in precision neutron transmuted indium doped silicon. In particular, these data suggest that the capture cross section of the 0.11 eV defect level (X-level[2-7]) is similar in magnitude to the indium level.

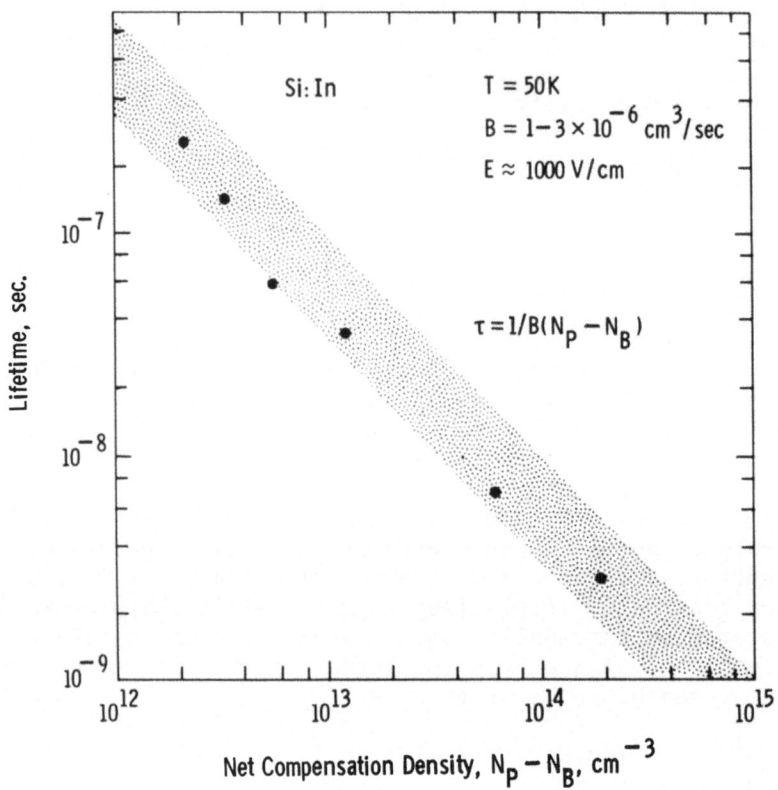

Fig. 17.   Measured lifetime at 50K as a function of net donor
           density compared with calculated lifetime for a re-
           combination coefficient of 1 - 3 x $10^{-6} cm^3$/sec.

     Owing to the highly uniform phosphorus concentration in
neutron compensated Si:In material, improved uniformity of response
can be expected. Figure 18 shows the signal current from discrete
detector elements in 10 element line arrays fabricated on a wafer
of float zone neutron compensated Si:In ($N_P/N_B \simeq 2$). Seven arrays,
spanning the diameter of the wafer, were measured and the photo-
current at 50K was found to be uniform to within ± 10% across the
wafer. Within a given array, the uniformity was considerably
better. If the outermost array near the left-hand edge of the
wafer is excluded, the average photocurrent uniformity of the
individual arrays was 5%.

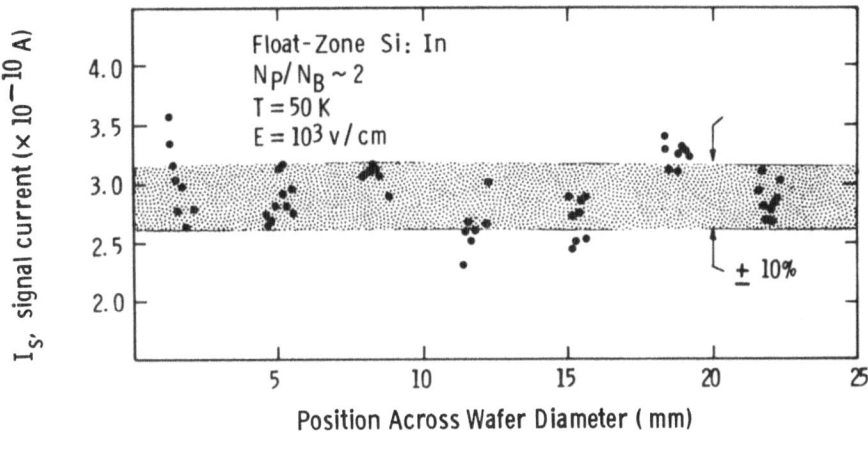

Fig. 18. Photocurrent uniformity of neutron transmuted indium
doped silicon.

5. SUMMARY

The neutron transmutation of silicon into phosphorus has been
utilized to achieve precise compensation of residual acceptor
impurities in the development of high lifetime, high responsivity
extrinsic silicon infrared detector material. Net compensation
densities as low as 2×10^{12} cm^{-3} have been obtained in high
purity, heavily indium doped, float zone grown extrinsic silicon.
In this material the very high measured responsivities of 100 A/W
at 860 V/cm and 50K corresponding to photocarrier lifetimes of
260 ns and mobility-lifetime products of 2×10^{-3} cm^2/V are a
direct consequence of the low net compensation densities which
were achieved by neutron transmutation.

ACKNOWLEDGEMENTS

We would like to thank Drs. D. K. Schroder and H. C. Nathanson
for several valuable discussions. We are particularly indebted to
Mr. J. A. Baker of Dow Corning Corporation for his willing cooperation in performing the neutron irradiations. Finally, the
excellent technical assistance of R. R. Papania, L. L. Wesoloski,
D. P. Nebel, and W. E. Bing is gratefully acknowledged.

REFERENCES

*   Supported by DARPA Contract DAAG53-76-C-0170 and monitored
      by Night Vision and Electro-Optics Laboratories, Fort
      Belvoir, VA 22060.

1)   R. N. Thomas, T. T. Braggins, H. M. Hobgood, W. J. Takei, and
      H. C. Nathanson, paper presented at 25th IRIS Detector
      Specialty Group Meeting, March 1977, Colorado Springs, CO.

2)   R. N. Thomas, T. T. Braggins, H. M. Hobgood, W. J. Takei,
      and H. C. Nathanson, J. Appl. Phys. $\underline{49}$, 2811 (1978).

3)   W. G. Pfann, Zone Melting, 2nd Edition (John Wiley and Sons,
      New York, 1966).

4)   G. Lucovsky, Solid State Commun. $\underline{3}$, 299 (1965).

5)   R. A. Messinger and J. S. Blakemore, Phys. Rev. $\underline{B4}$, 1873
      (1971).

6)   H. J. Hrostowski and R. H. Kaiser, J. Phys. Chem. Solids,
      $\underline{4}$, 148 (1958).

7)   R. Baron, M. H. Young, J. K. Neeland, and O. J. Marsh, Appl.
      Phys. Lett. $\underline{30}$, 594 (1977).

STUDY OF THE SPATIAL CHARACTERISTICS OF THE BREAKDOWN PROCESS IN SILICON PN-JUNCTIONS*

Gerald C. Huth

University of Southern California, Marina del Rey, CA

Boris L. Hikin and Vladimir Rodov

International Rectifier Corp., El Segundo, CA

ABSTRACT

The reduction in the magnitude of resistivity fluctuations in n-type silicon as a result of the neutron transmutation doping process (henceforth the "NTD" process) has had, and will have in the future, significant impact on silicon device technology. Its primary effect should be on high field devices such as high voltage power rectifiers, thyristors, internal gain or "avalanche" photo-detectors and nuclear particle detectors. Secondly, minimization of spatial variations in minority carrier lifetime should improve performance of silicon vidicon and image storage devices and, perhaps, improve production yields in the microcircuitry area. Depending upon the degree of enhanced spatial electric field uniformity provided by NTD, such concepts as silicon imaging devices with internal gain may become feasible for the first time. We begin by examining statistically the problem of achieving a high degree of dopant uniformity through NTD. Then, using the tool of a scanned, finely focused, optical beam, we indicate how the spatial area and general characteristics of breakdown can be visualized in silicon PN-junctions fabricated from NTD and other types of silicon. We will present scans of the pre-breakdown and breakdown characteristics of striated, pre-NTD silicon and indicate what we have found to be the physical character of the high voltage breakdown process in that type of silicon.

1.   INTRODUCTION

In the evolution of silicon PN-junction technology a succession of discrete advances have been made all of which lead ultimately to the high structural perfection junction that conceivably will breakdown over its entire area. The first of these was probably development of the Czochralski solution growth process followed somewhat later by the low oxygen content "floating zone" growth method. Then followed a lengthy period where surface effects limited junction performance. The surface was brought under control in high voltage junctions by the discovery of "beveling" or surface contouring and in low voltage devices by the oxide passivated "Planar" process. With the institution of these surface control methods, PN-junction performance became limited by internal bulk effects, specifically those affecting the then attainable high electric field regime. The next limitation then became the resistivity fluctuations or striations that seemed unavoidably characteristic of solution growth silicon crystal. This limitation has now been overcome with the development of the neutron transmutation (NTD) process for elegantly distributing dopant uniformly in silicon. Subsequently, silicon PN-junctions have reached a degree of structural perfection never attained before. It will be the purpose of this paper to show that in spite of the increased resistivity uniformity provided by the NTD process avalanche breakdown in silicon will still be initiated locally. With increasing reverse bias the number of these localized avalanching areas increases very rapidly. The rate of increase of these avalanching areas (and thus total area of breakdown) is a characteristic of the degree of resistivity uniformity of the material.

We will then consider this question both theoretically and experimentally. Views of microscopic breakdowns are presented which were obtained by the optical scanning technique. We believe that these may be the "last frontier" in the succession of developments in silicon junction technology.

2.   THEORETICAL

We begin by estimating how the spatial uniformity of resistivity (or density of impurity N) influences the voltage-current characteristic of a PN-junction in the high field or avalanche region.

Considering a silicon wafer with an area of S cm^2 containing an average impurity concentration N_0 cm^{-3}, the localized deviation from this average will be $\pm \gamma N_0$. Thus, taking all possible measurements of impurity concentration (or resistivity) across area S, one obtains $N_0 \pm \gamma N_0$. (We will consider, of course, a case where the concentration is exactly $N_0 + \gamma N_0$ on a point of the considered area.) It is clear that each act of measurement has associated with it a characteristic size ℓ. In the common four point probe this is the distance between probes and in single point measurement it is the length of current spreading.

Let us assume that the impurity distribution is purely random and the probability of finding a particular resistivity value is Gaussian. We note here that the parameter γ is a function of ℓ and it is clear that γ decreases when ℓ increases.

The probability of finding M impurity atoms in a rectangular volume with side ℓ is:

$$P(M) = \frac{1}{\sqrt{\pi}\sigma} \exp\left(-\frac{(M-M_0)^2}{\sigma^2}\right) \tag{1}$$

where $M_0 = N_0\ell^2 W_0$ with W_0 being the thickness of the silicon wafer when $W_0 < \ell$, and $W_0 = \ell$ when $W_0 > \ell$. The probability of finding a deviation from the average number of impurity atoms equal to $\gamma N_0 \ell^2 W_0$, must equal $\frac{\ell^2}{S}$ or

$$P \quad (N - N_0 > \gamma N_0\ell^2 W_0) = \frac{\ell^2}{S} \quad . \tag{2}$$

Or, using Eq. (1) we get

$$\frac{1}{\sqrt{\pi}\sigma}\left[\int_{-\infty}^{-\gamma N_0\ell^2 W_0} \exp\left(\frac{-(M-M_0)^2}{\sigma^2}\right) dM + \int_{\gamma N_0\ell^2 W_0}^{\infty} \exp\left(\frac{-(M-M_0)^2}{\sigma^2}\right) dM\right] = \frac{\ell^2}{S} \quad . \tag{2'}$$

Using Eq. (2'), σ can be calculated as follows

$$\sigma = \frac{\gamma N_0\ell^2 W_0}{erfc^{-1} \dfrac{\ell^2}{2S}} \quad . \tag{3}$$

In order to consider the influence of resistivity fluctuations on junction breakdown, we shall use Shockley's method[1] of using cubes with side W = the width of the space charge region. The probability of finding impurity concentration N in a volume W^3 will be according to Eq. (1):

$$P(N) = \frac{W^3}{\sqrt{\pi}\sigma} \exp\left(\frac{-W^6 (N-N_o)^2}{\sigma^2}\right) \tag{4}$$

Now substituting the expressions for abrupt junction voltage for N;

$$N = \frac{E_{crit.}\,\varepsilon\varepsilon_o}{2qV}, \quad \text{and} \quad W = \frac{2V}{E_{crit.}}$$

thus

$$(N-N_o) = \frac{(V-V_o)E_{crit.}\,\varepsilon\varepsilon_o}{2qV^2} \quad .$$

The voltage distribution across the junction is thus:

$$P(V) = \frac{1}{\sqrt{\pi}\sigma_v} \exp\left(-\frac{(V-V_o)^2}{\sigma_v^2}\right) \tag{5}$$

$$\text{where } \sigma_v = \gamma V_o \frac{\left(\dfrac{\ell^2 W_o}{W^3}\right)}{erfc^{-1}\left(\dfrac{L}{2S}\right)}$$

and V_o is the average breakdown voltage.

From Eq. (5) it follows that the current conducting area termed S_{on} in a PN-junction under reverse bias voltage V is:

$$S_{on} = \frac{S}{2} \, erfc\left(\frac{V_o - V}{\sigma_v}\right). \tag{6}$$

The minimum voltage at which the first part of the junction starts to conduct can be given by the expression:

$$W^2 = \frac{S}{2} \, erfc\left(\frac{V_o - V_{min}}{\sigma_v}\right) \quad . \tag{6'}$$

Equation (6') follows immediately from Eq. (6) when one considers that the minimum conducting area is of the order of W^2.

From Eq. (6') we can obtain

$$V_o - V_{min} = \sigma_v \, \text{erfc}^{-1} \left(\frac{2W^2}{S} \right) . \qquad (6'')$$

Let us now consider the voltage current characteristic of a breakdown starting from $V = V_{min}$ and assuming that each successive newly turned on area adds the same amount of current I_0. Then:

$$I = I_o S_{on}$$

or

$$I = I_o \, \exp \left(\frac{V - V_{min}}{\bar{V}} \right) \qquad (7)$$

In Eq. (7),

$$\bar{V} = \frac{\sigma_v}{2 \, \text{erfc}^{-1} \left(\frac{2W^2}{S} \right)} = \frac{\gamma V_o \, \frac{\ell^2 W_o}{W^3}}{2 \, \text{erfc}^{-1} \left(\frac{\ell^2}{S} \right) \text{erfc}^{-1} \left(\frac{W^2}{S} \right)} .$$

Also, in obtaining Eq. (7), it was assumed that $\frac{S_{on}}{S} \ll 1$, erfc X $\ll 1$ for X > 1, and erfc X $\approx \frac{1}{X\sqrt{\pi}} \, e^{-X^2}$.

From Eq. (7), the voltage current characteristic of an avalanching PN-junction is exponential. Figure 1 plots curves according to Eq. (7) for different values of ℓ and γ. Similar characteristic curves have been independently measured experimentally.[2] It was also noted in Eq. (2) that $\bar{V}$ does not depend on the area S which is in agreement with Eq. (7). It is interesting to note that if $\ell = W$, and γ is strictly associated only with the statistical fluctuation of an ideal, uniformly doped semiconductor, then all formulas derived herein lead to the result obtained by Shockley (Ref. 1). The voltage current characteristic for that case is also exponential, however, $\bar{V}$ is approximately two orders of magnitude less (see Fig. 1) and does not depend on the absolute value of the avalanche breakdown voltage.

Also plotted in Fig. 1 is the curve which approximately describes the current state-of-the-art with neutron doped silicon. It can be seen that with increasing doping uniformity, S_{on} is growing very rapidly or, in other words, the differences between avalanche breakdown voltages in neighboring areas is very small. Avalanche then will always be initiated at localized, discrete

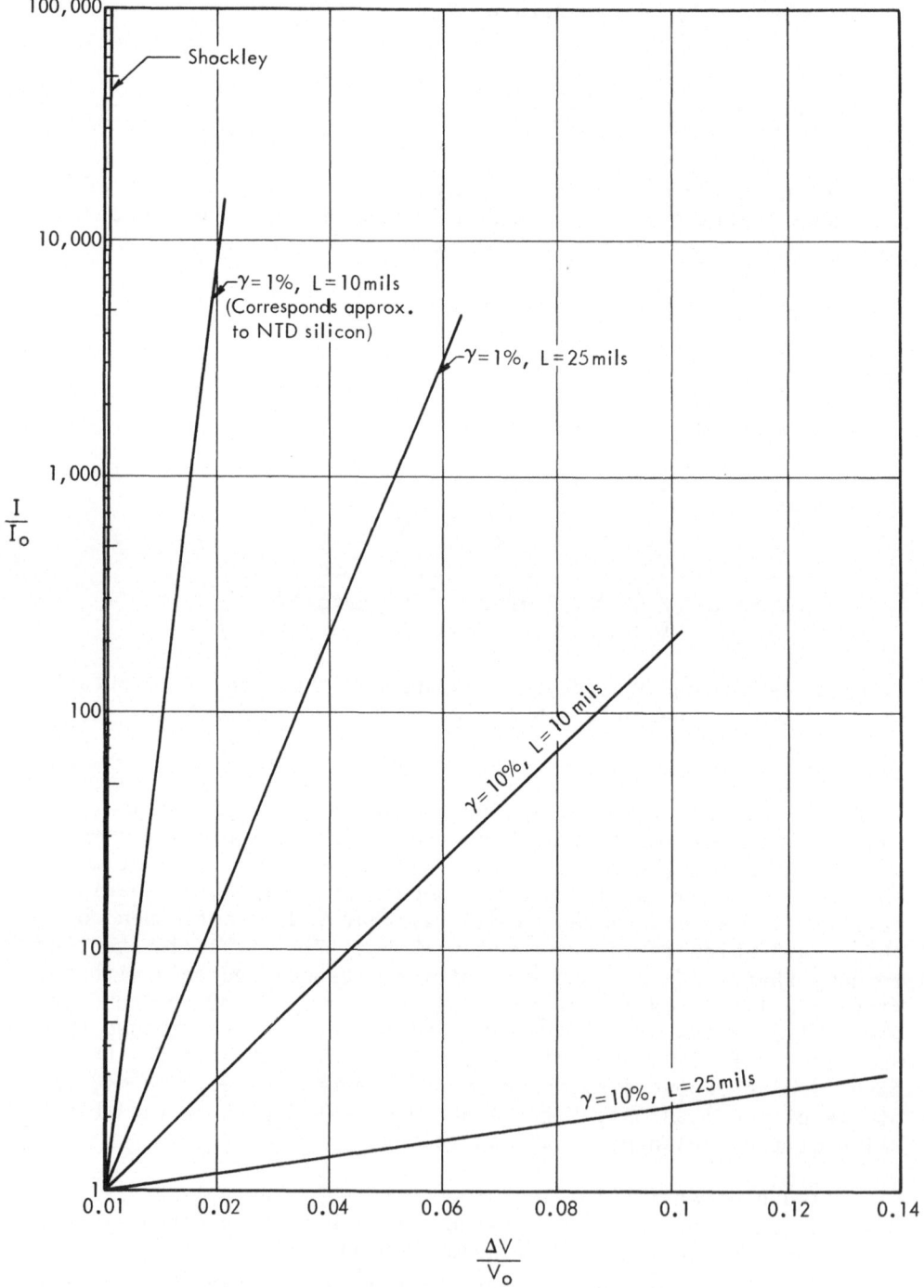

Fig. 1. Normalized junction current vs. incremental normalized
         bias voltage.

areas. Even when low magnification indicates large areas in
avalanche breakdown, higher magnification will still indicate a
multitude of small, discrete breakdown areas. As resistivity
uniformity increases, interactions between adjacent breakdown areas
will become important. At this point even small numbers of exchanged electrons would be sufficient to bring an adjacent area
into breakdown.

3.   STRUCTURAL PERFECTION AND ITS EFFECT ON THE ULTIMATE BREAK-
DOWN SITE IN SILICON PN-JUNCTIONS

What factors intervene before the previously discussed statistical limitation in dopant distribution becomes the limiting factor
on the lateral spread of breakdown area and the achievement of
total area breakdown in silicon PN-junctions? It is probable that
effects associated with physical imperfections in the silicon
crystal (dislocations, stacking faults, etc.) and the localized
impurity concentrations that are associated with them will intervene. The association between the ubiquitous and rapidly diffusing
metals--copper, iron, etc.--and dislocations create what can probably
be described as discontinuous conductors or lengths of micro "wires"
tending to shunt the junction space charge region. At the extreme
of this condition when the metal-dislocation defect becomes contin-
nous, the "bulk poisoned" voltage-current characteristic results
with its "soft from the origin" behavior. At the other extreme,
as when quenching from diffusion temperature is too rapid, the
metals can agglomerate locally and create conductive occlusions in
the space charge region. Each of these conditions has been described and even visualized (with near-infrared sensitive imagery)
in the technology of "decoration" of dislocations and other defects
in silicon.

The two somewhat slower diffusing elements--oxygen and carbon--
are present in silicon in quantitites great enough to cause
potential effects in attempts to achieve highly perfect junctions.
The agglomeration, precipitation and donor formation effects of
oxygen in silicon have been well studied.[3] Shockley long ago
discussed the potential effect of a field perturbing precipitate
that he assumed to be SiO_2 in silicon junction breakdown. Indeed
this precipitate, the source of which is inward diffusion of oxygen
in the "planar" process or residual oxygen from the silicon crystal
growth process, is probably the physical reason for the classical
'microplasma' breakdown observed in this space charge region in
low voltage silicon PN-junctions.

A potential relationship of the junction diffusant itself to
structural perfection must be taken into account. Interactions
between some junction forming impurities with crystal defects are
known to occur. Such interactions lead to irregular and non-
uniform junction fronts. The formation of "pipes" in certain

instances with phosphorus as a diffusant comes to mind. It has been our feeling from considerable study using the optical scanning measurement technique to be discussed in the following, that deeply diffused gallium probably provides the highest degree of structural perfection in PN-junctions in silicon. It is certain that this junction provides breakdown characteristics that are determined by base resistivity and associated crystal structural effects and which are not, to the first order, associated with the junction diffusant.

In order to assess the character and spatial distribution of junction breakdown, a number of investigators have studied imagery of the escaping, near-infrared, portion of the recombination radiation resulting from high current flow (generally localized) in the junction breakdown condition. This is a relatively insensitive method and additionally is not able to visualize the important pre-breakdown electric field configuration. We have constructed a series of optical "flying spot scanning" type systems for break-down studies. In this method, which visualizes spatial photo-current, it is possible to study what is, in reality, spatial electric field distributions of PN-junctions diffused into entire wafers and over the entire reverse biased voltage region. The first system constructed that used synchronized cathode ray tubes is shown in Fig. 2. A diagrammatic representation of the system operation in shown in Fig. 3. Care was taken to reduce the optical spot size on the junction interrogating CRT to a minimum-- about 1 mil. This is reduced further by use of lenses inserted between the CRT and the junction to view small breakdown areas. To the limit of optical resolution it is thus possible to use the system as a microscope viewing and magnifying very small areas. The system in current operation uses scanning mirrors and a helium-neon laser as a light source.

Preparation of silicon junctions for scanning is not difficult. In deeply diffused (75 microns) gallium junctions approximately 25 microns must be removed (by flat lapping and etching) to allow visible wavelength excitation from the yellow phosphor of the interrogating cathod ray tube. Use of longer optical wave-lengths minimizes this requirement. The junction, which might comprise an entire wafer, must of course be beveled and etched so that the high bulk field condition can be attained for study.

In study of many types of silicon single crystal and junction diffusion processes it has become apparent that a successive number of obstacles must be overcome before attainment of large area breakdown. These can be categorized into general areas as follows:

I.   The PN-junction diffusant and its effect (or effects) on the silicon lattice.

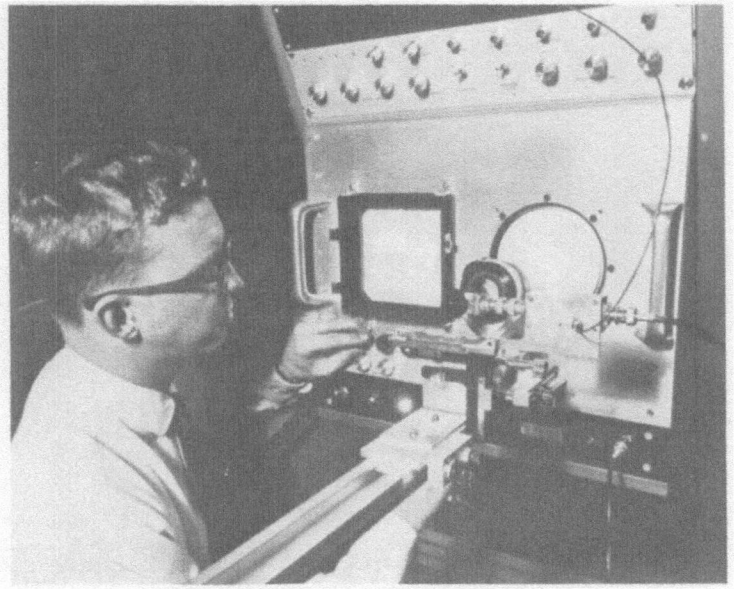

Fig. 2. View of flying spot scanner constructed for measurement of
gain distribution in avalanche detectors.

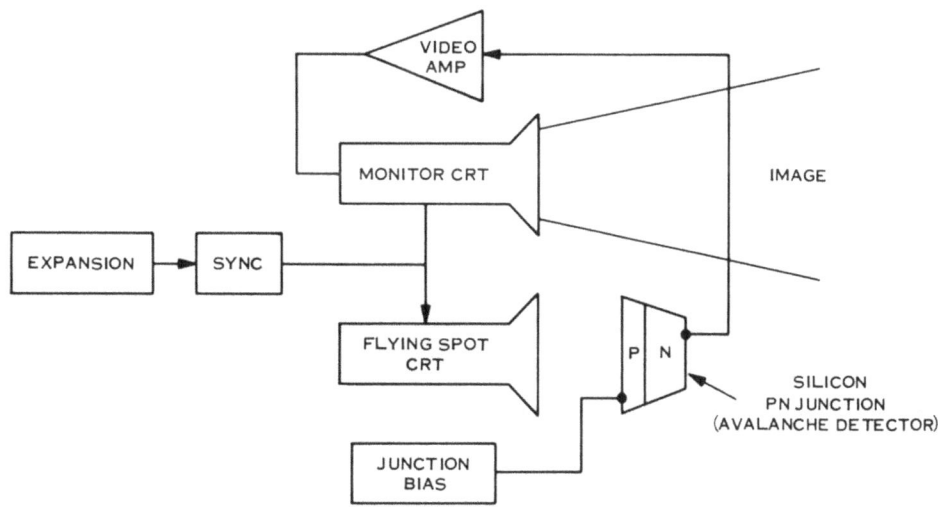

Fig. 3. Diagrammatic representation of operation of flying spot
scanner system.

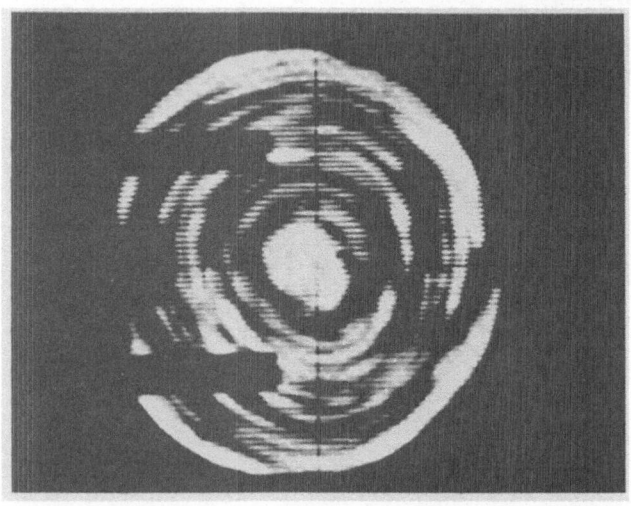

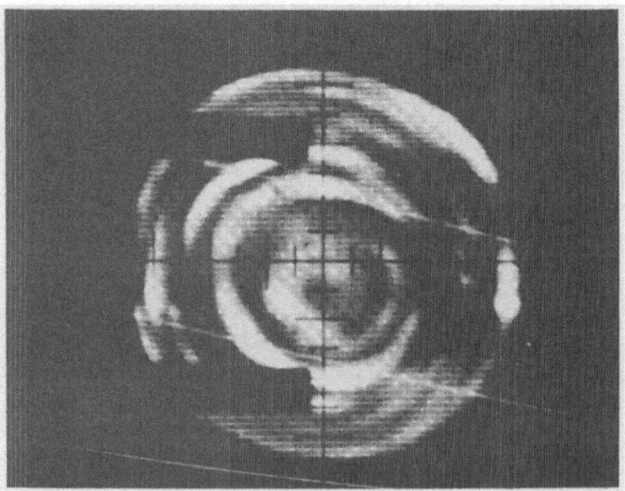

Fig. 4.   Scans of two types of conventionally doped silicon (wafers
          5 cm in diameter).

II.   Dopant (and thus base resistivity) striations and
        fluctuations in the silicon crystal.
III.   "Macro" defects (dislocations, stacking faults, etc.)
        inherent in the grown silicon crystal.

As stated above, it is our belief that the deeply diffused
gallium junction does not, at least to the first order, contribute
to non-uniform breakdown. There are probably two reasons for this:
1) the atomic size of the gallium atom produces minimum structural
damage to the silicon lattice, and 2) an effect of deep diffusion
in "smoothing" the junction front thereby minimizing the possibility of surface irregularity replicated, field perturbances in the
junction shape. At the other extreme, relative to junction
diffusants, one worries about the extreme damage produced by shallow
diffused boron and phosphorus species in solar cell applications.

It is Category II effects--base resistivity striations and
fluctuations--that neutron transmutation doped silicon has hope-
fully overcome. The "pre-NTD" situation is shown in optical scans
representative of two types of conventionally doped silicon as
shown in Fig. 4. The upper wafer shows a "cored" pattern with
outer, completely closed, and often very narrowly striated, rings.
This pattern was characteristic of one method of growing silicon
single crystal. The lower wafer shows the broad banded, "swirl"
pattern of striations characteristic of another manufacturer's
crystal growth method.

Figure 5 shows a magnification study of breakdown in a small
section cut from a larger wafer. This square section, 5.1 mm on
a side, is viewed through a circular aperture in the figure. The
overall "striped" character of electric field is seen in this
striated, pre-NTD material (of the type shown in the bottom photo
of Fig. 4). In the scan at 2473 V immediately preceding breakdown,
note how the intensity of the dark central band has decreased.
Breakdown is occurring in the upper band with consequent electric ·
field reduction in the center. The middle scan in the figure
results when the intensity of the light scanning beam is reduced.
Breakdown structure of Category type II or III now becomes evident
in the previously "overdriven" upper left section. The lower scan
is an optical magnification of this area comprising about 2.0 mm
on a side. Somewhat evident in the figure, but more obvious when
viewing directly, are localized breakdown areas arranged on
"threads" paralleling the resistivity striation bands.

Figure 6 shows yet higher magnifications--about 1 mm on a
side--of the same breakdown. We would characterize these breakdown sites as Category III, the ultimate, at least visually,
observable sites of the breakdown process. Two different photo-
graphic exposures are presented in the figure. At these magnifi-

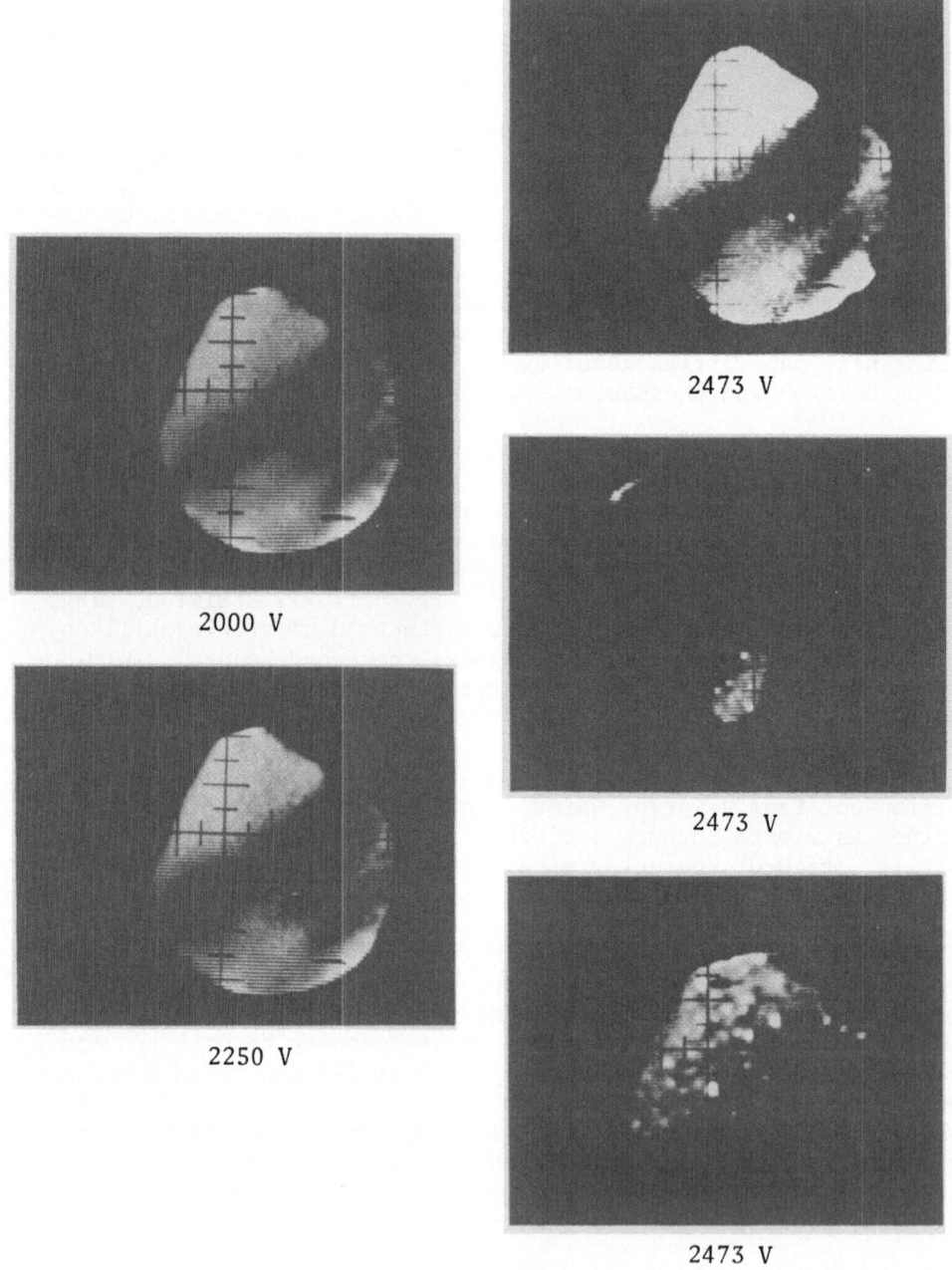

Fig. 5.   Magnification study of breakdown in a section of silicon
          wafer.   Scans at 2473 V are immediately before the onset
          of breakdown (light beam intensity decreased).

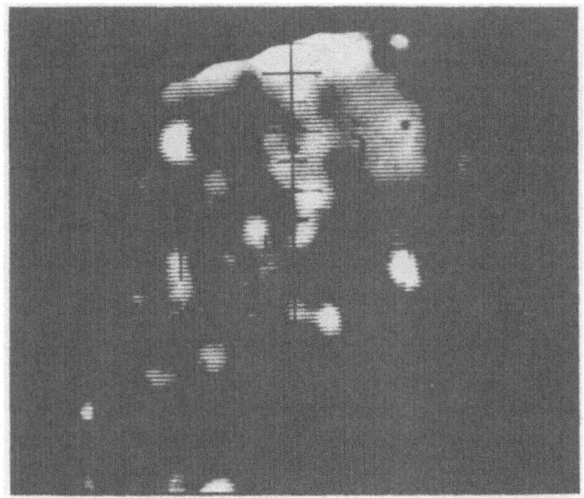

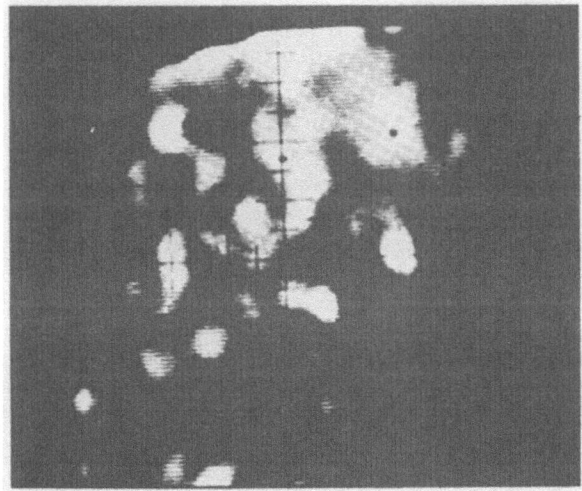

Fig. 6. Higher magnification of breakdown areas (1 mm on a side)
         showing angular character.

cations optical depth of field becomes a problem because of the long focal length lenses used. What becomes apparent, however, is that each breakdown area has an angular character and that these breakdowns seem to be oriented at different depths in the junction field. By varying the focus in these views it is possible to bring different breakdown spots into focus. In every case an approximately 60° angular character is observed.

Figure 7 shows scans of another silicon junction of approximately one centimeter diameter indicating what we would term Category III defects as discussed above. Large "macro" defects such as this that are not seemingly associated with any junction forming process do sometimes occur in silicon crystal. Again in this case the junction is biased to just prior to the onset of breakdown. Breakdown in this junction is seen to be characterized by very large angular areas (of approximately 60° internal angle) as indicated in the top scan of the figure. The lower scan is at higher magnification indicating again the discrete, specular nature characteristic of the breakdown. The left hand photos in Fig. 8 show a still higher magnification (view encompasses about 1 mm^2) of one "leg" of the angular breakdown region. Photos in the right side of the figure are the highest magnification possible (about 0.7mm^2). The same depth oriented, angular, character of these smallest visualizable breakdown sites is observed.

We show in Fig. 9 another instance of visualizable breakdown and pre-breakdown behavior that can be correlated with structural and impurity properties of silicon crystal. These scans are of the same high perfection gallium junctions diffused into silicon grown by the Czochralski rather than the float zone method. (The scans to the right are simply a pseudo three dimensional presentation of the left hand images). The rather uniformly distributed "texture" of localized breakdown sites seems characteristic of this material. One is tempted to think that these result from oxygen precipitates known to be associated with this material--but this has not been proven. This material does, however, tend to produce lower average peak inverse voltages than the float zone material used in the above discussed studies.

Our studies, and the results of many others in the semiconductor field using NTD silicon, have indicated that average inverse breakdown voltages are certainly higher with this material. This is undoubtedly the result of the minimization of Category II resistivity striations and fluctuations. There is also some indication, at least in early NTD work, that these fluctuations are not always completely masked. We have published one paper[4] reporting a result reflecting this improved condition in higher average internal "gain" obtainable in silicon avalanche type photodevices. Increased resistivity uniformity is translated into higher average pre-breakdown electric field which means increased "gain."

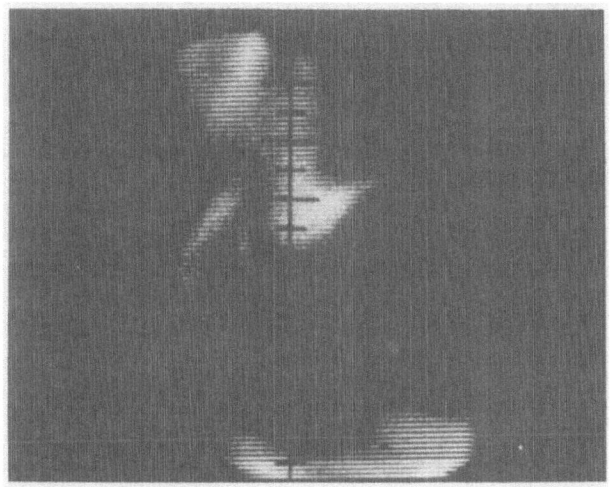

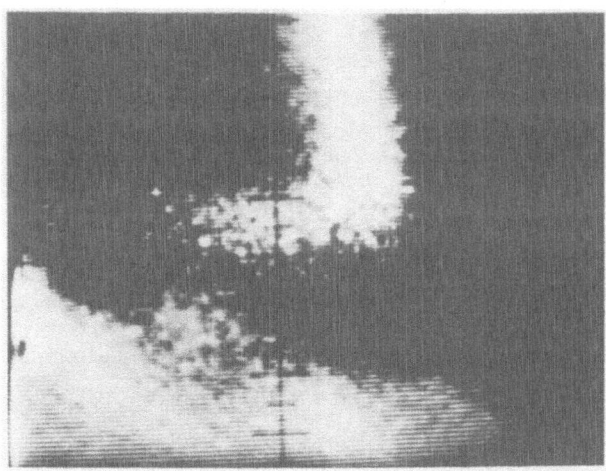

Fig. 7. Breakdown of silicon junction of one centimeter diameter.

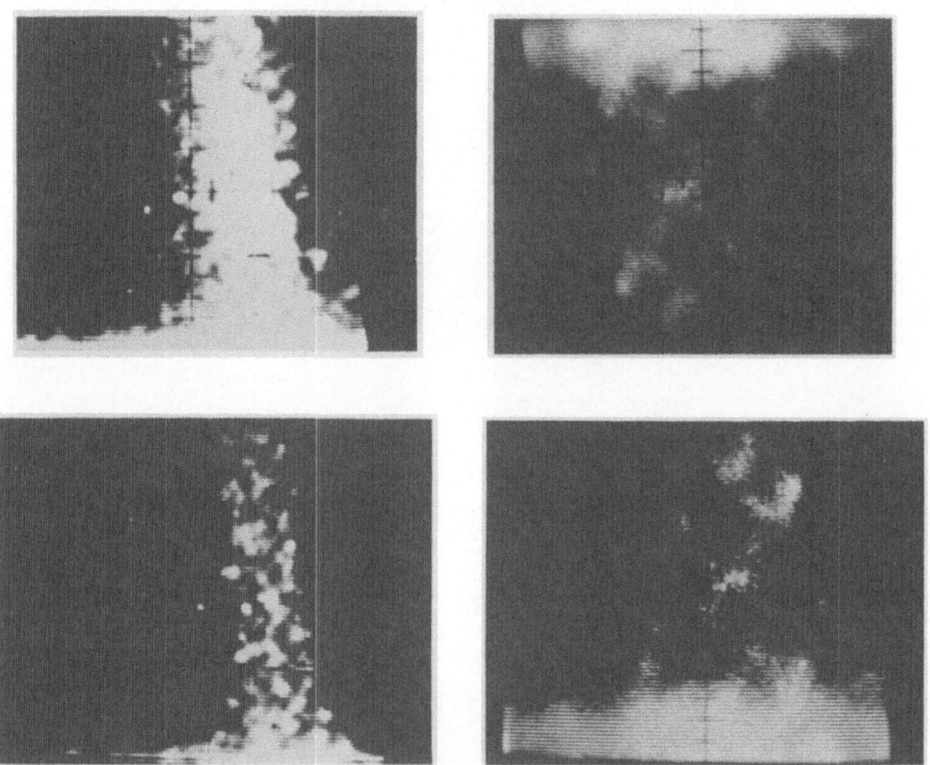

Fig. 8. Higher magnifications of breakdown areas.

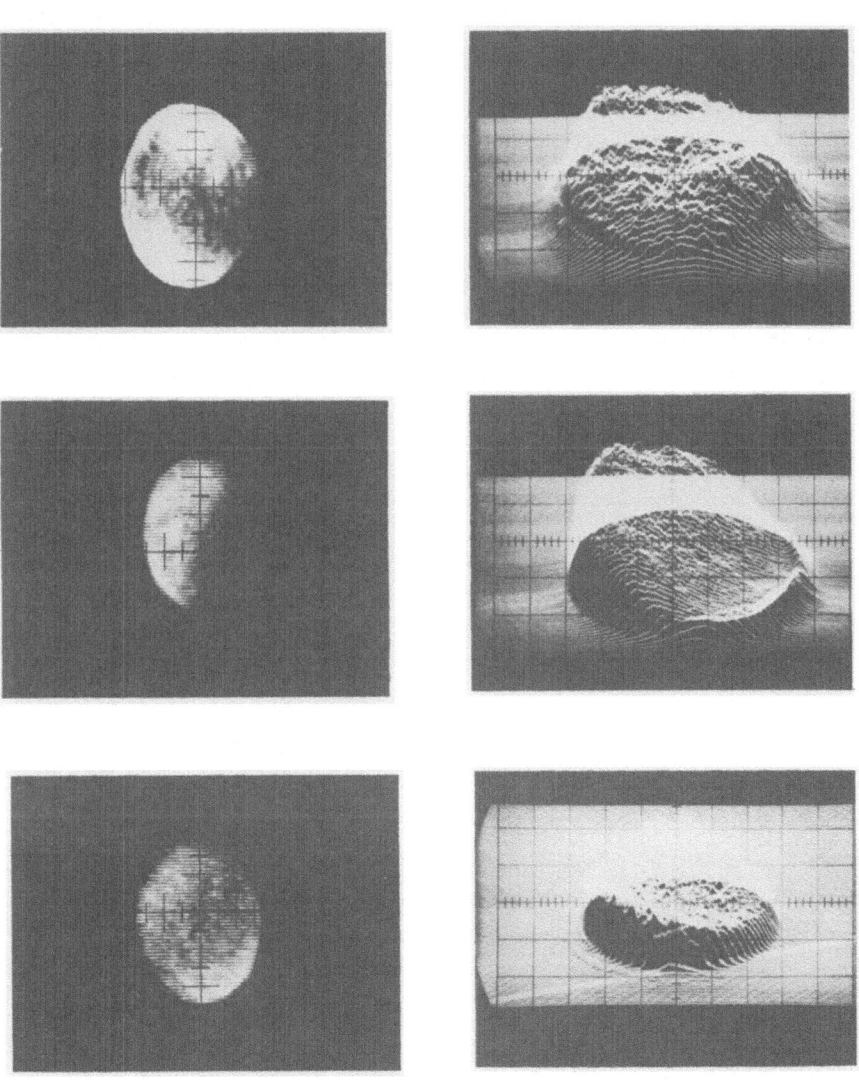

Fig. 9. Breakdown character of Czochralski silicon.

We have only recently begun to study the higher magnification breakdown characteristics of junctions diffused into NTD silicon. We expect, however, that crystalline structural imperfections and associated impurity effects will again intervene to limit attainment of the large breakdown areas that we have shown to be theoretically attainable. Perhaps the "next frontier" in silicon junction technology beyond NTD will be development of processes for eliminating stacking faults and other imperfections from diffused silicon.

REFERENCES

*   Work supported in part by the Department of Energy contract No. EY-76-S-03-0113.

1)   W. Shockley, Sol. State Electronics, 2, 35 (1961).

2)   I. V. Grekhov, N. N. Krukova, and V. V. Lebedev, Soviet Physics, Semiconductors, 6, N12 (1967).

3)   W. Kaiser, H. L. Frisch and H. Reiss, Phys. Rev. 112, 1546 (1958).

4)   V. L. Gelezunas, W. Seibt, and G. Huth, Appl. Phys. Letters, 30, 118 (1977).

RESISTIVITY FLUCTUATIONS IN HIGHLY COMPENSATED NTD SILICON*

J. M. Meese[+] and Paul J. Glairon[++]

University of Missouri Research Reactor

Columbia, MO 65211

ABSTRACT

Calculations of the resistivity fluctuation $\Delta\rho/\bar{\rho}$ as a function of compensation ratio in NTD-Si are presented which are valid near exact compensation. These calculations are compared with experimental data taken on silicon which has been compensated by the NTD process to resistivities as high as 100,000 Ω-cm. Calculations are also presented of the maximum possible mean resistivity obtainable before fluctuation induced type conversion occurs as a function of the initial starting material resistivity fluctuation.

1. INTRODUCTION

Janus and Malmros have given expressions to predict the effect of transmutation doping on the relative impurity concentration fluctuation.[1] They have shown that for n-type NTD-Si doped well beyond exact compensation, the relative concentration inhomogeneity decreases linearly with increasing fluence (doping factor). For many envisioned detector applications, it is desirable to use the NTD-process to compensate to very high resistivities near exact compensation. Several attempts to transmutation compensate p-type float zone silicon to very high resistivities for various detector applications have been reported.[2-4] The usefulness of this material is thought to depend on the final resistivity uniformity which can be achieved. It is clear that this uniformity is not linear with fluence near exact compensation due to the importance of the intrinsic carrier concentration. It is, therefore, desirable to obtain expressions for uniformity as a function of fluence or compensation ratio for p-type as well as n-type material near exact compensation. It is also desirable to express this uniformity in terms of resistivity rather than impurity concentration since the wafer uniformity is always determined experimentally by spreading resistance or four-point probe resistivity measurements, not impurity concentration. It will be shown that NTD compensation yields a significant improvement in resistivity uniformity compared to conventional compensation by melt doping with phosphorus in the range of compensation ratios useful for extrinsic silicon IR detector applications.

2. A RESISTIVITY FLUCTUATION MODEL

In general, the resistivity as measured by one of the usual probe techniques is a complicated function of the probe coordinates on the surface of the wafer. An exact definition of mean resistivity and deviation from this mean is therefore hopelessly complicated from an experimental point of view. An inspection of typical radial traces of spreading resistance fortunately shows that the periods and amplitudes of the observed fluctuations due to doping inhomogeneities are usually rather well defined. In fact, the obvious periodicity of these traces has lead Voltmer and Ruiz to apply Fourier analysis to these patterns.[5] They find that the Fourier transformed patterns contain relatively few dominant spatial frequencies which they have been able to relate to ingot pulling speed and rotation frequency.

Because of the amplitude uniformity usually found in spreading resistance traces, it is possible and desirable to define the mean of the maximum and minimum resistivities, $\bar{\rho}$, and the maximum fluctuation, $\Delta\rho$, as

$$\bar{\rho} = (\rho_{max} + \rho_{min})/2 \qquad (1)$$

and

$$\Delta\rho = (\rho_{max} - \rho_{min}) \qquad (2)$$

where these maximum and minimum resistivities are analogous to the minimum and maximum concentrations discussed by Janus and Malmros.[1] By calculating these two resistivities as a function of fluence (or compensation ratio), contact between experiment and theory can easily be made.

We will also define the fraction

$$\Delta\rho/\bar{\rho} = 2\left(\frac{\rho_{max} - \rho_{min}}{\rho_{max} + \rho_{min}}\right) \tag{3}$$

as a measure of the relative resistivity fluctuation. Once we have determined how the resistivity at a general point on the wafer changes with fluence Φ, i.e., determined the function $\rho(\Phi)$, the quantities defined by Eqs. (1) - (3) are easily calculated since all resistivity points must follow the same $\rho(\Phi)$ curve. We have presented a means of calculating $\rho(\Phi)$ previously,[4] therefore, we will only outline the principle results here.

From charge conservation, and using the equilibrium condition $np = n_i^2$, the usual expression for carrier concentration in an ionized p-type semiconductor is given by

$$p = \frac{1}{2}\left[(N_A - N_D) + \sqrt{(N_A - N_D)^2 + 4n_i^2}\right] \ .$$

If N_p is the concentration of phosphorus added by transmutation and K is the phosphorus production rate per unit fluence, Φ, then $N_p = K\Phi$ and

$$p = \frac{1}{2}\left[(N_A - N_D - K\Phi) + \sqrt{(N_A - N_D - K\Phi)^2 + 4n_i^2}\right] . \tag{4}$$

The production rate is about $K = 3.355$ ppb per $10^{18}n/cm^2$ or 1.676×10^{-4} P-atoms/cm^3 per n/cm^2.[4]

For exact compensation, the concentration of acceptors exactly equals the concentration of initial donors, N_D, plus those donors added by transmutation, $Np = K\Phi$. Therefore, the critical fluence, Φ_c, necessary to produce exact compensation for a particular point on the wafer with initial hole concentration $N_A - N_D$ is

$$\Phi_c = (N_A - N_D)/K. \tag{5}$$

From Eqs. (4) and (5), then

$$p = \frac{K\Phi_c}{2}\left[(1 - \Phi/\Phi_c) + \sqrt{(1 - \Phi/\Phi_c)^2 + (2n_i/K\Phi_c)^2}\right], \tag{6}$$

p-type.

Similarly, the carrier concentration as a function of fluence for n-type material is given by

$$n = \frac{K\Phi_c}{2} \left[-(1-\Phi/\Phi_c) + \sqrt{(1-\Phi/\Phi_c)^2 + (2n_i/K\Phi_c)^2} \right],$$ (7)

n-type.

Equations (6) and (7) can be substituted directly into expressions for resistivity as a function of carrier concentration

$$\rho(p) = p/[e\mu_p(bn_i^2 + p^2)], \text{ p-type}$$ (8)

or

$$\rho(n) = n/[e\mu_p(bn^2 + n_i^2)], \text{ n-type}$$ (9)

which have been derived from the usual expressions, $\rho=(ne\mu_n+pe\mu_p)^{-1}$, $np = n_i^2$, and $b = \mu_n/\mu_p$. Therefore, a direct calculation of $\rho(\Phi)$ is obtained from Eqs. (6) and (8) for p-type and Eqs. (7) and (9) for n-type.

The variation of resistivity as a function of thermal neutron fluence calculated from the above expression is shown in Fig. 1 for the region near exact compensation, i.e., the region where ρ is not proportional to the reciprocal of the carrier concentration.

Since the electron and hole mobilities are not equal in silicon, the resistivity reaches a maximum before the critical fluence for exact compensation, Φ_c. We have previously shown[4] that this resistivity maximum, ρ_H, is given by

$$\rho_H = (2en_i \sqrt{\mu_n\mu_p})^{-1}$$ (10)

and that the corresponding carrier concentration, p_H, is given by

$$p_H = \sqrt{bn_i^2}$$ (11)

where $b = \mu_n/\mu_p$ is the drift mobility ratio.

In silicon at room temperature, using typical values for n_i, μ_n, and μ_p,[4,6] it is found that $\rho_H = 270,488$ Ω-cm and that the resistivity at exact compensation is $\rho_i = 237,694$ Ω-cm. It can also be shown that[4]

$$\Phi_c - \Phi_H = \frac{b-1}{\sqrt{b}} \frac{n_i}{K}$$ (12)

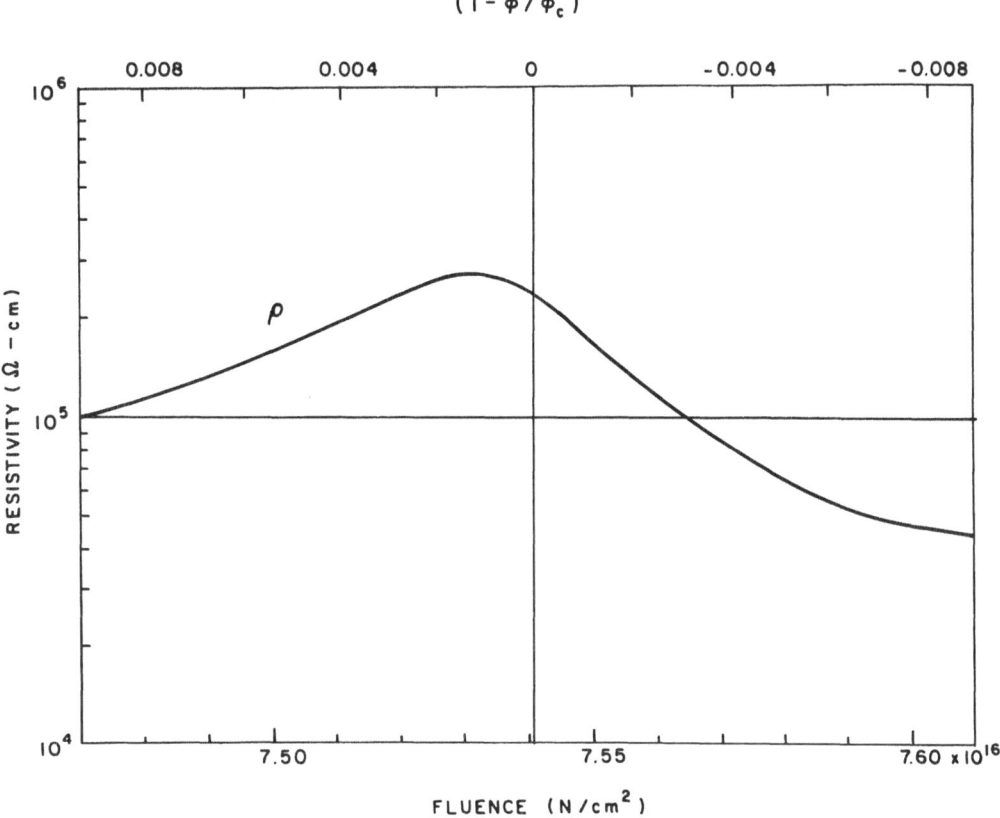

$$(1 - \phi / \phi_c)$$

Fig. 1. Variation of resistivity vs. fluence for NTD-Si calculated using the parameters $\rho_0 = 1000$ Ω-cm, $n_i = 1.391 \times 10^{10}$ cm^{-3}, and $\mu_n = 1396$ cm^2/V-sec, $\mu_p = 492$ cm^2/V-sec.

where Φ_c has been defined in Eq. (5) and Φ_H is the fluence to produce the maximum resistivity, ρ_H. Since this fluence difference is independent of starting resistivity, this suggests that the curve shown in Fig. 1 is a universal curve, an assertion easily proven using Eqs. (6) - (9).[4]

Now that $\rho(\Phi)$ has been determined on a universal curve, it is possible to calculate the behavior of ρ_{max}, ρ_{min} and $\Delta\rho/\bar{\rho}$ vs. fluence as shown on Fig. 2. We shall assume for convenience that the initial phosphorus concentration, P_0, is negligible compared to the boron concentration, B, and that the boron concentration remains constant throughout the NTD-compensation process. The first assumption is reasonable for multipass float zone Si

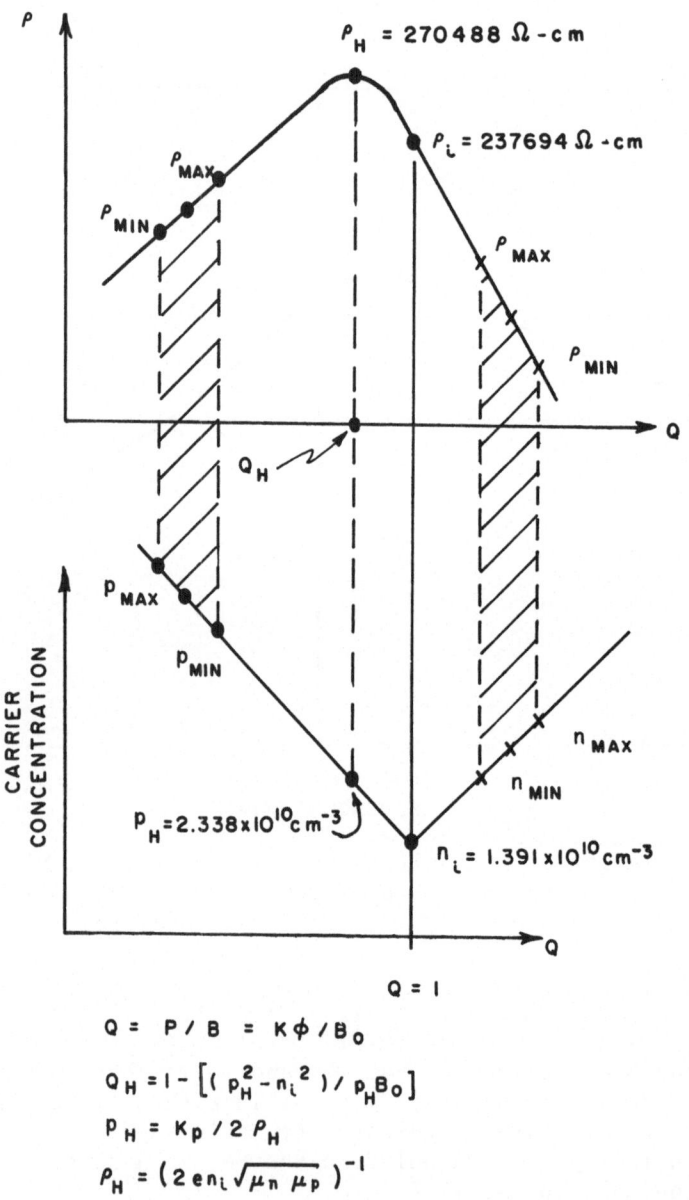

Fig. 2.   Schematic representation of resistivity and carrier concentration vs. phosphorus to boron ratio, Q, which is
proportional to fluence. The maximum and minimum values
of ρ and p are shown for the starting material and define
the shaded area on the left. These fluctuations are also
shown after transmutation compensation to a ratio Q which
represents an overcompensated case in the shaded area on
the right.

detector material since the segregation coefficient for boron
is near unity. After multipass zone refining, then, we would expect to obtain a boron doped sample with a very low degree of donor
compensation.[7-8] The second assumption is easily justified by
comparing the rate of ^{10}B burn-up (σ = 3837 barns, 20% abundant)
compared to the rate of transmutation of ^{30}Si (σ = 0.11 barns,
3% abundant). The other boron cross sections are too small to
consider. If we assume an initial boron concentration of typically
3 x 10^{12}cm^{-3} in multipass float zone and an initial ^{30}Si concentration of (0.03 x 5 x 10^{22}Si/cm^{3}), then using N = $N_0\sigma\Phi$ we obtain

$$\frac{d^{10}B}{d\Phi} = -\ 2.28 \times 10^{-9}\ \ ^{10}B/cm^3/n/cm^2$$

and

$$\frac{d^{31}P}{d\Phi} = \ 1.68 \times 10^{-4}\ \ ^{31}P/cm^3/n/cm^2.$$

Therefore, by taking the ratio of these two rates, we see that one
boron atom is lost for each 73,700 phosphorus atoms produced by
transmutation.

Since we have neglected the phosphorus concentration in the
starting material, the initial concentration fluctuation is a boron
fluctuation. We can define the quantities

$$\bar{B} = \frac{B_{max} + B_{min}}{2} \tag{13}$$

and

$$\Delta B = B_{max} - B_{min} \tag{14}$$

as the mean boron concentration and boron concentration fluctuation.
These are constants of the problem since the burn-up of boron is
negligible. These quantities can be calculated from the initial
resistivities ρ^{o}_{max}, ρ^{o}_{min} using

$$B_{min} = f(K_p/\rho^{o}_{max}) - \frac{n_i^2}{f(K_p/\rho^{o}_{max})} \tag{15}$$

and

$$B_{max} = f(K_p/\rho^{o}_{min}) - \frac{n_i^2}{f(K_p/\rho^{o}_{min})} \tag{16}$$

where

$$f(x) = \left(x + \sqrt{x^2 - 4bn_i^2}\right)/2,$$ (17)

$$K_p = (e\mu_p)^{-1}, \text{ and } b = \mu_n/\mu_p.$$

The above expressions can be derived easily. From Eq. (8),

$$\rho = \frac{pK_p}{bn_i^2 + p^2}$$

and solving for p, we find that

$$p = \tfrac{1}{2}\left[\left(\frac{K_p}{\rho}\right) + \sqrt{\left(\frac{K_p}{\rho}\right)^2 - 4bn_i^2}\right].$$ (18)

Likewise, from charge conservation ($N_D \cong 0$ according to our previous assumption for the starting material)

$$p = N_A + n = N_A + n_i/p$$

and solving for N_A,

$$B = N_A = p - \frac{n_i^2}{p}$$ (19)

Substituting Eq. (18) into (19), we obtain either Eq. (15) or (16) by assigning max, min subscripts to B and realizing that a minimum resistivity is the result of a maximum boron concentration, etc.

We define as a measure of the compensation ratio the quantity, Q as follows:

$$Q = P/\bar{B} = K\Phi/\bar{B}$$ (20)

where K is the phosphorus production rate $\sigma N(^{30}Si)$ and Φ the fluence. [This definition of compensation ratio is the usual one for Q<1 but for Q>1, Q^{-1} represents the usually defined compensation ratio.]

Since $\bar{B}$ and ΔB are constant, Q will be the only variable in our calculation. We now wish to find the net impurity concentration as a function of Q (or fluence). This is obviously given by

$$N_{A_{max}} - N_D = \bar{B} + \frac{\Delta B}{2} - P = \bar{B}(1 - Q + \tfrac{1}{2}\frac{\Delta B}{\bar{B}}) \qquad (21)$$

and

$$N_{A_{min}} - N_D = \bar{B}(1 - Q - \tfrac{1}{2}\frac{\Delta B}{\bar{B}}) \qquad (22)$$

for p-type material and the negative of the above equations for n-type. We have assumed here that transmutation doping is completely random and uniform so that $\Delta P = 0$. The maximum and minimum carrier concentrations are now easily determined as a function of Q using the expression given previously

$$p(Q)_{\substack{max \\ min}} = \tfrac{1}{2}\left[(N_{A_{\substack{max \\ min}}} - N_D) + \sqrt{(N_{A_{\substack{max \\ min}}} - N_D)^2 + 4n_i^2}\right] \qquad (23)$$

and

$$n(Q)_{\substack{max \\ min}} = \tfrac{1}{2}\left[-(N_{A_{\substack{max \\ min}}} - N_D) + \sqrt{(N_{A_{\substack{max \\ min}}} - N_D)^2 + 4n_i^2}\right] \qquad (24)$$

for p-type and n-type respectively.

Using Eqs. (23) and (24), we can calculate the minimum and maximum resistivities for any value of Q from Eqs. (8) and (9) as

$$\rho(Q)_{\substack{max \\ min}} = \begin{cases} K_p p(Q)_{\substack{max \\ min}}/(p^2(Q)_{\substack{max \\ min}} + bn_i^2), & \text{n-type} \\[2mm] bK_n n(Q)_{\substack{max \\ min}}/(bn^2(Q)_{\substack{max \\ min}} + n_i^2), & \text{p-type} \end{cases} \qquad (25)$$

where K_n has been defined previously as $K_n = (e\mu_n)^{-1}$. Knowing these resistivities, it follows that

$$\frac{\Delta\rho}{\bar{\rho}}(Q) = 2\left[\frac{\rho(Q)_{max} - \rho(Q)_{min}}{\rho(Q)_{max} + \rho(Q)_{min}}\right] \qquad (26)$$

using Eq. (3).

To summarize the calculation, the first step is to determine ρ_{max} and ρ_{min} in the starting material before NTD compensation.

Using Eqs. (15) and (16), B_{max} and B_{min} can be determined. From Eqs. (13) and (14), $\bar{B}$ and ΔB are now calculated and substituted into Eqs. (21) and (22) to obtain the net impurity concentration. Equation (21) will be used as given when Q≤1 but must be multiplied by -1 for Q>1. The maximum and minimum carrier concentrations can now be calculated using Eqs. (23) and (24). Some care is required in determining which equation to use in the compensation region where ρ_{max} and ρ_{min} cross the resistivity peak in Fig. 2 and the intrinsic resistivity at exact compensation. It can easily be shown that the value of Q at the resistivity peak is given by

$$Q_H = 1 - [p_H{}^2 - n_i{}^2)/p_H\bar{B}]. \qquad (27)$$

Solving $p = \left[(\bar{N}_A - N_D) + \sqrt{(\bar{N}_A - N_D)^2 + 4n_i{}^2}\right]/2$ for $\bar{N}_A - N_D$, we find that

$$\bar{N}_A - N_D = \frac{p^2 - n_i{}^2}{p^2} \quad .$$

From Eqs. (21) and (22), however,

$$\bar{N}_A - N_D = \bar{B}(1 - Q).$$

Therefore

$$(1 - Q) = \frac{p^2 - n_i{}^2}{p^2\bar{B}}$$

from which Eq. (27) follows immediately by the application of suitable subscripts.

Equation (27) is useful in deciding which of Eqs. (23) and (24) to choose as ρ_{max} and ρ_{min} pass through ρ_H and ρ_i. Also, at exact compensation $\rho = \rho_i$ and Q = 1.

Once the carrier concentrations are calculated from Eqs. (23) and (24), then (25) and (26) yield the desired result.

3. CALCULATED RESULTS OF FLUCTUATION MODEL

A program was written in BASIC to perform the calculations outlined in the previous section.[9] The results are shown in Fig. 3 for a resistivity before transmutation compensation of 2500 Ω-cm

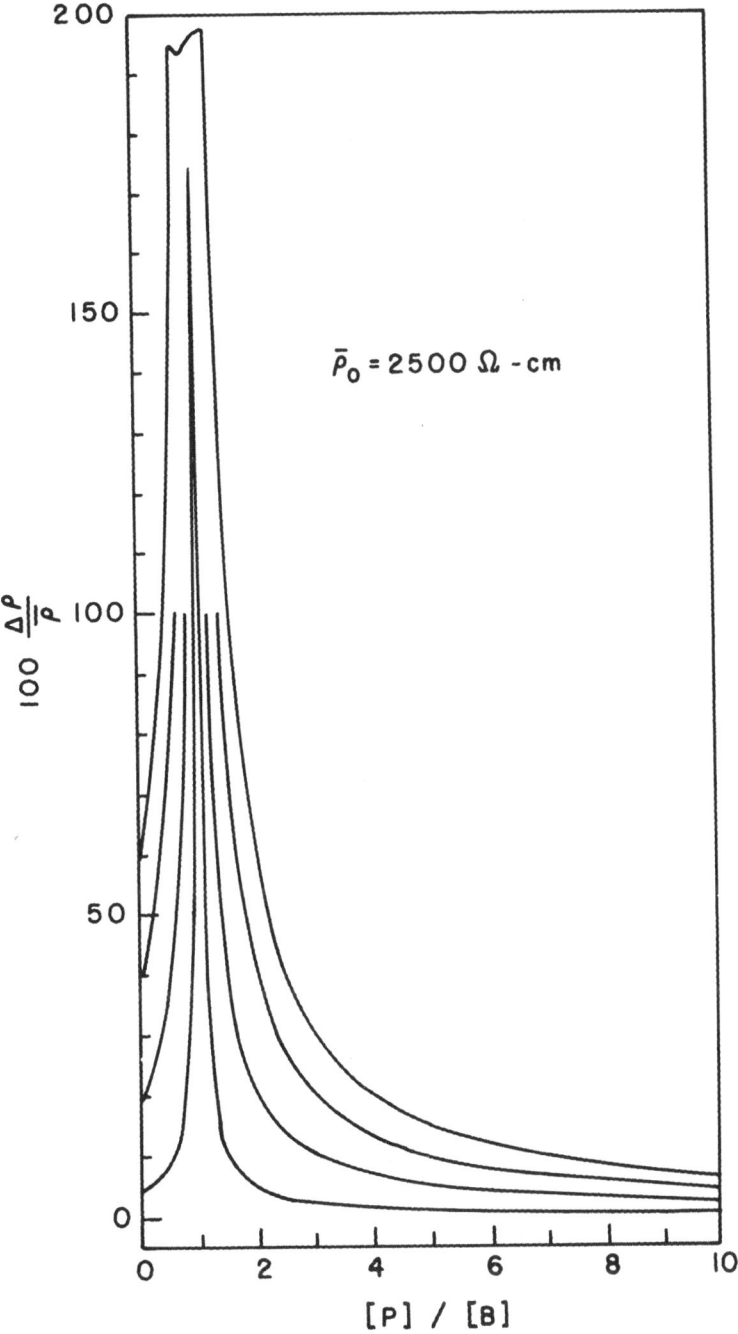

$\bar{P}_0 = 2500 \ \Omega$ - cm

Fig. 3. Calculated values of $100 \ \Delta\rho/\bar{\rho}$ vs. P/B = Q = KΦ/B for a starting material of 2500 Ω-cm resistivity, p-type. Four curves are represented for initial resistivity fluctuations of 5, 20, 40, and 60%.

and for initial resistivity fluctuations of 5, 20, 40 and 60%. Similar sets of curves were calculated for initial resistivities of 5000 Ω-cm and 10,000 Ω-cm. Although slight differences for these other resistivities were found numerically near $Q = P/B = 1$, the differences could hardly be identified when plotted on the scale shown in Fig. 3. We can conclude, therefore, that the curves shown in Fig. 3 are very nearly universal curves which are almost independent of the average initial resistivity before compensation.

It is apparent in Fig. 3 that increasing the compensation by transmutation doping on the p-type side of exact compensation ($Q<1$) always causes the resistivity fluctuation $\Delta\rho/\bar{\rho}$ to increase rapidly above its initial value. In fact, fluctuations greater than 100% are to be expected for all typical starting material which is nearly exactly compensated by the NTD process. It must be pointed out, however, that this is still a better situation than compensating by conventional means since no new impurity inhomogenity is added to the starting material in the NTD process. In fact, the effects of conventional compensation can be estimated from these curves as follows.

Assume, for illustration that two identical samples, each with initial starting resistivity fluctuations of 5% are compensated, one by the NTD process and one by more conventional means. The NTD sample would follow the 5% curve of Fig. 3. We might guess that the addition of phosphorus in the molten state would add a phosphorus concentration fluctuation ΔP of two or three times ΔB since boron is less sensitive to temperature fluctuations in the melt than phosphorus. Then if P is varied but ΔP is constant, the resistivity as a function of compensation ratio would most likely fall above our 20% curve shown in Fig. 3 for the conventionally compensated sample.

For a particular device resistivity uniformity requirement of $\Delta\rho/\bar{\rho} < 10\%$, a comparison of these two curves shows that the NTD material can easily meet these requirements for $0 < P/B < 0.5$ and for $P/B > 1.5$ while the conventionally compensated material could only meet the device requirement for $P/B > 3$ on the n-type side. A greater variety of detector devices is therefore possible, both n-type and p-type, using NTD-Si than by conventional melt compensation.

A second interesting feature is shown in Fig. 3 for the 60% curve. The small dip near the peak of this curve is a result of the fact that silicon has an absolute maximum resistivity defined by Eq. (10). This means that as ρ_{max} and ρ_{min} approach ρ_H in Fig. 2 with increasing compensation, $\Delta\rho$ will actually start to decrease once Q for ρ_{max} passes Q_H. This effect can be seen in the calculations of all the curves shown in Fig. 3 but does not

appear in most of the plots to the scale shown. It is doubtful
that this effect can be seen experimentally, however, since for a
very slightly larger fluence, ρ_{max} passes Q = 1 and the material
is of mixed type until ρ_{min} also passes Q = 1. We should therefore
expect to see mixed typing for samples compensated near Q = 1.
This will be amplified in a later section.

It should also be noted in Fig. 3 that for P/B ratios of
approximately 2, the resistivity fluctuation is roughly equal to
the initial resistivity fluctuation. This result is apparently
true for a number of starting material resistivities and resistivity fluctuations. This result is interesting since an empirical
rule of thumb for extrinsic IR detector compensation is developing
which suggests that the P/B ratio should ideally fall between
2 and 3.

One further comment should be made concerning Fig. 3. At very
high compensation ratios (P/B $\sim$ 10), the curves indicate that
$\Delta\rho/\bar\rho$ is rapidly approaching the accuracy of a spreading resistance
measurement if we assume this accuracy to be about 5%. This effect
can have dire consequences for the device manufacturer since he
will be unable to discover from these measurements which source of
NTD-Si had the better initial impurity concentration fluctuation.
(The P/B ratio for 100 Ω-cm material might be between 10 and 20).
On the other hand, intuition and various models for junction breakdown would suggest that the impurity concentration fluctuation, not
resistivity fluctuation controls breakdown. Perhaps this explains
the rather varied opinion today in the power device industry as to
whether the NTD process improves breakdown voltage. We tend to
believe that the ideal junction will still demand excellent starting
material and that the NTD process will produce the best junction
breakdowns when the starting material is p-type and has the best
uniformity. Further experimentation will be required to investigate
this hyphothesis.

4.   EXPERIMENTAL VERIFICATION OF MODEL

In this section we will present experimental measurements of
resistivity fluctuations for different compensation ratios in NTD
wafers produced at MURR. Multipass float zone silicon with an
initial p-type resistivity of about 19,000 Ω-cm and an initial
resistivity fluctuation of between 4 and 7% was used in this experiment.

Resistivity traces across the diameters of the wafers were
measured before and after transmutation doping using a traveling
sample stage and a four point probe with a probe spacing of 0.0625".
Samples were lapped in 400 grit SiC and water. Sample surfaces

were then allowed to age slightly to stabilize any drifts in the contact potential between the probes and the sample.

An offset voltage was observed on the high resistivity samples due to a contact potential which appeared in the absence of any current through the current probes. Although this potential varied with each new placement of the probes on the surface and the voltage was sometimes as large as the resistivity voltage, it was found by taking I-V plots that the slope of the current vs. voltage was not affected by this contact potential. It was therefore possible to obtain reproducible resistivities by averaging the apparent resistivity values obtained for forward and reverse current provided care was taken not to heat the sample by using too high a current. This could be determined by reducing the current in steps until no further increase in resistance was observed.

Samples were irradiated to appropriate fluences in a Cd ratio of about 30:1 and annealed to 850°C in a Spectrosil quartz tube in flowing argon for 30 min. The samples were then allowed to cool slowly to room temperature as the furnace cooled. All samples were intentionally overcompensated so that the n-type side of the $\Delta\rho/\bar{\rho}$ vs. P/B curves could be investigated. Emphasis was placed on n-type material since it is thought that for IR extrinsic detectors the boron should be slightly overcompensated.

Figure 4 shows selected sections of the normalized resistivity vs. probe distance across the wafer for five samples compensated to different P/B ratios. We have selected regions of the trace which show both the maximum and minimum resistivity values observed.

Figure 5 is a plot of the experimental data taken from Fig. 4 where $\Delta\rho/\bar{\rho}$ was obtained using Eq. (3) and the P/B ratio calculated from $\bar{\rho}$ assuming no initial compensation in the sample. The two solid lines are calculated from the model presented above for initial $\Delta\rho/\bar{\rho}$ values of 4.23% and 7.12%. These limits represent the maximum and minimum values of $\Delta\rho/\bar{\rho}$ found for the five wafers before NTD-doping. These calculated curves are similar to those of Fig. 3 except that only the n-type side (P/B > 1) is shown here. The model and experimental data agree reasonably well, however the experimental points tend to fall slightly outside the expected limits. This may be the result of our neglecting the initial compensation of the samples. An initial donor concentration would shift the experimental points to the right.

Also shown on this figure is one data point for a conventionally melt-compensated sample where the compensating impurity is again phosphorus. Clearly the NTD material is superior with respect to uniformity for a P/B ratio of 2. $\Delta\rho/\bar{\rho}$ for the NTD sample was

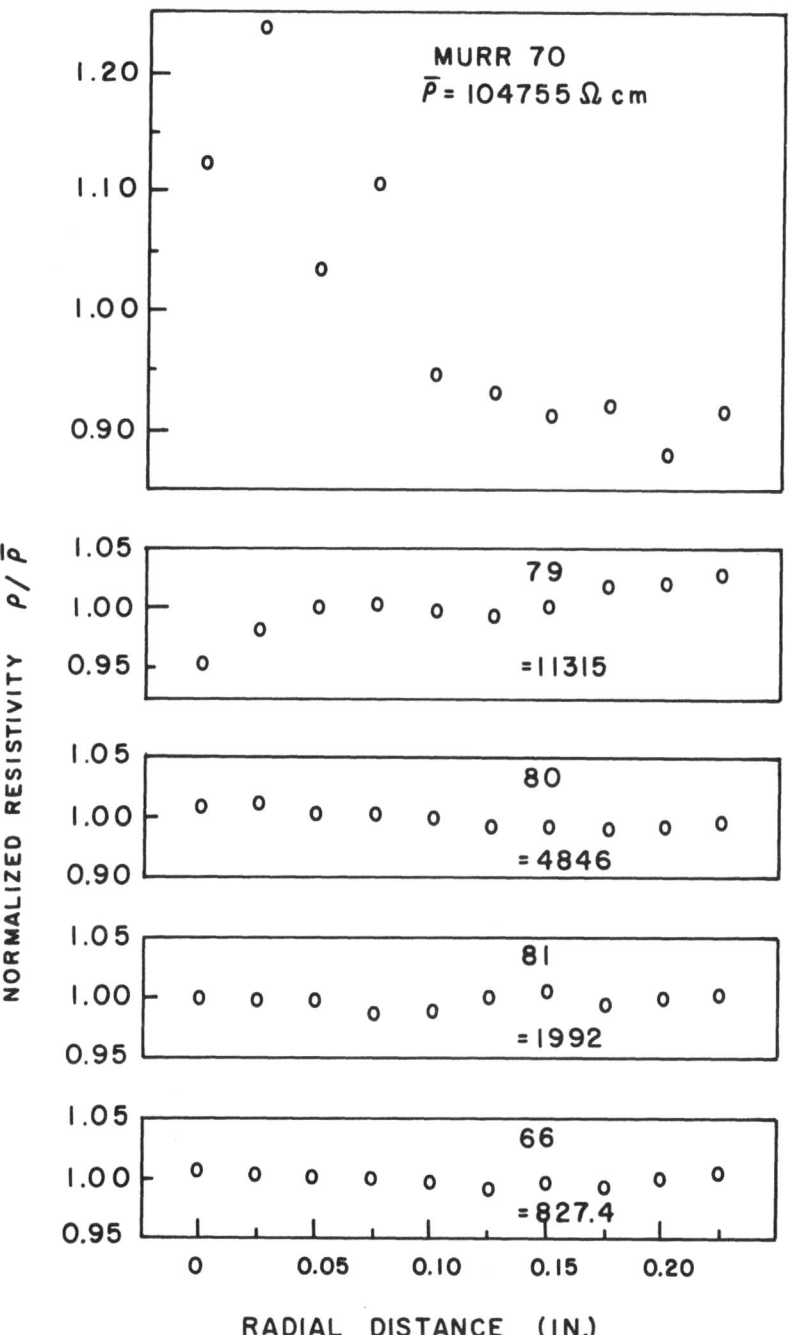

Fig. 4. Normalized resistivity vs. radial probe distance for four-point resistivity measurements on 5 wafers compensated to different n-type resistivities by NTD processing.

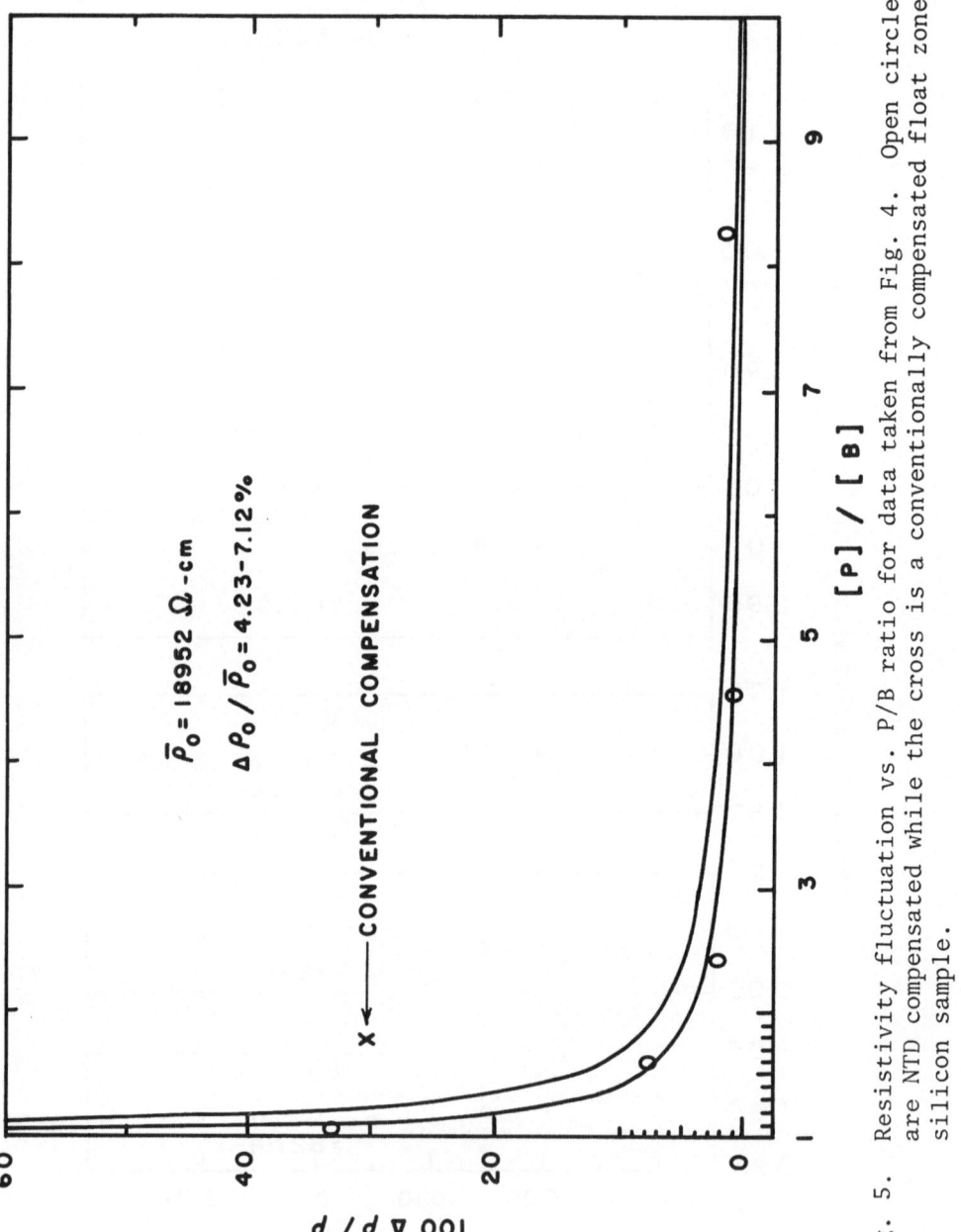

Fig. 5.   Resistivity fluctuation vs. P/B ratio for data taken from Fig. 4.   Open circles are NTD compensated while the cross is a conventionally compensated float zone silicon sample.

experimentally about 5.5% while it was 30% for the conventionally compensated sample.

It is not clear how $\Delta\rho/\bar{\rho}$ varies for conventional compensation vs. compensation ratio, however, several authors[1,10-11] have reported that $\Delta\rho/\bar{\rho}$ is between 10 and 30% for 100 Ω-cm material (a P/B ratio of between 10 and 20). We therefore suspect that the value of $\Delta\rho/\bar{\rho}$ obtained for conventional compensation is relatively independent of compensation ratio.

5. MAXIMUM RESISTIVITY WITHOUT MIXED TYPING

As a last topic, we will use the equations of the previous model to calculate the highest resistivity which can be obtained by NTD-compensation before mixed typing (part of sample n-type and part p-type) occurs. An inspection of Fig. 2 shows that as the fluence increases both ρ_{max} and ρ_{min} move to the right over the peak at ρ_H. When the Q value for the first point, ρ_{max} equals unity, then $\rho_{max} = \rho_i$ and any further compensation will produce mixed typing. This mixed typing will continue until sufficient irradiation causes the Q value for ρ_{min} to pass Q = 1. At this point the crystal becomes entirely n-type and the mixed typing no longer exists.

We can easily picture from Fig. 2 that increasing the initial value of $\Delta\rho/\bar{\rho}$, which corresponds to the width of the shaded region increasing, will lower the average resistivity obtainable as it is defined by Eq. (1) when ρ_{max} is set equal to ρ_H. We have calculated this effect and show the results on Fig. 6. If the initial value of $\Delta\rho/\bar{\rho}$ is zero, then the maximum resistivity for p-type material is ρ_H but the maximum n-type resistivity is clearly ρ_i as seen on Fig. 2. This explains the discontinuity seen in Fig. 6 at the initial fluctuation value of $\Delta\rho/\bar{\rho} = 0$.

The procedure to calculate the rest of the curves is slightly different for n-type than for p-type. For p-type material of an initial $\bar{\rho}$ and $\Delta\rho/\bar{\rho}$, the initial values of ρ_{max} and ρ_{min} are easily determined as

$$\rho_{max} = \bar{\rho} + \frac{\Delta\rho}{2}$$

and

$$\rho_{min} = \bar{\rho} - \frac{\Delta\rho}{2} \quad .$$

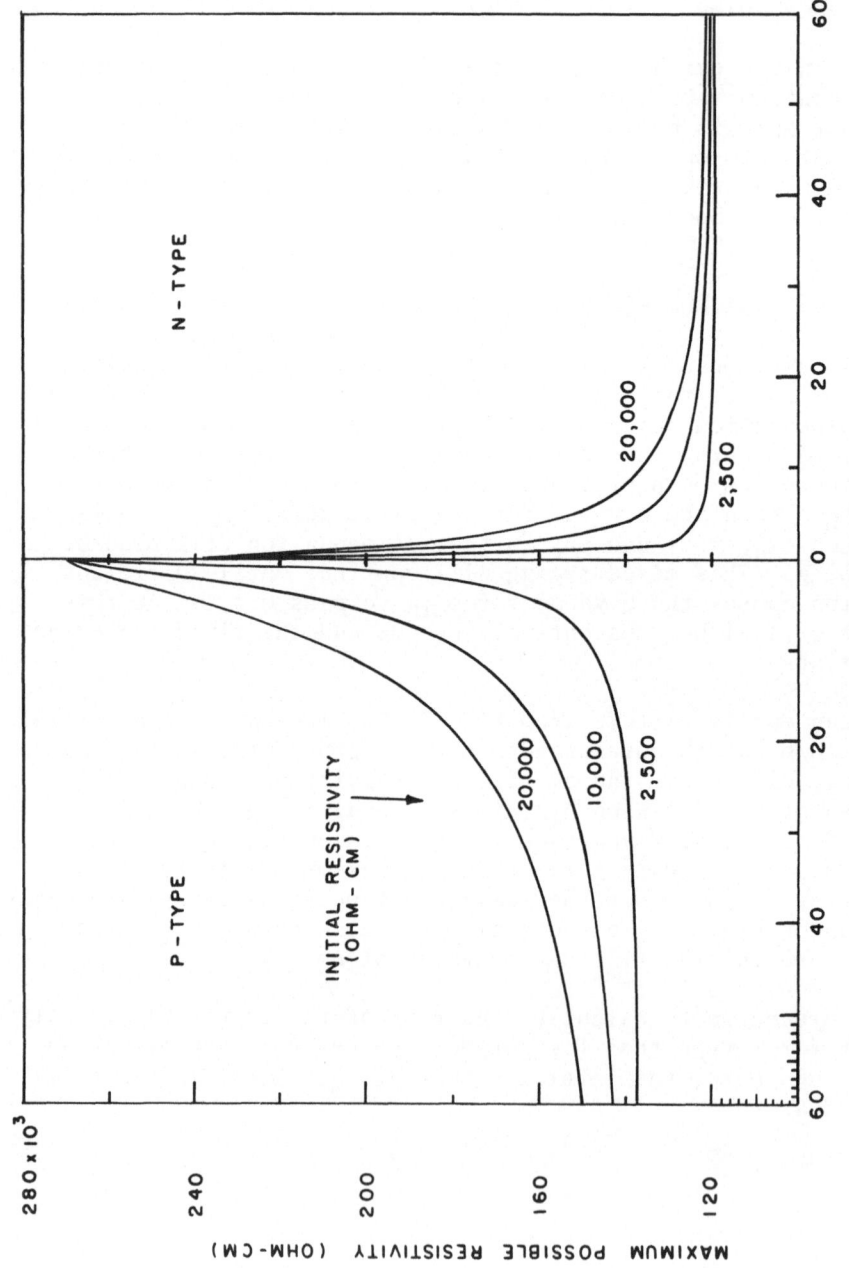

Fig. 6.   Maximum possible resistivity by transmutation compensation before mixed typing
          occurs as a function of the initial resistivity fluctuation before transmutation
          doping.

From Eqs. (15) and (16), the values of B_{max} and B_{min} can now be determined. From these $\bar{B}$ and ΔB are calculated as before.

Now for p-type material, the highest value of $\bar{\rho}$ occurs when $\rho_{max} = \rho_H$ in Fig. 2. This occurs at $Q = Q_H$ which can be calculated from Eqs. (11) and (27). It is clear that

$$N_{A_{min}} - N_D = \bar{B}(1 - Q_H)$$

and that

$$N_{A_{max}} - N_D = \bar{B}(1 - Q_H + \Delta B/\bar{B}).$$

Note that the whole fluctuation ΔB is used now to obtain $N_{A_{max}}$. These equations serve the same purpose that Eqs. (21) and (22) served before. Therefore, substituting these last two equations into Eqs. (23) and (24), and Eqs. (23) and (24) into Eq. (25), the new values of $\rho_{max} = \rho_H$ and ρ_{min} can be determined from which the largest mean resistivity possible before type conversion occurs can be determined by using Eq. (26).

The procedure is similar on the n-type side except that ρ_{max} is now fixed at ρ_i ($Q = 1$) and the fluctuation places ρ_{min} at a lower resistivity and higher Q.

An inspection of the results of Fig. 6 shows that for an initial resistivity fluctuation of 10%, the highest p-type average resistivity possible is just over 200,000 Ω-cm for the best case and less than 200,000 Ω-cm for lower starting resistivities. On the n-type side, there seems little hope of achieving average resistivities above 140,000 Ω-cm without mixed typing. We caution that with typical starting material available today, claims of achieving higher resistivities than these by NTD-compensation suggest that either mixed typing or incomplete annealing of the radiation damage has occurred.

ACKNOWLEDGEMENTS

We wish to thank M. Arst of Rockwell International for supplying the float zone silicon for this experiment and R. Nelson of Rockwell for asking the questions which led to this work. We also wish to thank R. S. Spry of Air Force Materials Laboratory for his continued interest and support.

REFERENCES

+   Also, Department of Physics, University of Missouri-Columbia.
++   Also, Department of Electrical Engineering, University of
       Missouri-Columbia.
*   Sponsored in part by Air Force Materials Laboratory, Air Force
       Systems Command, United States Air Force, Wright-Patterson
       AFB, Ohio 45433 under Contract No. F33615-76-C-5230.
1)   H. J. Janus and O. Malmros, I.E.E.E. Trans. on Electron
       Devices ED-23, 797 (1976).
2)   J. Messier, Y. le Coroller and J. M. Flores, I.E.E.E. Trans.
       Nuc. Sci. NS-11, 276 (1964).
3)   R. N. Thomas, T. T. Braggins, H. M. Hobgood and W. J. Takei,
       J. Appl. Phys. 49, 2811 (1978).
4)   J. M. Meese, Silicon Detector Compensation by Nuclear
       Transmutation, Technical Report AFML-TR-77-178. Air Force
       Materials Laboratory, Wright-Patterson AFB, Ohio 45433,
       (Feb. 1978).
5)   F. W. Voltmer and H. J. Ruiz, N.B.S. Special Pub. 400-10,
       Spreading Resistance Symposium, Gaithersburg, Md. (June 1974),
       p. 191.
6)   H. Y. Fan, Solid State Physics 1, 283 (1955).
7)   F. A. Trumbore, Bell Syst. Tech. J. 39, 205 (1960).
8)   M. L. Schultz, Infrared Phys. 4, 93 (1964).
9)   A listing of this program is available upon request of the
       authors for the price of postage plus listing time.
10)   H. A. Herrmann and H. Herzer, J. Electrochem. Soc. 122, 1568
       (1975).
11)   W. E. Hass and M. S. Schnoller, J. Electronic Mat. 5, 57
       (1976).

TRANSISTOR GAIN TRIMMING IN I^2L INTEGRATED CIRCUITS USING THE NTD PROCESS

E.J. Caine, P.J. Glairon, E.J. Charlson and E.M. Charlson

Dept. of Electrical Engineering, U. of Missouri

Columbia, MO 65211

ABSTRACT

Current injection logic (I^2L) is a linear compatible process using standard bipolar transistors in the inverse mode. This process involves careful control of both base and substrate doping levels to insure that inverse betas are larger than the maximum number of collector outputs (typically three).

This paper concerns application of NTD to alteration of the base and substrate doping to optimize inverse characteristics of I^2L bipolar transistors. Calculations of doping profiles before and after irradiation will be compared to measurements of Gummel numbers and the consequent substrate and base doping levels using the Moll-Ross equations.

$$I_c = \frac{q\, n_i^2 A_j}{(Q_B/D_B)} \exp\left(\frac{q\, V_{BE}}{kT\cdot}\right) \; ; \quad \frac{Q_E}{D_E} = \left(\frac{Q_B}{D_B}\right)\beta$$

Using best estimates of diffusion coefficients, lifetimes in substrate regions are calculated.

Results show that a tenfold increase in substrate concentration increased inverse beta by a factor of approximately ten. Minority carrier diffusion length in the substrate was reduced from approximately 9 to 3 microns as a result of the irradiation. Overall results will be compared with I^2L made on epitaxial substrates with comparable doping.

1. INTRODUCTION

First introduced in 1972, integrated injection logic (I^2L),[1] also known as merged transistor logic (MTL),[2] has become an increasingly popular bipolar logic because of its relatively high speed, high packing density and low power dissipation. The University of Missouri Solid State Laboratory has constructed several logic circuits of I^2L using conventional bipolar fabrication techniques in order to become better acquainted with this new technology.

The basic operation of I^2L can be modeled with a multi-output inverter gate. Figure 1(a) shows the appropriate block diagram, arranged so that if input B is at logic "1", all outputs will be at logic "0" and vice versa. A three collector transistor, together with a constant current source from emitter to base, Fig. 1(b) and (c), make up the I^2L version of the same gate. Note that a lateral PNP transistor is used for the current source and that certain areas of the lateral are common to the NPN multicollector device. Indeed, the collector of the lateral transistor is the base of the NPN vertical transistor and the base of the lateral transistor is the emitter of the vertical transistor. This multiple use of doped regions gives rise to the name of merged transistor logic.

The switching operation of the gate is achieved by forward biasing the base-emitter junction of the lateral to a constant value of current. Unless this current is drawn away from the base of the vertical device where it is entering, by applying a logic "0" to input B, (corresponding to 0.2 volts from the saturated collector of another transistor), the transistor will turn on, causing the base-emitter junction to rise to 0.7 volts (a logic "1"). All output collectors will then try to sink current and drop to 0.2 volts.

Figure 1(d) shows the device layout and cross section. The most important thing to note is that the collectors of the multi-output structure are located on the top of the wafer creating a transistor operated in the inverse manner. Standard bipolar devices have the substrate as the collector and the n+ diffusion as the emitter.

By forward biasing the base-emitter junction of the lateral, holes are injected into the N substrate and some are collected by the P region of the vertical device causing its base-emitter junction to be forward biased. This latter junction will then inject holes back into the substrate and electrons forward into the base of the vertical transistor. Assuming the inverse transistor beta is larger than unity (designated beta inverse because of the collector location), more electrons will be injected from the substrate into the base than holes injected into the substrate.

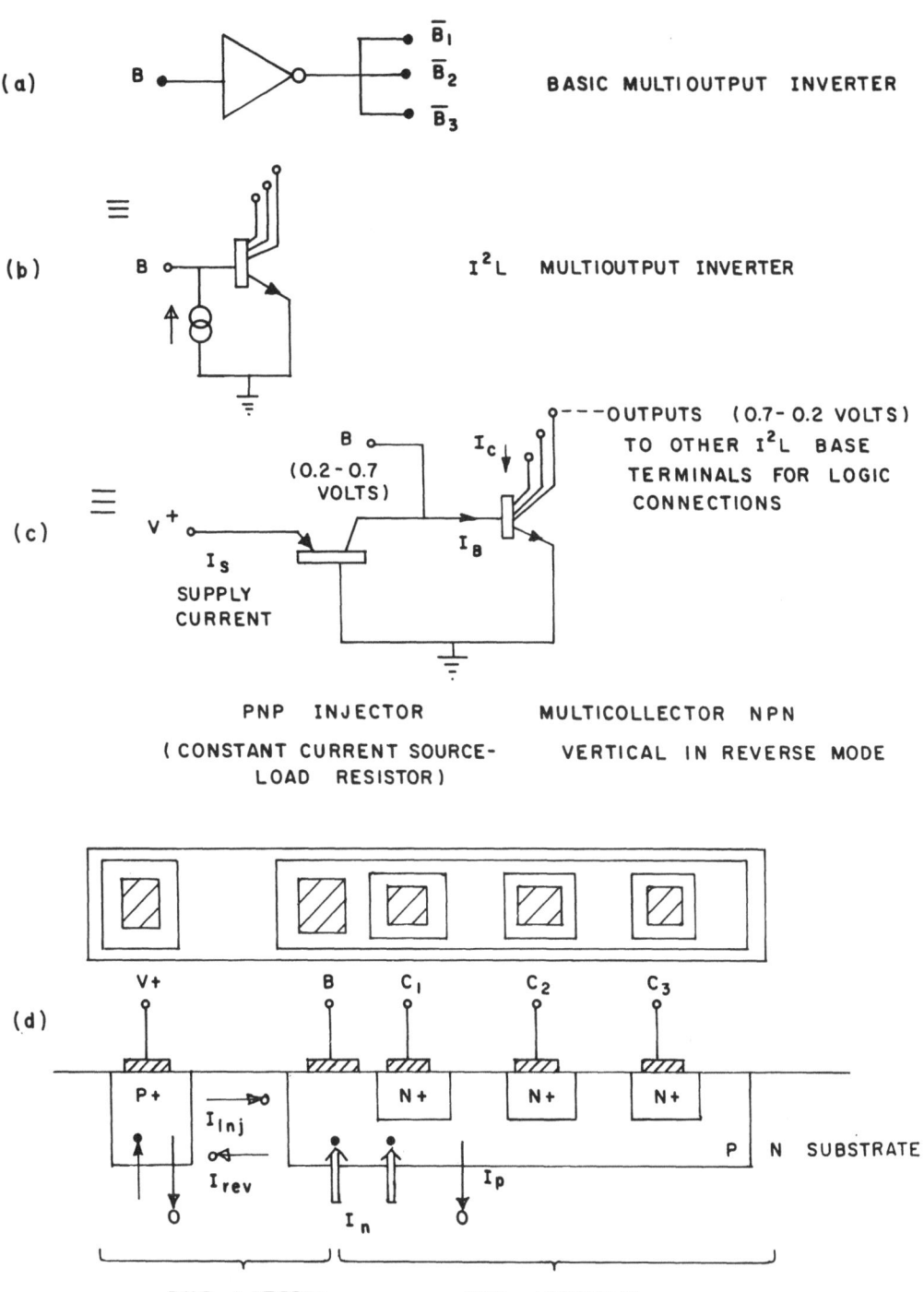

Fig. 1. Schematic and circuit layout of I²L inverter gate.

2. ELECTRICAL REQUIREMENTS FOR I^2L

To change logic states of the inverter, each collector of the vertical device must be able to sink all of the current from a base (input) to which it is connected. Thus, beta inverse must be larger than the maximum number of collectors (fanout) or:

$$\beta_{inv} = \frac{I_C}{I_B} > \text{ maximum number of collectors.}$$

To explain the relationship of beta inverse to process parameters, it is necessary to review some parameters of transistor performance.

Common base current gain:

$$\alpha = \gamma\beta_T$$

Common emitter current gain:

$$\beta = \frac{\alpha}{1-\alpha} \tag{1}$$

For an NPN device such as the multicollector vertical:

γ = Injection efficiency

= $\dfrac{\text{Electron carrier current injected at base-emitter junction}}{\text{Total base-emitter junction current}}$

β_T = base transport factor

= $\dfrac{\text{Electron current at collector junction}}{\text{Electron current at emitter junction}}$

Ideally, γ and β_T should be unity. But γ is less than one because of hole injection across the base-emitter junction and recombination in the base-emitter depletion region. Recombination of injected electrons across the base region causes β_T to be less than one.

Recent studies[3-4] have shown that γ is the limiting factor in α and not β_T as had been previously thought. (Experimental results given later will add support to this claim.) Because of this:

$$\alpha = \gamma = \frac{1}{1 + \dfrac{Q_B/D_B}{Q_E/D_E}} \tag{2}$$

where $Q = \int_0^{x_j} n(x)\, dx$ is the Gummel number[5] and D is the

minority carrier diffusivity.

In fabricating the first samples of I^2L using standard bipolar technology, it was found that inverse beta was typically 0.05. In order to increase beta inverse, two things can be done according to Eq. (2); one, reduce Q_B and two, increase Q_E.

3.   BACKGROUND FABRICATION PROCEDURE

The standard bipolar processing applied to an I^2L structure is listed below:

1.   Diffuse p⁺ regions (25 $\Omega/\square$) to form emitter of injection device (injection bar) and diffuse P regions (200 $\Omega/\square$) to form collector of lateral and base of vertical device. Substrate material is 5 Ω-cm, (111), n-type Czochralski silicon.

2.   Diffuse shallow n⁺ regions (10 $\Omega/\square$) to form collectors of vertical devices and guard rings around P base regions to suppress sidewall injection. (The latter guard rings are not shown in Fig. 1(d) for simplicity.)

3.   Open oxide for contact holes and apply aluminum for inter-connections.

A theoretical diffusion profile has been calculated for the vertical NPN structure using erfc and Gaussian distributions and is shown in Fig. 2(a). Again, note the emitter, base and collector locations from the wafer surface at 0.0 μ depth. This device is labeled Normal because standard processing techniques were used.

4.   NTD PROCESSING

In order to reduce Q_B and/or increase Q_E, it was decided to use neutron transmutation doping (NTD). In this manner, N region doping could be uniformly increased in concentration through the addition of phosphorus and P region doping accordingly decreased through acceptor compensation.

One wafer was stripped of its surface aluminum and placed in the University of Missouri Research Reactor. The NTD data is given as follows:

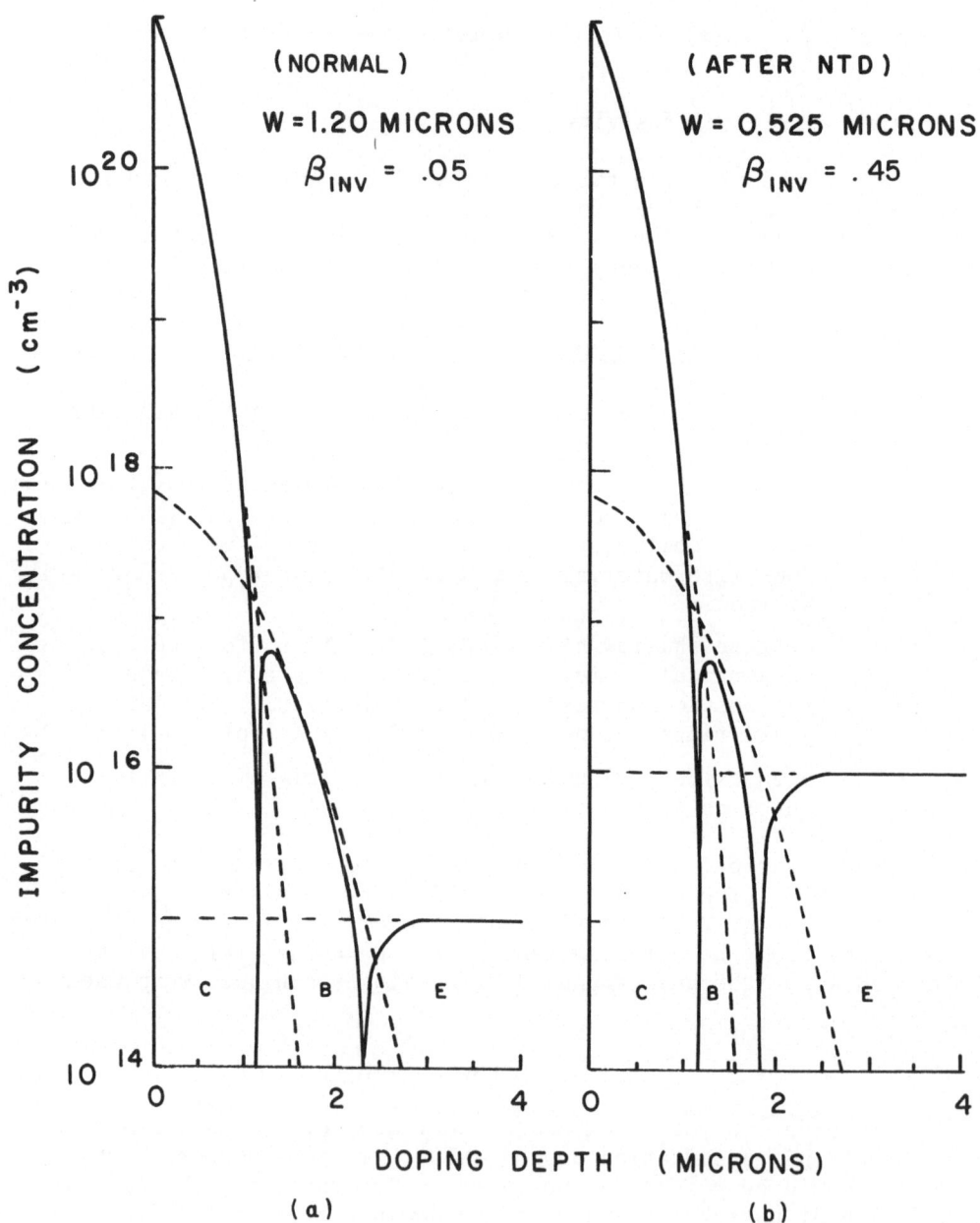

Fig. 2.   Vertical NPN transistor doping profiles.

1.   300 hour irradiation at a 5 x 10^{13}n/cm²sec flux.

2.   9.06 x 10^{15}/cm³ additional phosphorus content.

3.   Sample was then annealed for 15 minutes at 850°C in an argon ambient. Cool-down to room temperature was done slowly over a two and one-half hour period.

Surface aluminum was reapplied for connections, followed by a 15-minute anneal at 400°C in a nitrogen ambient.

Figure 2(b) gives the same vertical transistor diffusion profile as in (a) except for the addition of 9 x 10^{15} phosphorus atoms per cm³ due to NTD. Substrate (emitter) concentration theoretically increased to 10^{16}/cm³ while the base width was reduced from 1.2 μ to 0.525 μ. The ramification of these two changes is that the inverse beta increased from 0.05 with normal processing to typically 0.45 after NTD.

A third version of I²L was also fabricated during this same period whose substrate material was a 5 μ epitaxial layer (1 x 10^{16}/cm³) on a highly doped n⁺ substrate (0.01 Ω-cm). Identical diffusions were used. It should be pointed out that the semiconductor industry normally uses this type of epitaxial wafer as starting material.

5. EVALUATION OF NORMAL, NTD PROCESS AND EPITAXIAL DEVICES

In order to quantitatively explain the improved beta inverse, the Moll-Ross equation was used:[6]

$$I_C = \frac{qn_i^2 A_j}{Q_B/D_B} \; \exp\left(\frac{qv_{BE}}{kT} \right) \tag{3}$$

where:

q =   electronic charge

n_i =   intrinsic carrier concentration for silicon at 300°K

A_j =   base-emitter junction area

k =   Boltzmann constant

T =   absolute temperature

v_{BE}=   base-emitter junction voltage

Q_B =   Gummel number for base region

D_B =   diffusivity of minority carriers in base region

This equation shows that under active bias, transistor collector current is exponentially related to base-emitter voltage.

Several voltage-current points were measured for vertical NPN devices on each of the normal, NTD and epitaxial I^2L and are shown in Figs. 3 and 4. A straight line on the log-linear graph for approximately two decades of current verifies the exponential relationship. Deviations at higher currents are due to bulk material resistive voltage drops.

With these data lines extrapolated to zero voltage points (v_{BE} = 0) a saturation collector current can be determined. Then from Eq. (2), Q_B/D_B can be found. Also, it can be shown that by using Eqs. (1) and (2):

$$\frac{Q_E}{D_E} = \left(\frac{Q_B}{D_B}\right) \beta \tag{4}$$

After determining Q_B/D_B and beta from data, Q_E/D_E can be calculated using Eq. (4). Assuming a bulk diffusion contant of D_E = 12.93 cm^2/sec, emitter Gummel number, Q_E, diffusion length for minority carriers in the emitter, L_E, and lifetime for emitter minority carriers, τ_p, can be determined. Figure 5 summarizes the experimental results.

Results of Q/D for devices before and after irradiation follow predictions. For the base region, Q_B/D_B decreased to approximately one-third that of the non-irradiated sample and in the emitter region, Q_E/D_E increased to approximately 3.5 times that of the non-irradiated sample. The emitter Gummel number found by multiplying Q_E/D_E by a best estimate of bulk material hole diffusivity, shows a four-fold increase. Then using this same diffusion constant for both normal (non-irradiated) and NTD (irradiated) I^2L, one obtains a decrease in emitter diffusion length and minority carrier lifetime after irradiation.

Operating the multicollector vertical transistor in a non-I^2L mode (collector and emitter interchanged), enabled the forward beta to be measured. Little difference is seen between samples, indicating that a decrease in base width and most likely injected minority recombination, did not play a major role in determining gain of the devices. So β_T, which is dependent on this recombination, is not the primary gain parameter.

Epitaxial I^2L with its characteristics is shown for comparison. Its inverse beta was typically 1.7, indicating that lower resistivity substrate material is a key consideration. Although still not large enough for proper circuit operation, it is believed that beta inverse can be increased with certain device geometry changes.

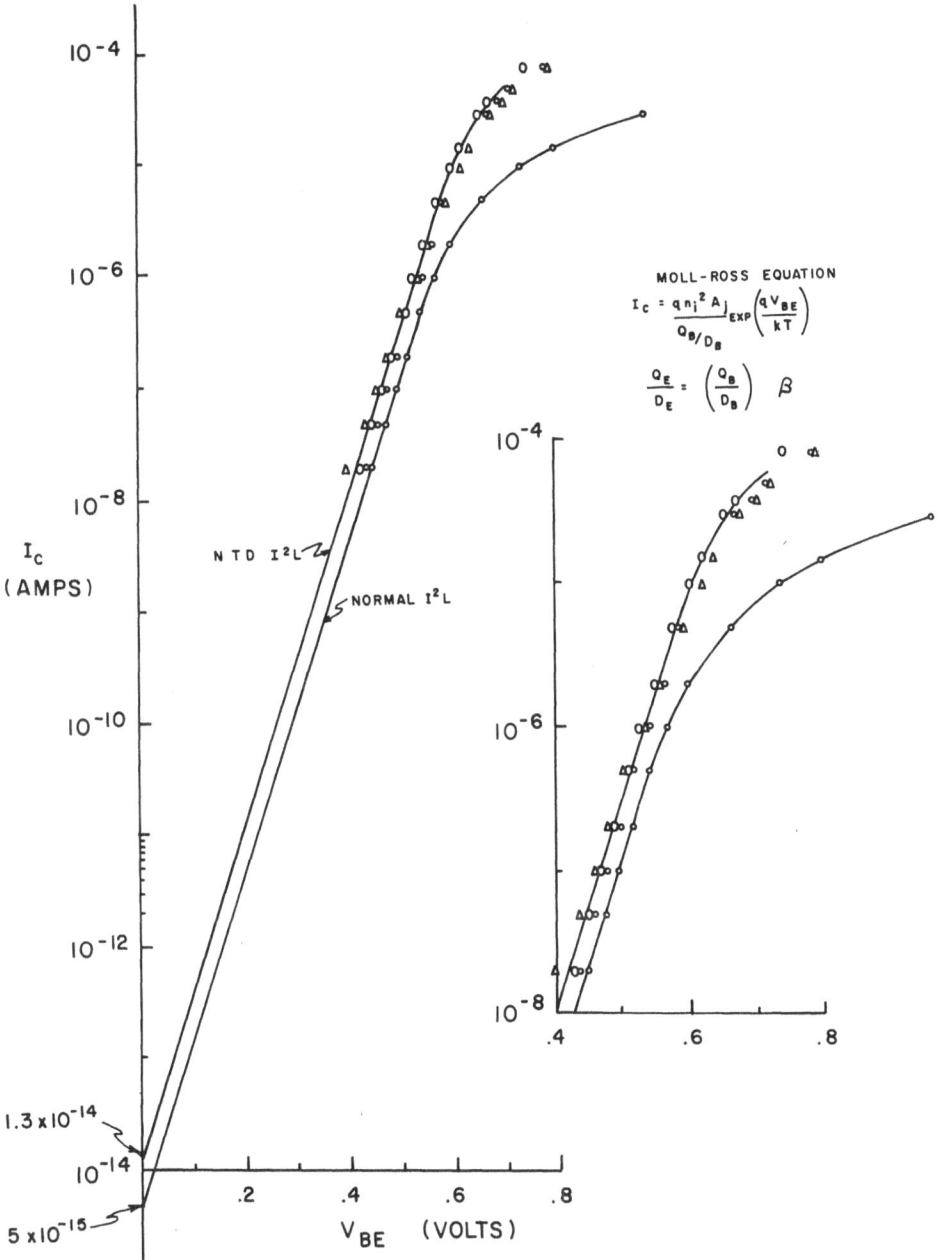

Fig. 3.   Normal and NTD I²L current-voltage curves.

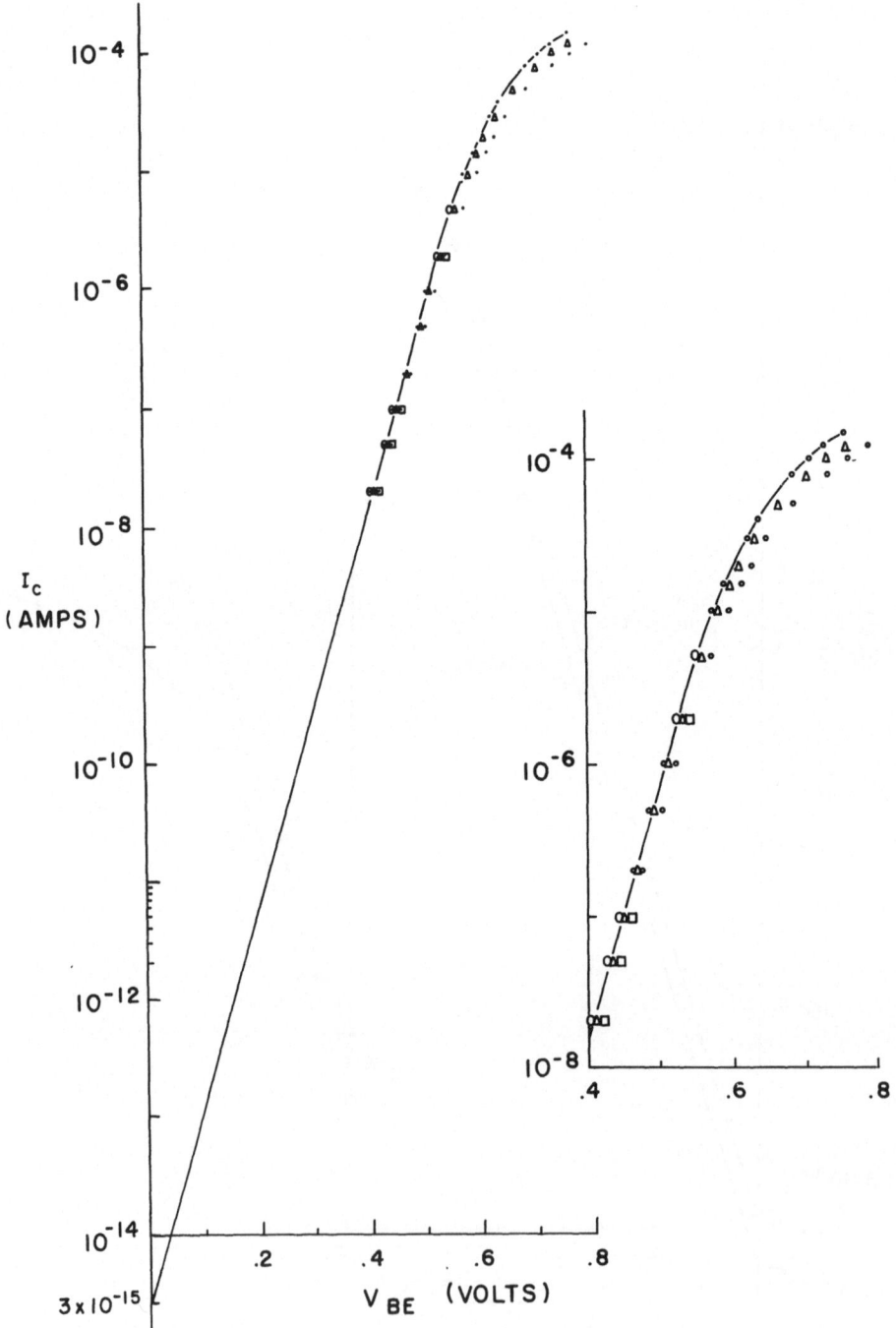

Fig. 4.   EPI I^{2}L current-voltage curves.

	Normal I^2L	NTD I^2L	Epi I^2L
$\dfrac{\text{Base Gummel No.}}{D_B} = \dfrac{Q_B}{D_B} \left(\dfrac{\text{sec}}{\text{cm}^4}\right)$	1.77×10^{12}	6.77×10^{11}	2.93×10^{12}
$\dfrac{\text{Emitter Gummel No.}}{D_E} = \dfrac{Q_E}{D_E} \left(\dfrac{\text{sec}}{\text{cm}^4}\right)$	8.83×10^{10}	3.05×10^{11}	4.98×10^{12}
*Emitter Gummel No. $= Q_E$ (cm^{-2})	1.15×10^{12}	3.95×10^{12}	6.46×10^{13}
Diffusion Length in Emitter $= \dfrac{Q_E}{N_E} = L_E$ (μ)	11.45	3.95	51.6
Forward Beta $= \beta$	50	$48\text{-}55$	$100\text{-}210$
Inverse Beta $= \beta_{INV}$	0.05	0.45	1.7
Emitter Minority Carrier Lifetime $= \tau_p$ (sec)	1.01×10^{-7}	1.207×10^{-8}	2.06×10^{-6}

*Bulk diffusion constant assumed $= 12.93 \dfrac{\text{cm}^2}{\text{sec}}$

Fig. 5. Experimental results.

Typical common emitter output characteristic curves are shown in Fig. 6 for both forward and inverse beta configurations. In the forward mode, devices actually improved by increasing the collector current level at which gain exhaustion appeared. The increased beta can be seen for the inverse mode with a decrease in access resistance.

6. SUMMARY

The results of this paper can be summarized as follows:

1. NTD can predictably be used in modifying a semiconductor device after standard processing steps giving the designer an added dimension in flexibility.

2. Little degradation is seen in device characteristics.

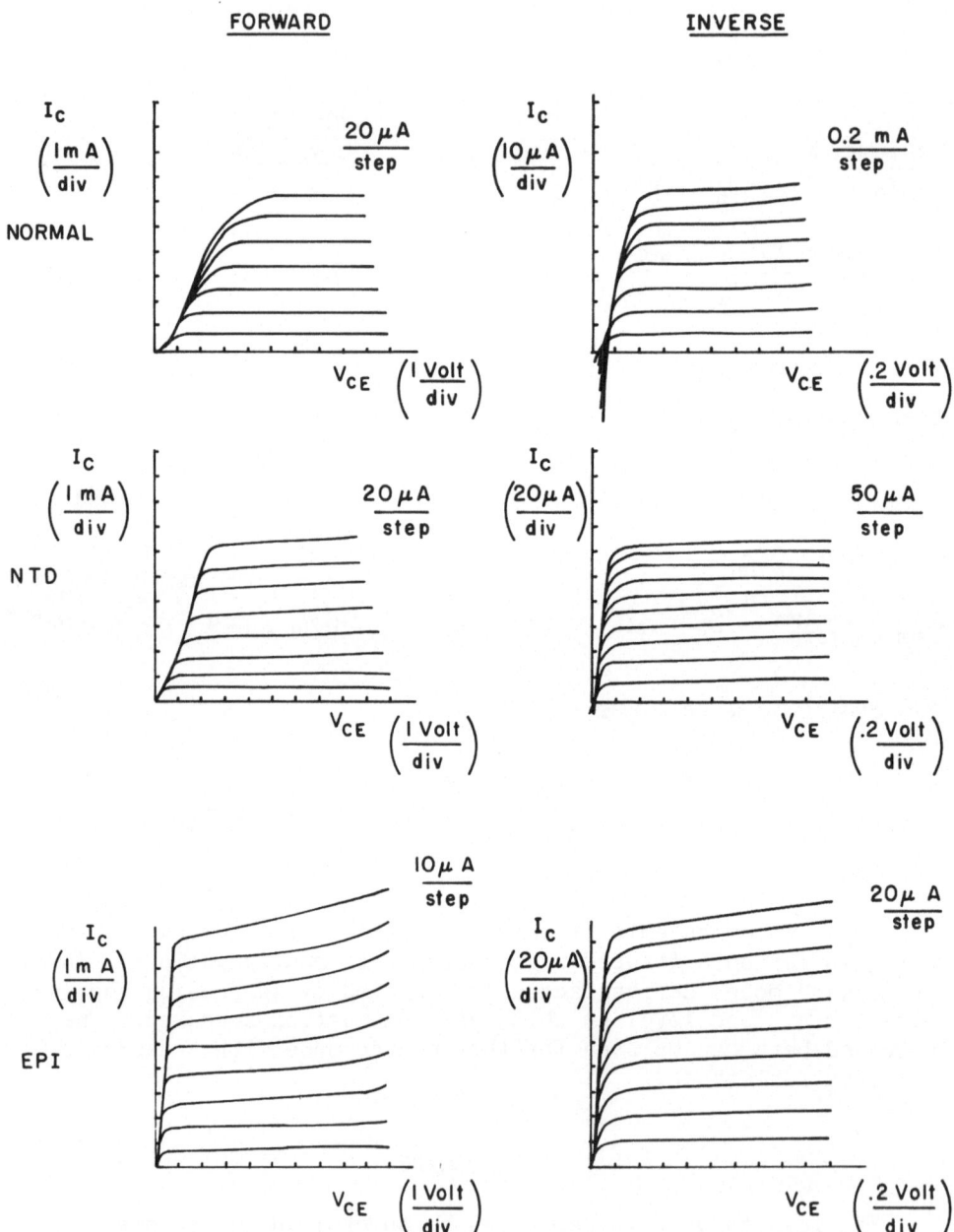

Fig. 6.   Common emitter output curves.

3. NTD could possibly be used in other device modifications,

    a. Turn-on voltages of MOSFETS.

    b. Junction characteristics:

        Voltage breakdown
        Capacitance vs. voltage.

At present, the I²L integrated circuit layout is being modified to increase beta inverse primarily by increasing the collector diffusion areas. Future work will include building I²L on 5 Ω-cm float zone starting material to facilitate easier post-irradiation study and using 0.5 Ω-cm non-epitaxial starting material to compare results with NTD I²L already seen here.

7. ACKNOWLEDGMENT

The authors wish to thank Woo Joong Youn for the preliminary I²L integrated circuit layout used for this work and Western Electric at Lee's Summit, Missouri for the epitaxial material.

REFERENCES

1) K. Hart and A. Slob, IEEE J. Solid-State Circuits, SC-7, 346 (1972).

2) H. H. Berger and S. K. Wiedmann, IEEE J. Solid-State Circuits, SC-7, 340 (1972).

3) P. J. Kannam, IEEE Trans. Electron Devices, ED-2, 845 (1973).

4) R. P. Mertens, et al., IEEE Trans. Electron Devices, ED-20, 772 (1973).

5) H. K. Gummel, Proc. IRE, 49, 834 (1961).

6) J. L. Moll and I. M. Ross, Proc. IRE, 44, 72 (1956).

DETERMINATION OF THE NEUTRON FLUX AND ENERGY SPECTRUM AND CALCULA-
TION OF PRIMARY RECOIL AND DAMAGE-ENERGY DISTRIBUTIONS FOR MATERIALS
IRRADIATED IN THE LOW TEMPERATURE FAST-NEUTRON FACILITY IN CP-5*

M. A. Kirk and L. R. Greenwood

Argonne National Laboratory

Argonne, IL 60439

ABSTRACT

We have determined the absolute differential neutron-energy
spectrum for the low temperature fast-neutron irradiation facility
in the CP-5 reactor by means of a 20-foil activation technique. This
technique employs the most recent version of the SAND-II computer
code, which iteratively unfolds the neutron spectrum by fitting the
foil activities. A Monte Carlo routine was also employed to calculate standard-deviation errors in each neutron-energy group. Using
this differential neutron spectrum we have calculated for numerous
elements, total recoil cross sections, detailed primary recoil group
distributions, total damage-energy cross sections, damage-energy
distributions, and an error analysis based on the uncertainties in
the neutron spectrum. The significance of this information with
respect to the interpretations of various neutron radiation damage
experiments is discussed. A general observation about neutron radiation damage in silicon is also suggested.

1. INTRODUCTION

As the field of neutron radiation damage in materials has
matured, the need has arisen for more detailed knowledge of the
neutron flux and energy spectra in which experiments are performed.
Impetus has been provided by the current need to correlate experiments performed at different neutron sources, especially fission
and 14-MeV neutron facilities; the necessity to simulate high-dose
neutron bombardment by ion bombardment; and recent advances in
theoretical and computer treatments, for which a closer comparison
with experiments is desirable. In the past, the results of many

neutron-damage experiments have been less than precise. Agreement
among the results of experiments performed at different facilities,
and between experiment and theory, has often been no better than
within a factor of two or three. This has usually been a result of
experimental difficulties inherent in these kinds of studies, especially uncertainties in neutron fluence and energy spectra. As the
experiments and their interpretations become more sophisticated,
more accurate determinations of neutron fluence and spectrum details
will permit more exact physics to be done. It is for these reasons
that we have performed the measurements and calculations described
in this paper.

The Neutron Radiation Damage Group, Materials Science Division,
Argonne National Laboratory, is concerned with basic studies on the
interaction of neutrons with various materials. Recent work has
included studies of metals, alloys and superconductors bombarded by
thermal neutrons, fast neutrons and fission fragments. Changes in
properties such as resistivity, magnetic saturation, volume, critical
superconducting current and temperature, and neutron sputtering have
been measured during or following bombardments at temperatures between 5 and 100°K and near room temperature. These property changes
have been related to various damage production mechanisms, annealing
and clustering of defects, void formation and flux pinning. The
irradiation facilities used in these studies are located at the CP-5
reactor. A recent addition to these facilities is a moderately high
temperature thimble for irradiation by thermal neutrons or fission
fragments in a controlled atmosphere and at controlled temperatures
between approximately 500 and 900°K. Past emphasis, however, has
been on irradiation at lower temperatures, down to $\sim$5°K. Two
different cryogenic facilities exist for irradiation with either a
highly thermalized flux or a slightly degraded fission-neutron flux.
Both are capable of controlled irradiation temperatures between
about 5 and 400°K. The cryogenic thermal-neutron facility has a
nominal flux of 6 x 10^{16}n/m^2/sec and a cadmium ratio of approximately 400. The cryogenic fast-neutron facility, VT53, has a flux
of 2.2 x 10^{16}n/m^2/sec for neutron energies > 0.1 MeV. The measurement of the flux and neutron-energy spectrum in this latter
facility is the subject of the next section.

2. NEUTRON FLUX AND ENERGY SPECTRUM

Figure 1 shows a simplified schematic of VT53, the cryogenic
fast-neutron irradiation facility at CP-5 (a more detailed description is given in Ref. 1). This vertical thimble is located in the
graphite reflector region surrounding the D$_2$O moderator tank, 1.2 m
from the center of the reactor core of radius 0.3 m. The main
elements displayed in this schematic are the Zircaloy-2 + 15 wt%
^{235}U fuel cylinder (the source of fission neutrons), boron carbide

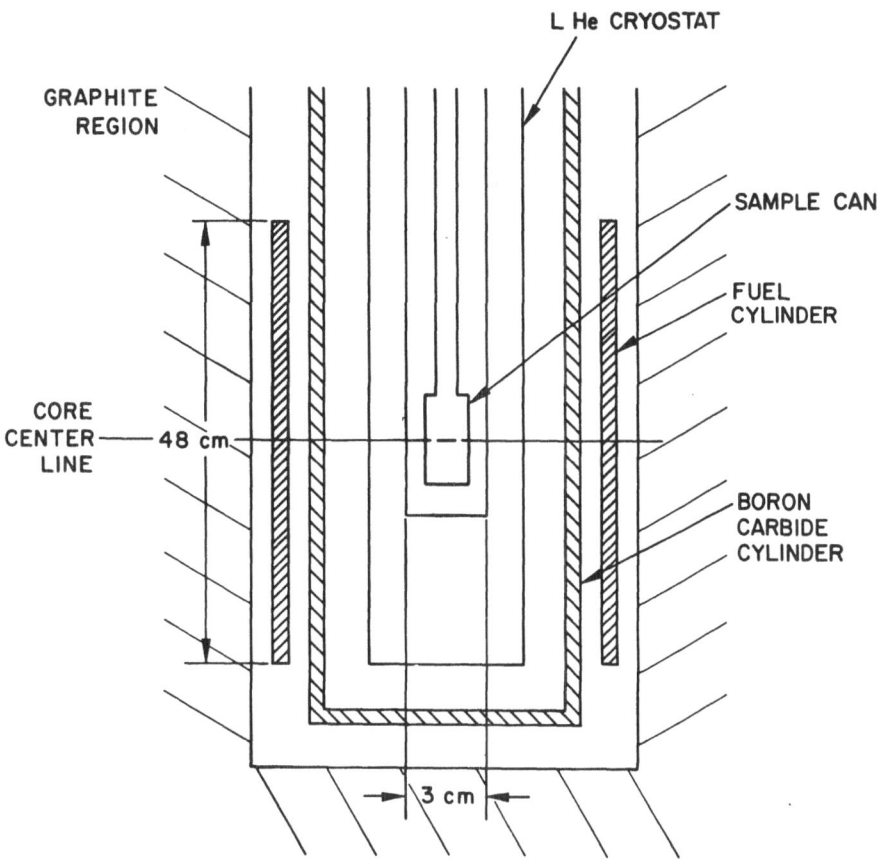

Fig. 1. Simplified schematic of the cryogenic fast-neutron irradiation facility in CP-5 (VT53).

cylinder (the thermal-neutron shield), simplified liquid helium
cryostat, and typical sample location.

A multiple foil activation method was used to determine the
neutron flux and energy spectrum in the illustrated sample location.
The SAND-II computer code[2] was used to iteratively unfold the neutron
spectrum by fitting the foil activities. This method calculates
activities per target nucleus of a set of n foil reactions by means
of the expression

Table 1. Foil neutron reactions used in neutron energy spectrum
          fitting with the SAND-II computer code and deviations
          measured from calculated activity.

Foil Reaction	Deviation (%)
Au-197 (n, γ) Au-198	5.98
Au-197 (n, γ) Au-198 (Cd-covered)	-6.74
Co-59 (n, γ) Co-60	-5.11
Co-59 (n, γ) Co-60 (Cd-covered)	4.53
Sc-45 (n, γ) Sc-46	-0.172
Sc-45 (n, γ) Sc-46 (Cd-covered)	0.29
Fe-58 (n, γ) Fe-59	-0.01
U-235 (n, F) Fission products	-4.44
U-238 (n, F) Fission products	5.64
Np-237 (n, F) Fission products	-0.27
U-238 (n, γ) U-239	6.74
Fe-54 (n, p) Mn-54	2.01
Al-27 (n, α) Na-24	-6.04
Ti-46 (n, p) Sc-46	5.67
Ti-47 (n, p) Sc-47	-9.65
Ti-48 (n, p) Sc-48	7.95
Co-59 (n, p) Fe-59	-6.68
Ni-58 (n, p) Co-58	2.71
Ni-58 (n, 2n) Ni-57	0.00
Nb-93 (n, 2n) Nb-92m	-0.86

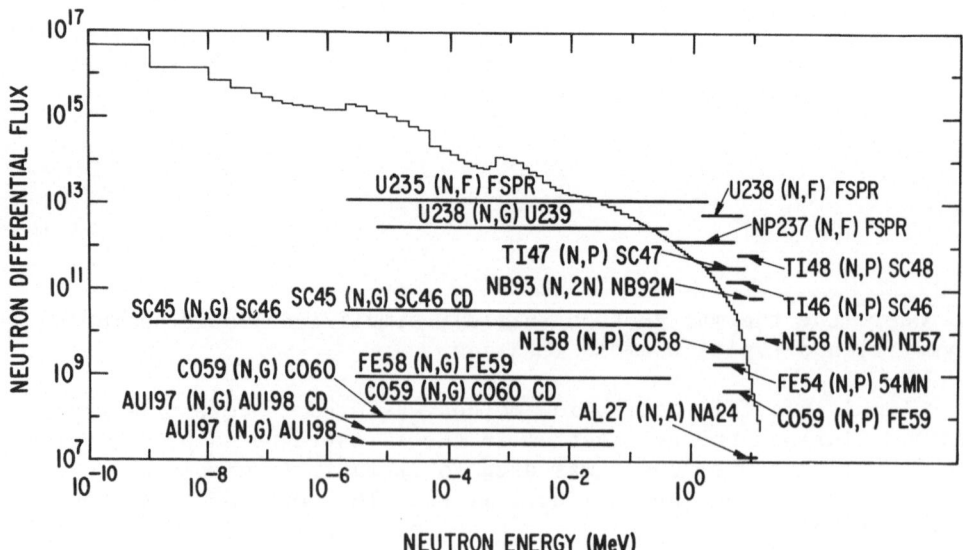

Fig. 2. SAND-II solution of the VT53 differential neutron flux energy
         spectrum. Horizontal bars indicate the energy ranges over
         which 90% of the various reaction activities are produced.

$$A_i^{(K)} = \int_0^{18 \text{ MeV}} \sigma_i(E) \phi^{(K)}(E) dE \tag{1}$$

where

$A_i^{(K)}$ = the saturated activity of the i'th foil calculated at the Kth iteration,

$\sigma_i(E)$ = the energy dependent cross section for the pertinent (n,γ), (n,p), (n,α), (n,n'), n,2n), (n,f), etc. reactions in the i'th foil,

and $\phi^{(K)}(E)$= the Kth iterative differential neutron flux.

The input spectrum, $\phi^{(o)}(E)$, was taken from an earlier approximate determination in the same facility.[3] The energy dependent cross sections were taken from ENDF/B-IV. Foil activities were measured by the Dosimetry Group of the Analytical Chemistry Laboratory with Ge(Li) detectors over several decay-γ half-lives for each reaction. Peak integrations and Compton background subtractions were done by means of computer programs in routine use by the Dosimetry Group. Similar activity measurements have been shown to agree within 1.5% with measurements made at other laboratories in the ILRR program. Cross section corrections for self-shielding and cover foils were made for foil and wire geometries in an isotropic flux prior to spectrum unfolding. The SAND-II program compared the calculated activities with the measured activities following each iteration of Eq. (1). It then adjusted the differential neutron spectrum (100 energy groups) after each iteration until the standard deviations of measured-to-calculated activity ratios became stable to within less than one percent change per iteration. This was accomplished after 17 iterations. The spectral changes are smoothed over energy during each iteration to maintain smooth spectra and avoid resonance effects. Table 1 gives the 20 foil reactions employed with the deviations of measured from calculated activities after 17 iterations. No foil reactions were discarded. Figure 2 shows the solution of the differential neutron flux spectrum along with the energy ranges over which 90% of the various reaction activities are produced.

An error analysis procedure was performed with the new code SANDANL. This code employs a Monte Carlo routine to simultaneously vary uncertainties in the measured activities, cross sections and input spectrum. A standard deviation error is thus produced in each neutron energy group. Figure 3 shows the mean, plus and minus one standard deviation, neutron energy spectra in lethargy units. Some portions of the spectrum, notably between 0.1- and 1.0-MeV neutron energies, have relatively large errors because of a lack of sensitivity of foil reactions to neutrons in this energy region (see Fig. 2). Unfortunately, no neutron reactions have any significant

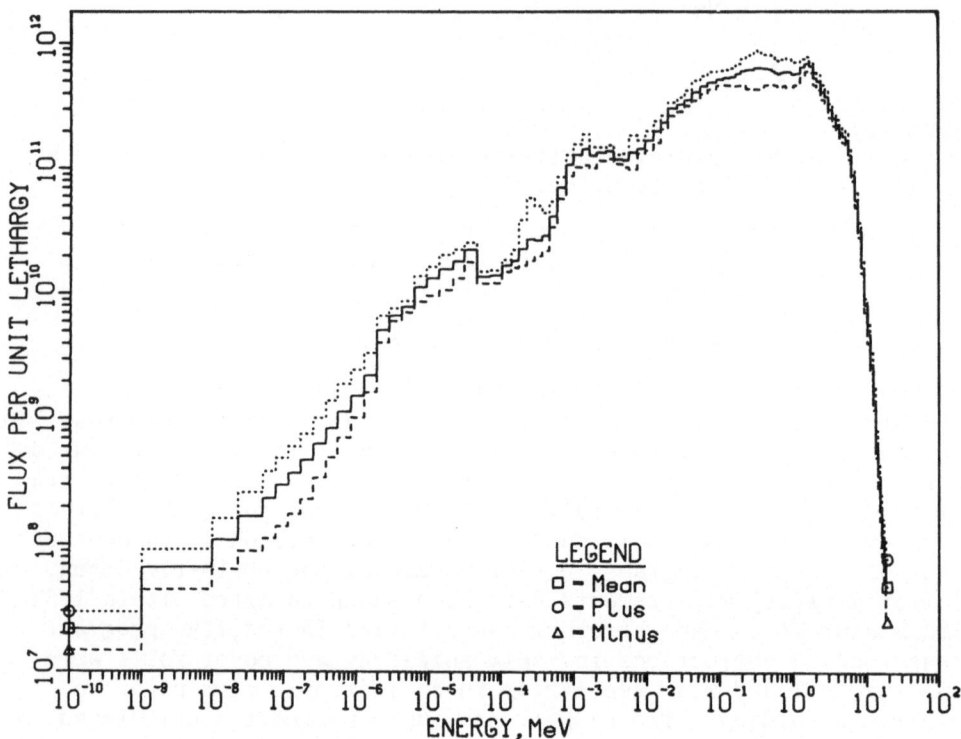

Fig. 3. SANDANL results for the mean, plus and minus one standard
        deviation, neutron energy spectra in lethargy units.

response in any element within this region. The bump in the spectrum
near the 10^{-3} MeV neutron energy region is possibly due to fission
neutrons from the fuel cylinder, reflected back into the cryostat by
the surrounding graphite.

Finally the integrated flux for neutron energies greater than
0.11 MeV is 2.2 x 10^{16}n/m^2/s (± 13%) with the reactor operating at
5 MW (full power). This neutron flux and energy spectrum is one of
the most accurately determined in any facility to date.

3. DAMAGE PARAMETER CALCULATIONS

An accurate description of the neutron flux and energy spectrum
is not very useful by itself in helping the experimentalist to
understand neutron radiation damage effects. However, in the past
few years several computer programs[4-6] have been developed to calculate radiation damage parameters of interest using the experimen-

tally determined neutron energy spectrum. One of these programs (DISCS), taken from the work of Odette and Doiron,[4] is used here. In addition, the program SPECTER was created to calculate spectrum averaged quantities of interest. These two programs were used to calculate, for 21 elements, the differential and group primary recoil distributions and the spectrum averaged total primary recoil cross sections. The elastic, inelastic, and three nonelastic contributions to the total recoil distributions were tabulated. For the present degraded fission neutron energy spectrum, the spectrum averaged recoil cross sections are dominated by the elastic contribution (> 95% for all 21 elements). The contribution of (n, γ) recoil processes is not included in these recoil spectra. They can be approximately included[6] based on the calculated average (n, γ) recoil energies[7] for various elements. However, they are of minor significance in the present neutron spectrum where the ratio of subcadmium to epicadmium flux is about 0.001 (Cd ratio of 1.001).

The elements silicon, nickel, niobium, and gold were selected to illustrate the results of these calculations. Table 2 gives the group distributions (partially collapsed from the original, more detailed group structure) of primary recoils and the total primary recoil cross sections+ for these four elements. Also displayed in this table (for gold only) is an error of one standard deviation for recoils in each group, based on the covariant error matrix generated during the neutron spectrum error analysis by the SANDANL code. To facilitate comparison among the four elements, Fig. 4 shows the integral distribution of primary recoils.

Using the Robinson analytical approximation[8] to the Lindhard et al.[9] theory of electronic energy losses, it is also possible to calculate the damage energy distribution and spectrum averaged total damage energy cross sections. The number of Frenkel defects (interstitial and vacancy pairs) produced by a primary recoil of energy T is generally proportional to the damage energy available from this recoil, which is just the total recoil energy, T, minus the electronic energy losses at this recoil energy. The distribution of damage energy over the recoil energy groups thus gives a good indication of how the Frenkel defects are distributed with primary recoil events. As an example, Table 3 gives the distributions of damage energy in recoil energy groups and the spectrum averaged damage energy cross sections for the same four elements shown in Table 2. Figure 5 graphically illustrates the corresponding integral damage energy distributions.

4. DISCUSSION

The integrated neutron flux determined by this work (2.2×10^{16} n/ m^2/s $\pm 13\%$, for E_n > 0.1 MeV) for the low temperature fast neutron

Table 2. Group distributions of recoils and spectrum averaged cross sections in gold, niobium, nickel, and silicon irradiated in the VT53 fast neutron energy spectrum.

	Si	Ni	Nb	Au	
Primary Recoil Energy Group	Primary Recoil Distributions (%)				
					+/-
0-5 eV	2.4	4.0	2.6	11.8	1.4
5-10 ↓	0.4	1.7	1.2	5.1	0.9
10-20	0.6	2.9	1.9	5.6	1.0
20-50	1.5	6.2	4.3	6.7	0.6
50-100	1.9	6.8	3.9	5.4	0.3
100-200	2.4	8.7	4.9	7.2	0.6
200-400	2.7	10.1	6.1	9.4	0.9
400-600	1.8	8.0	4.7	6.4	0.7
600-800	1.5	6.8	4.1	4.7	0.6
800-1000	1.3	4.4	3.5	3.7	0.6
1-1.5 keV	2.8	5.4	7.2	6.6	1.4
1.5-2 ↓	2.4	3.9	5.5	4.2	1.1
2-3	3.9	5.3	8.4	5.4	1.5
3-5	6.7	5.8	11.5	5.8	1.3
5-7	5.5	3.6	7.2	3.0	0.5
7-10	6.9	3.9	6.7	2.6	0.4
10-20	18.8	5.7	9.0	3.9	0.4
20-40	18.1	3.3	4.7	2.1	0.1
40-60	6.7	1.3	1.5	0.3	0.02
60-80	3.5	0.8	0.6	0.08	0.004
80-100	2.2	0.5	0.3	0.04	0.003
> 100	6.0	0.9	0.3	0.2	---
	Spectrum Averaged Cross Sections (barns)				
$\sigma_{elastic}$	2.86	8.82	6.77	9.35	±12%
$\sigma_{inelastic}$	0.05	0.12	0.40	0.80	± 9%
$\sigma_{n,\ 2n}$	0.0001	0.001	0.001	0.002	±21%
$\sigma_{n,\ p}$	0.002	0.02	0.0001	--	
$\sigma_{n,\ \alpha}$	0.001	0.001			
σ_{Total}	2.91	8.96	7.17	10.15	±11%

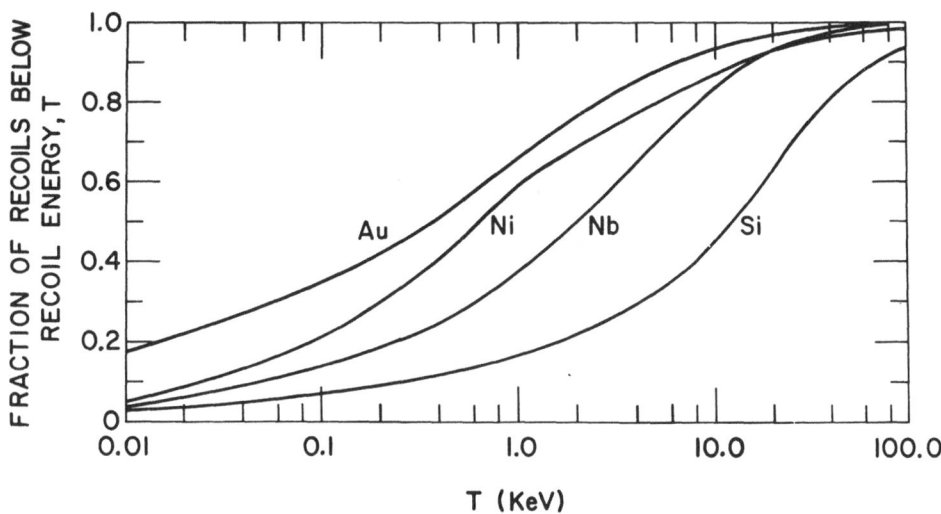

Fig. 4. Integral distributions of primary recoils in Au, Nb, Ni, and Si irradiated in VT53.

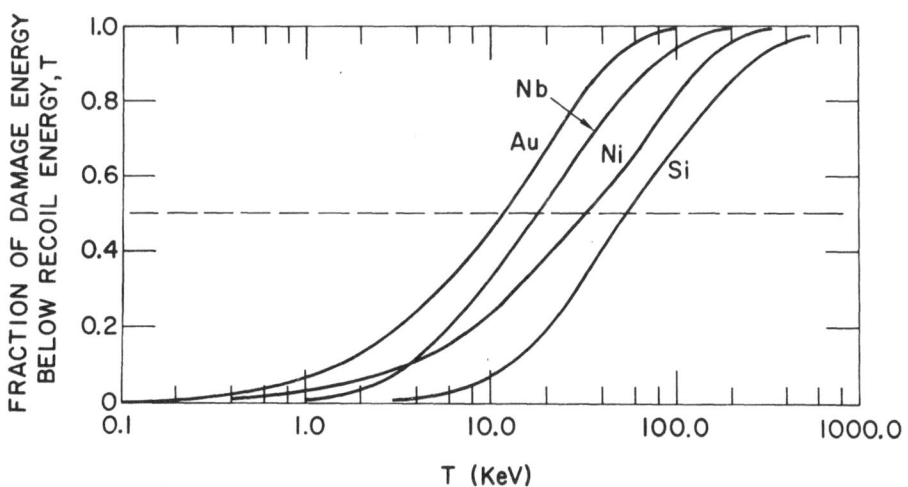

Fig. 5. Integral damage energy distributions for Au, Nb, Ni, and Si irradiated in VT53.

Table 3. Group distributions of damage energy and spectrum averaged
          damage energy cross sections ($<\sigma T_d>$) in gold, niobium,
          nickel and silicon irradiated in the VT53 fast neutron
          energy spectrum.

	Si	Ni	Nb	Au
Primary Recoil Energy Group	Damage Energy Distribution (%)			
0-5 eV	0.00	0.001	0.001	0.01
5-10 ↓	0.0002	0.003	0.002	0.02
10-20	0.0005	0.01	0.006	0.04
20-50	0.003	0.05	0.03	0.11
50-100	0.008	0.1	0.05	0.2
100-200	0.02	0.3	0.14	0.5
200-400	0.04	0.6	0.3	1.2
400-600	0.05	0..8	0.4	1.4
600-800	0.06	1.0	0.5	1.4
800-1000	0.06	0.8	0.6	1.5
1-1.5 keV	0.2	1.4	1.6	3.6
1.5-2 ↓	0.2	1.4	1.7	3.2
2-3	0.5	2.6	3.6	5.7
3-5	1.3	4.3	7.6	9.5
5-7	1.6	4.1	7.1	7.4
7-10	2.8	6.2	9.2	8.9
10-20	12.7	14.7	20.7	22.6
20-40	21.4	16.2	20.4	22.3
40-60	12.6	10.7	10.9	6.2
60-80	8.9	9.5	6.3	2.0
80-100	6.6	6.8	3.5	1.2
> 100	31.1	18.5	5.4	0.9

Spectrum Averaged Damage Energy Cross Sections (keV-barns)

	Si	Ni	Nb	Au
$< \sigma T_d >$	42.8	36.0	33.9	19.5 (±8.7%)

facility in CP-5 is 70% greater than that determined less accurately
in 1964.[3] The amount of [235]U burnup in the fuel cylinder over this
period of time is not known with certainty. However, based on a
comparison of resistivity damage rate measurements in copper made
over a comparable period of time,[10-11] the burnup is about 5% for
an 8 year time period. Thus, it is reasonable to assume that the
flux (corrected for burnup) in this facility has remained constant
in time within the uncertainty of the present measurement. It is
indeed found that the major cause of the difference between the
flux measurements is the improvement in accuracy of the cross section

data on which these flux determinations are based. Therefore, any
use of data from previous experiments in this facility will employ
the presently determined neutron spectrum and integrated flux
values. One caution that is perhaps obvious should be noted. In
comparing experimental data from different neutron irradiation
facilities, one must be careful when using integrated flux values
that have been determined at different times, since cross sections
have changed with time.

The calculations of damage parameters in various materials
based on the neutron energy spectrum, as displayed in the previous
section, offer a great advantage in the interpretations of experiments. Both the spectrum-averaged quantities and quantities
distributed with respect to recoil energy can be used to make
detailed comparisons with experiments at other neutron facilities,
experiments that employ other types of irradiating particles, and
theories of radiation damage phenomena. Two examples of this can
be found in recent analyses of the sputtering of gold by fast
neutrons[12] and the changes in superconducting properties of Nb_3Sn
caused by various types of irradiation.[13] A "typical" recoil
energy for a particular material irradiated in the present neutron
energy spectrum is best described as the recoil energy at the median
of the damage energy distribution. This "typical" recoil is useful
in interpretations of measurements that are primarily sensitive to
total Frenkel-defect production (e.g., changes in resistivity during
low temperature irradiation). However, the detailed distribution
of recoils or damage energy is useful in interpreting experiments
that measure property changes which are due to a damage mechanism
or a distribution of defects that are dependent on recoil energy.
One example of this is the irradiation disordering of Ni_3Mn, where
the mechanism of disordering at high recoil energies differs considerably from the low energy mechanism.[14] Another example is the
interpretation of experiments designed to measure the concentration
of freely migrating defects during high temperature irradiation of
order-disorder alloys in the partially ordered state.[15] In this
case, the fractional contribution of the low energy recoils to the
concentration of freely migrating defects is of the same order of
magnitude as that of the high energy recoils.

An application of the above ideas more pertinent to this conference on Neutron Transmutation Doping of Silicon is displayed in
Table 4. In a thermal reactor, the neutron energy spectrum can be
approximately expressed as the superposition of a degraded fission
neutron spectrum, a 1/E spectrum and a Maxwellian thermal neutron
distribution. The recoils in Si caused by the 1/E portion of the
neutron spectrum may be neglected as they contribute only a small
portion of the recoil distribution; thus, the relative contributions
of the fast and thermal neutrons to the fraction of recoil and
damage energy can be roughly calculated as a function of cadmium

Table 4.   (n, γ) recoil and damage energy fractions in silicon for
            a mixed neutron spectrum approximated by a superposition
            of thermal neutron flux on the VT53 fast neutron spectrum.

Cd Ratio	(n, γ) Recoils (% of Total)	(n, γ) Damage Energy (% of Total)
1	0	0
10	35	1.5
100	84	12
1000	98	56

ratio. The construction of Table 4 assumes a superposition of the
present degraded fission neutron energy spectrum and various values
of thermal neutron flux to yield the cadmium ratios in the first
column. The average silicon (n, γ) cross section used is 0.16
barns.[16] The average (n, γ) recoil energy is taken to be 474 eV[7]
and is corrected for electronic energy loss by Robinson's approximation[8] to the Lindhard et al. theory.[9] An obvious result of
Table 4 is that even for cadmium ratios near 1000, > 40% of the
radiation damage is in the form of Frenkel defects created by the
fast neutron component of the total flux. Furthermore, the majority
of these defects created by the fast neutron component will be
clustered in high energy defect cascades created by primary recoil
events with a "typical" energy of 55 keV (see Fig. 5). Thus, any
investigation of radiation damage effects in silicon irradiated
in a highly thermalized flux must take into account this contribution to the distribution of defects in the material.

5.   ACKNOWLEDGEMENTS

The advice of T. H. Blewitt and the assistance of R. R. Heinrich and R. L. Malewicki is warmly acknowledged.

REFERENCES

*   Work supported by the U. S. Department of Energy.
+   We would like to express our appreciation to A. Gabriel of
     ORNL and D. Parkin of LASL for performing calculations of
     primary recoil distributions with our neutron spectrum. This
     was done for the purpose of comparison with the DISCS-SPECTER
     codes. The results of all three computer codes agreed quite
     well.
1)   A. C. Klank, T. H. Blewitt, J. J. Minarik and T. L. Scott,
     Bull. Inst. Int. Froid., Suppl. 5, 373 (1966).
2)   W. N. McElroy, S. Berg, T. B. Crockett and R. J. Tuttle, Nucl.
     Sci. Eng. 36, 15 (1969).

3)   A. D. Rossin, published by T. H. Blewitt and T. J. Koppenaal, in Radiation Effects, edited by W. F. Sheely (Gordon and Breach, New York, 1966) p. 561.

4)   G. R. Odette and D. R. Doiron, Nucl. Tech. 29, 346 (1976).

5)   D. M. Parkin and A. N. Goland, Rad. Effects 28, 31 (1976).

6)   T. A. Gabriel, J. D. Amburgey and N. M. Green, ORNL/TM-5160.

7)   R. R. Coltman, Jr., C. E. Klabunde, D. L. McDonald and J. K. Redman, J. Appl. Physics 33, 3509 (1962); and R. R. Coltman, C. E. Klabunde and J. K. Redman, Phys. Rev. 156, 715 (1967).

8)   M. T. Robinson, in Nuclear Fusion Reactors, ed. by J. L. Hall and J. H. C. Maple (British Nuclear Energy Society, London, 1970), p. 364.

9)   J. Lindhard, V. Nielsen, M. Scharff and P. V. Thomsen, Mat. Fys. Medd. Dan. Vid. Selsk. 33, No. 10 (1963).

10)   J. A. Horak published in Ref. 3, later published as J. A. Horak and T. H. Blewitt, Phys. Stat. Sol. 9, 721 (1972).

11)   B. S. Brown, T. H. Blewitt, T. L. Scott, and A. C. Klank, J. Nucl. Mat. 52, 215 (1974).

12)   M. A. Kirk, R. A. Conner, D. G. Wozniak, L. R. Greenwood, R. L. Malewicki and R. R. Heinrich, submitted to Physical Review for possible publication.

13)   B. S. Brown and T. H. Blewitt, submitted to J. Nucl. Mat. for possible publication.

14)   M. A. Kirk, T. H. Blewitt and T. L. Scott, J. Nucl. Mat. 69/70, 780 (1978).

15)   T. H. Blewitt, A. C. Klank, T. L. Scott and W. Weber, in Radiation Induced Voids in Metals, ed. by J. W. Corbett and L. C. Ianniello (USAEC, 1972).

16)   F. W. Walker, G. J. Kirouac and F. M. Rourke, Chart of the Nuclides, Knolls Atomic Power Laboratory, 12th edition (1977).

NEUTRON DOPING OF SILICON IN THE HARWELL* RESEARCH REACTORS

T. G. G. Smith

A. E. R. E. Harwell

Oxfordshire, England

ABSTRACT

The irradiation of silicon, for the purpose of transmutation doping with phosphorus, has been carried out in the research reactors, DIDO and PLUTO since 1975. Ton quantities of material have been irradiated for various customers and the high utilization of the two reactors, with their offset operating cycles, guarantees production throughout the entire year.

The present facilities offer a capacity of some 15 tons of medium resistivity material per year under conditions of good thermal to fast neutron ratios. Flux profile modifications have been made to provide uniformity of dopant along crystal length; the methods adopted and the results are illustrated and discussed.

The irradiations are carried out on a commercial contract basis and the specification of target resistivity, distribution, variations and guarantee of accuracy with an indicative pricing structure are discussed. The methods applied for Quality Assurance and Control are described.

Measurements of decay rates of induced radioactivity and the system of clearance and certification such as to allow the silicon to be internationally transported as "Exempt Material," as defined in IAEA Regulations, are dealt with.

Plans are at an advanced stage for the provision of additional facilities, specifically designed to produce material in the 5-10 ohm-cm range, each capable of producing about 20 tons of doped silicon per year.

1.   INTRODUCTION

Irradiation of silicon for the purpose of phosphorus doping has been carried out in the Harwell research reactors since 1975. Initially the quantities were small, but built up rapidly to tonnes per year. At the present time the capacity is approximately 15 tonnes per year of medium range (40-50 Ω-cm) resistivity material but we plan to increase our capacity, initially to 30 tonnes per year, with provision for further substantial increases in the future. In increasing the capacity we have provided for production of low resistivity (5-10 Ω-cm) material in considerable quantity.

2.   HARWELL IRRADIATION FACILITIES

At Harwell silicon is irradiated in the twin, heavy-water, materials-testing reactors DIDO and PLUTO. Both reactors operate continuously throughout the year and each achieves an availability of greater than 86% of calendar time. Lost operating time arises only from routine monthly refueling and maintenance shutdowns, and as the shutdowns of the reactors are staggered, relative to each other, a continuous irradiation service is guaranteed.

As the reactors are D_2O moderated and cooled the irradiation conditions are particularly good for silicon doping. The irradiation positions used for this work are arranged in a circle at a radius of 110 cm from the core center lines. They are situated in an annulus of graphite (reflector) immediately outside the reactor heavy-water tank, where the axial and radial neutron flux gradients are small.

The ratio of thermal to fast neutrons is in excess of 1000:1, which minimizes the damage which has to be removed by annealing. The majority of the irradiation positions are of 10 cm (4 in.) diameter but there are also 15 cm (6 in.) and 25 cm (10 in.) diameter facilities.

3.   DOPING ACCURACY

In considering the accuracy which can be achieved in the neutron doping process it is necessary to consider the neutron flux profiles and gradients. Figure 1 shows a typical unperturbed flux profile for an irradiation position over a length of 50 cm spaced about the maximum flux value. It will be noted that the maximum and minimum flux values differ by 8-12% of the maximum, and the gradient at the lower end is particularly steep. To reduce this variation and to smooth the profile, flux flatteners or neutron screens, in the form of stainless steel tubes are fitted

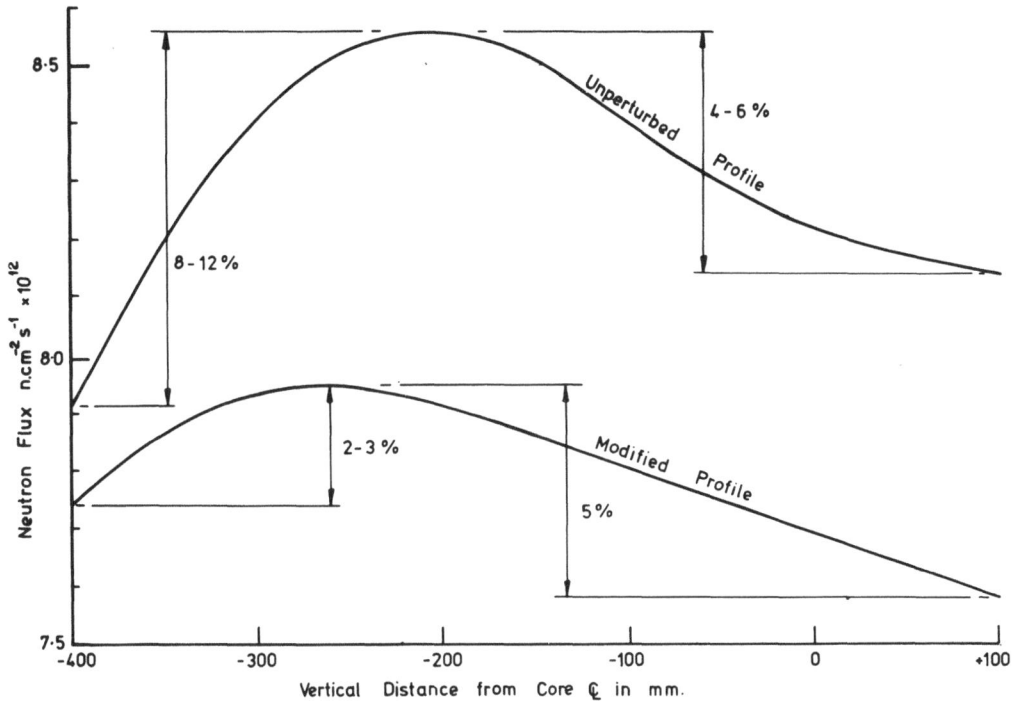

Fig. 1. Typical neutron flux profiles.

to the facility liners with the result shown as "modified profile" in Fig. 1. The severe gradient arising from the 8 - 12%
variation has been reduced to 2 - 3% and the overall variation
reduced to 5%.

    The "modified profile" can now be examined in more detail, and
this is shown in Fig. 2. The profile can be considered in terms
of "resistivity distribution" which is of interest to the customer
and can form the basis for the technical specification of the
product. The "Average Target Resistivity" or A.T.R.50 is, as the
name suggests, the average of the resistivity over a length of 50
cm. Achievement of this value is subject to variations arising
from the irradiation timing, the measurement of the mean flux, and
the distribution shape; therefore we apply a tolerance of ± 5% to
the A.T.R.50 value. The exact shape of the "distribution" is also
subject to variations due to disturbances in the reactor such as
control-absorber movements and other irradiations and experiments;
tolerances are therefore also applied to the "Resistivity Distribution" of 5% maximum greater, and 10% maximum less, than A.T.R.50.

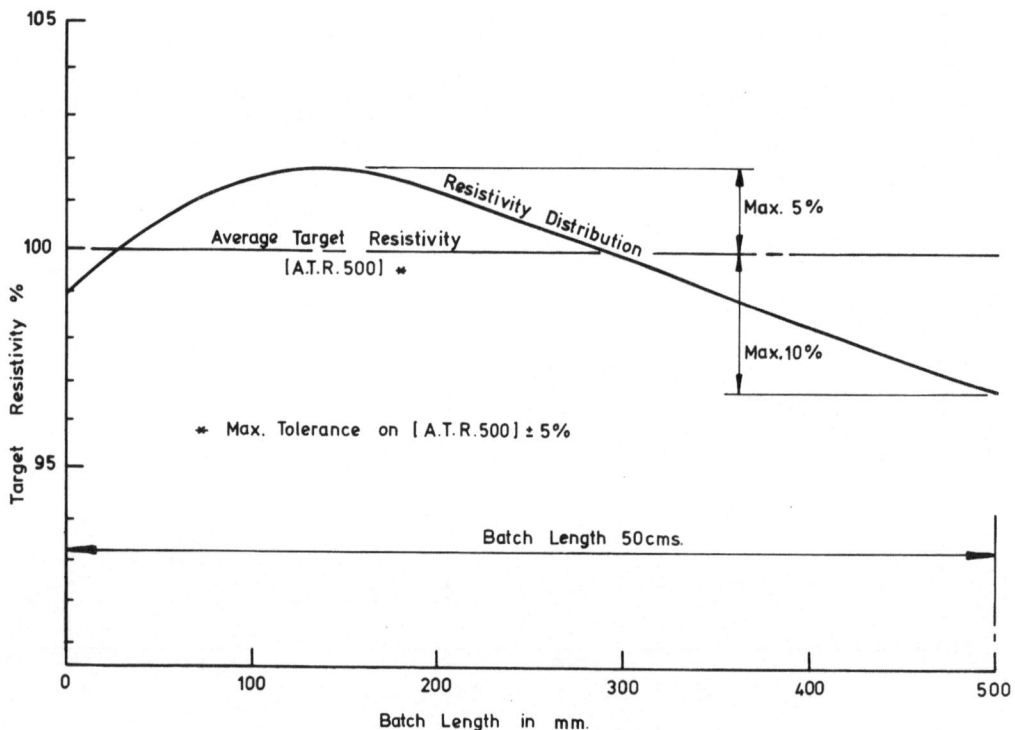

Fig. 2.   Resistivity variations.

As we guarantee that these tolerances will not be exceeded they contain a margin over and above the variations which occur, and feedback of information from our customers indicates that we regularly achieve an accuracy of less than half the specified tolerances. For example, a statistical survey of the average results from 100 batch irradiations reveals that:

$$\text{Average} \quad \frac{\text{Measured Resistivity}}{\text{Target Resistivity}} = 1.01 \pm 0.015$$

Although radial gradients are small, crystals are rotated during irradiation, and a maximum variation of ± 1% on a diameter of 10 cm (4 in.) is guaranteed. In practice variations are less than can be measured within the accuracy of a conventional four-point probe.

4. CONTRACTUAL TERMS AND PRICING

We offer irradiation by volume. Figure 3 illustrates this and shows the volume which for convenience we describe as a "batch." It is a cylindrical volume of 90 mm diameter and 500 mm in length. The customer is at liberty to fill this volume with one crystal or a number of crystals, but crystals in excess of three per batch attract a small handling surcharge for each additional crystal.

The price which is charged per batch is dependent on dose and quantity, and we prefer to negotiate an annual agreement with each of our customers for a fixed basic price per batch and an annual average dose or resistivity per batch. The average annual dose is typically $6.5 \times 10^{17} \text{n cm}^{-2}$, which corresponds to a resistivity of 35 Ω-cm. This averaging of the dose allows the customer freedom to produce material over a wide range of resistivities without negotiating individual prices.

5. CERTIFICATION AS "EXEMPT MATERIAL"

In many of the literature references on the neutron doping of silicon one finds the statement that "three days, or at most a week, after irradiation, silicon is safe to transport and to handle." This is, of course, a relative statement and it is necessary to define what is meant by "safe." In the I.A.E.A. publication "Regulations for the Safe Transport of Radioactive Materials (1977)" it states that to qualify as "Exempt" or safe material the following conditions must be met:

para 302: The radiation level at any point on the external surface of the package shall not exceed 0.5 mRem/h.

and in

para 303: The non-fixed radioactive contamination of any external surface shall not exceed $10^{-4} \mu\text{Ci cm}^{-2}$. This level is permissible when averaged over any area of 300 cm^2 of any part of the surface.

These are internationally accepted standards for the transport of packages, and we comply with them. However, we are aware that few, if any, silicon manufacturers possess the complex equipment necessary to measure the non-fixed contamination of the crystals or to determine the contamination levels which could arise from silicon particles liberated from the heart of the crystal during machining or polishing operations. As the producers of the irradiated material we consider that we are legally and morally obligated to ensure that the material will meet all safety criteria for the workplace and the operators in all future processes and

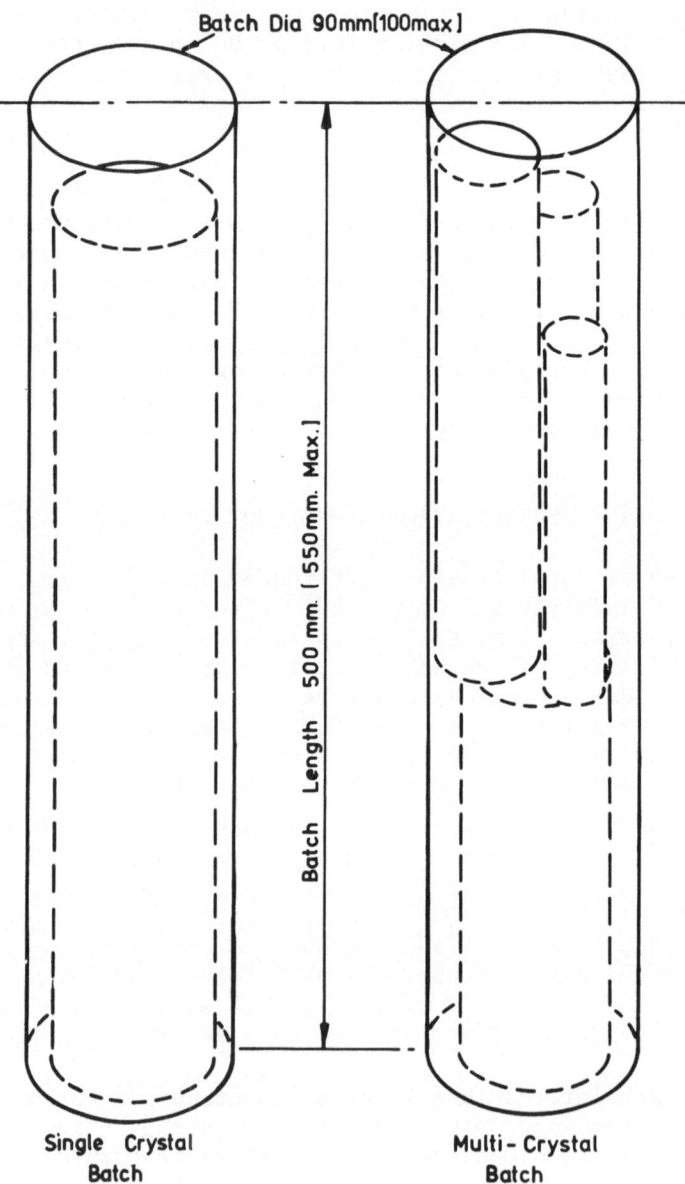

Fig. 3.   Batch dimensions.

handling. To meet this obligation we ensure that each crystal, or
piece, of silicon complies with the following criteria before
certification for dispatch:

> Radiation:　　　Less than 0.1 mRem/h, a factor of 5 less than
> that necessary for the package according to
> I.A.E.A. Regulations.

and

> Contamination: Less than $5 \times 10^{-5}\mu Ci\ cm^{-2}$, a factor of 2 less
> than that necessary for the package.

As a result of the application of these low levels which we consider to be essential only three days delay prior to shipment is
not always possible, particularly for material irradiated down to
low resistivities.

The certification for shipment is carried out, following an
independent check, by the health physicists of the Environmental
and Medical Sciences Division at Harwell.

6.　FUTURE FACILITIES AND CAPACITY

From discussions with our customers and from other knowledge-
able sources in the silicon industries we have concluded that, in
the foreseeable future, the demand for neutron doped silicon will
increase and that there will be an increasing requirement for a
product in the resistivity range 5-10 Ω-cm.

To meet these requirements we are planning to bring a new
facility into operation in 1979 which will have a capacity of
about 20 tonnes per annum of mixed medium- and low-resistivity
material. The large capacity is achieved by an increase in thermal
neutron flux of approximately 10 times that in our present
facilities and by an increase in the "batch" diameter and length.
The length will be 60 cm (24 in.) and the diameter 11.5 cm (4.5
in.). The ratio of thermal to fast neutrons will be approximately
100:1, but tests on material from this irradiation position show
that this does not cause a problem when annealing. Further identical facilities can be added should the demand arise. We shall
continue to utilize the existing facilities.

Costs of neutron doping will be comparable with our present
prices, and each customer will have an option of obtaining up to
25% of material in the 5-10 Ω-cm range.

---

*　　The Atomic Energy Research Establishment of the United Kingdom
Atomic Energy Authority.

SILICON IRRADIATION FACILITIES AT THE NBS REACTOR

N. A. Bickford and R. F. Fleming

National Bureau of Standards

Washington, DC 20234

ABSTRACT

A program of silicon irradiation is being carried out at the National Bureau of Standards 10 MW, heavy water moderated reactor. The facility, which operates on a 40 day round-the-clock cycle, can provide a wide range of neutron fluxes with an equally wide range of cadmium ratios.

1. DESCRIPTION OF FACILITIES

A plan view of the NBS reactor core showing several of the irradiation locations is shown in Fig. 1. A set of five pneumatic rabbit tubes, useful for irradiating silicon chips to analyze for impurities or to study irradiation damage, provide a range of thermal fluxes from 2×10^{11}n cm^{-2}sec^{-1} (copper-cadmium ratio of 3400) to 6×10^{13} (copper-cadmium ratio of 46).[1] Irradiation times of up to 24 hours and sample sizes up to 20 grams can be accommodated.

Those researchers interested in long term silicon doping irradiations can currently use two vertical facilities designed G2 and G4. Both facilities are D_2O filled and are completely isolated from the reactor coolant. Since they are isolated, encapsulation of the silicon is unnecessary and only an aluminum harness is needed to hold the sample. The G2 tube will accept samples up to 1.6 inches in diameter and has a neutron flux at the core midplane of 1.1×10^{14} (copper-cadmium ratio of 55).[1] A vertical flux profile of this facility is shown in Fig. 2. Irradiation of

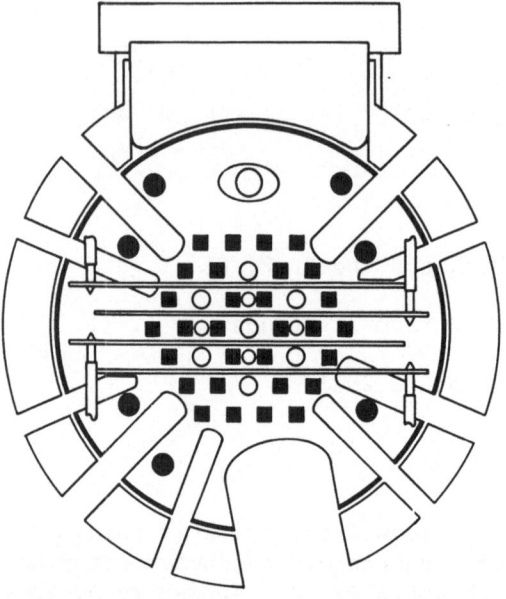

■ Fuel Element
● 4 inch Experimental Thimble
○ 3½ inch Experimental Thimble
o 2½ inch Experimental Thimble

Fig. 1. Plan view of NBS reactor core and irradiation facilities.

samples in G2 for periods of one day to six weeks has been done for Oak Ridge National Laboratory (ORNL). The predicted phosphorus doping rate of 7.5 x 10^{13}atoms $cm^{-3}hr^{-1}$ yielded a concentration in excellent agreement with that measured by ORNL.

The G4 tube is located at the center of the reactor core and will accept samples up to 3 inches in diameter. Its neutron flux has roughly the same shape as that in G2 but is about 28% greater. A one kilogram silicon sample has been irradiated in G4 to a measured phosphorus concentration of 1.4 x 10^{17}atoms cm^{-3}.

Several additional vertical tube facilities are available for development. In addition to 3.5 inch diameter incore positions with fluxes similar to G2, several 4 inch reflector positions are available with somewhat lower fluxes but higher cadmium ratios.

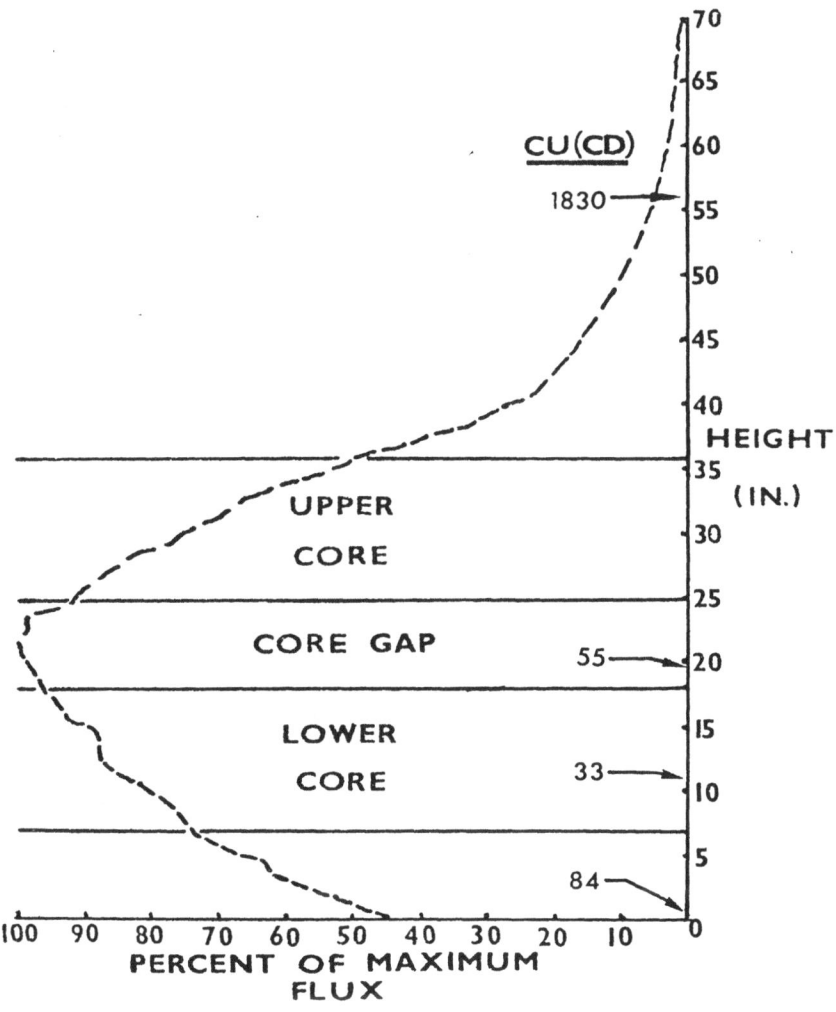

Fig. 2. Vertical flux profile of position G2.

2. CHARACTERIZATION OF IRRADIATION FACILITIES

When comparing irradiation facilities for transmutation doping of silicon, two quantities, in addition to the spatial uniformity of the neutron flux, are important: 1) the ^{31}P production rate and 2) the rate of radiation damage. The ^{31}P production rate is determined by the magnitude of the thermal neutron flux while the radiation damage rate has three components:

1. The unavoidable damage resulting from the $^{30}Si(n,\gamma)^{31}Si$ reaction itself, which is essentially proportional to the ^{31}P production.

2. The damage resulting from high energy neutrons whose energies are well above any appreciable cross section for the $^{30}Si(n,\gamma)^{31}Si$ reaction.

3. The damage resulting from the intense gamma ray field.

In order to establish some measured quantities which can be used to characterize irradiation facilities we should consider the several radioactivities that are produced when pure silicon is irradiated in a reactor flux. When the 2.62 hour half life ^{31}Si decays there is a 1266.1 keV gamma ray emitted with an intensity of about 0.07%.[2] The specific activity of this gamma line is of course directly proportional to the ^{31}P doping rate. The high energy neutrons always present in any reactor spectrum produce three other activities in silicon via the reactions $^{28}Si(n,p)^{28}Al$, $^{29}Si(n,p)^{29}Al$ and $^{30}Si(n,\alpha)^{27}Mg$. Both the ^{28}Al and the ^{27}Mg activities are also produced by thermal neutron reactions on any aluminum or magnesium impurity that may exist in the silicon. The 6.52 minute ^{29}Al activity is unique and results in a gamma ray at 1273.3 keV. This suggests that a simple characterization of a silicon irradiation facility can be accomplished by irradiating high purity silicon and then gamma counting the ^{31}Si and the ^{29}Al activities produced.

The corrected count rate, A_o, for any particular activity is defined as

$$A_o \equiv \frac{\lambda N e^{\lambda t_1}}{(1-e^{-\lambda\tau})(1-e^{-\lambda\Delta})} \left[\frac{e^{\lambda\delta}-1}{\lambda\delta} \right]$$

where

λ = $\dfrac{\ell n\ 2}{\text{half life}}$ = decay constant

N = net counts in the photo peak

t_1 = decay time from end-of-irradiation to start-of-count

τ = irradiation time

Δ = counting live time

δ = counting dead time

The quantity in square brackets is a correction[3] to account for variable dead time during the counting of an activity of short half life. The measured quantity, A_o, is proportional to the reaction rate per nucleus and to the sample mass, m.

$$A_o = K \, m \int \sigma(E) \; \phi(E) dE$$

The constant K includes detector efficiency, branching ratio, isotopic abundance, etc.

It follows that the total number of ^{31}P atoms produced in a sample during an irradiation of duration τ is just

$$M_P = \frac{\tau \, A_o(^{31}Si)}{\epsilon \; \Gamma\gamma}$$

where
$\Gamma\gamma$ = number of 1266 keV gammas per decay of ^{31}Si

ϵ = detector efficiency at 1266 keV averaged over sample volume.

Unfortunately, the measured value of $\Gamma\gamma$ for ^{31}Si is poorly known[2] and needs to be carefully redetermined. However, the quantity $A_o(^{31}Si)$ by itself provides a <u>relative</u> measure of the ^{31}P production rate in a facility.

We propose the ratio $A_o(^{31}Si)/A_o(^{29}Al)$ as a measure of the ratio of phosphorus production to neutron damage in an irradiation facility. Although the silicon damage function has a significantly different energy dependence than the $^{29}Si(n,p)^{29}Al$ cross section, both are sensitive to the flux of high energy neutrons. The fact that the gamma ray energies of the two activities are only 7 keV apart means that the difference in the Ge(Li) detector efficiencies can be ignored and no detector calibration is necessary. The measurement consists of counting the 1273.3 keV line of ^{29}Al and then, after it and the 1267.9 keV single escape line from ^{28}Al have decayed away, counting the 1266.1 keV ^{31}Si line.

3.   RESULTS FOR THE NBS FACILITIES

The measurement of $A_o(^{31}Si)$ and of the ratio $A_o(^{31}Si)/A_o(^{29}Al)$ were carried out on the vertical tubes G2 and G4 and on the pneumatic tube RT4. The results are shown in Table 1. We see that G4 not only has a higher ^{31}P production rate than G2, but that the production-to-damage ratio is somewhat improved. Notice also that at 6 inches above core midplane in G2, which is in the fueled region, the production-to-damage ratio has dropped significantly. The results in RT4 show that if high ^{31}P production rates are not necessary it is possible to obtain doping with greatly reduced fast neutron damage.

Table 1. Activity of ^{31}Si and ratio of activities of ^{31}Si to ^{29}Al.

IRRADIATION FACILITY	$A_0(^{31}Si)$**	$\dfrac{A_0(^{31}Si)}{A_0(^{29}Al)}$
G2*	1.0	0.74
G2 (2" ABOVE)*	0.99	-
G2 (6" ABOVE)*	0.89	0.40
G4*	1.28	0.83
RT4	0.086	9.51

*CORE GAP CENTER LINE
**NORMALIZED TO G-2 CENTERLINE VALUE

4. SUMMARY

We have proposed two measured quantities, $A_0(^{31}Si)$ and $A_0(^{31}Si)/A_0(^{29}Al)$, which can be used to characterize irradiation facilities for silicon doping. They provide a measure of ^{31}P production rate and a figure-of-merit for the production-to-damage ratio. This characterization has been applied to the facilities at the NBS reactor.

REFERENCES

1) D. A. Becker and P. D. LaFleur, J. Radioanalytical Chemistry 19, 149 (1974).
2) W. S. Lyon and J. J. Manning, "Radioactive ^{31}Si," Phys. Rev. 93, 501 (1954).
3) D. DeSoete, et al., Neutron Activation Analysis (John Wiley, New York, 1972), p. 490-496.

GENERAL ELECTRIC TEST REACTOR NTD SILICON DEVELOPMENT PROGRAM

J. E. Morrissey and T. Tillinghast

General Electric Vallecitos Nuclear Center, Pleasanton, CA

A. P. Ferro and B. J. Baliga

General Electric Corporate R & D Center, Schenectady, NY

ABSTRACT

The General Electric Test Reactor (GETR) was designed and constructed to provide large irradiation volumes outside the reactor
pressure vessel in a surrounding water pool. The thermal neutron
flux available for silicon irradiation spans four decades, 10^{11} to
$> 10^{14}$nv. The large irradiation volume permits the inclusion of
flux flattening and spectral softening devices if desired.

General Electric's development program is planned as three
phases: Phase 1 - calibration of a prototypical facility and
irradiation volume, Phase 2 - limited pilot scale irradiations of
selected test specimens under precisely controlled conditions, and
Phase 3 - optimize the process and scale up for commerical irradiation of production quantities.

Phase 1 was successfully completed early in 1977. The results
provided a calibrated flux monitor for use in subsequent irradiations and detailed neutron spectra and spatial variation in the
prototypical irradiation volume.

The temporary shutdown of the GETR has delayed the Phase 2
program which will utilize the calibrated flux monitor and spectral
data from Phase 1, coupled with two additional degrees of control,
rotation and time, to irradiate three inch diameter silicon samples.
Relatively long irradiation time is favored to reduce the error in
absolute nvt to << 5%. Phase 2 results will be used to optimize
the engineering design of a production facility incorporating in-

situ nvt monitors, dry irradiation space and demand charge/discharge
operations for variable diameter silicon specimens throughout the
desired resistivity range.

1. INTRODUCTION

NTD silicon offers significant technical advantages over chemically doped silicon. In particular, NTD silicon has a more uniform
phosphorus concentration across the radius of an ingot or wafer
than chemically doped material. The uniformity could approach 1%
for a three inch diameter wafer activated in the 50MW General
Electric Test Reactor (GETR). Availability of such uniform NTD
silicon would make it possible to manufacture higher power density
thyristors for high voltage applications. In addition this technology may open new areas of application dependent on uniform
resistivity.

General Electric's Irradiation Processing Operation (IPO) located at the Vallecitos Nuclear Center (VNC) near San Francisco,
California, operates the United States' largest privately owned test
reactor which produces over thirty commercial radioisotopes for
medicine and industry. The VNC consists of two operating nuclear
reactors, hot cells, fuels, chemistry and metallographic laboratories and a staff of over 500 engineers, scientists and skilled
technicians. VNC is the company's corporate center for the handling of nuclear materials used in research and development programs.

The GETR is a uniquely versatile irradiation facility for the
production of NTD silicon. The thermal neutron flux available for
silicon irradiation spans four decades, 10^{11} to 10^{14}nv. The irradiation volume potentially available for silicon doping exceeds
that available from other domestic facilities. These two capacity
related factors, coupled with the Center's experienced staff and
commercial operation make the GETR ideally suited for NTD silicon
production.

General Electric's participation in NTD silicon is planned in
three phases: Phase 1 - characterization of a prototypical facility
and irradiation volume, Phase 2 - limited pilot scale irradiations
of selected test specimens at precisely controlled conditions and
Phase 3 - commercial irradiation of production quantities. Phase 1
was successfully completed December 1976 and was a joint effort
of several company components including the Corporate Research and
Development Center, Schenectady, New York and the Static Power
Component Operation, Collingdale, Pennsylvania. Phase 2 work scope
has been established but the development program has been deferred
because of the temporary shutdown of the GETR due to the alleged
presence of a new earthquake fault. It is hoped the GETR will
return to operation in mid-1978.

2. PHASE 1 - CHARACTERIZATION OF IRRADIATION VOLUME AND CALIBRATION
OF FLUX DETECTOR

The objectives of Phase 1 were:

o Calibrate a neutron flux detector to permit its use in
control of subsequent silicon irradiations.

o Determine the neutron flux spectra and measure the radial,
axial and tangential flux gradients in regions of the
pool where subsequent silicon irradiations would be performed.

o Determine the radial variation in silicon doping across
a typical irradiation facility.

To meet these objectives, a facility (Fig. 1) was designed,
built and successfully operated in the GETR in late 1976. The
facility contained 18 small discs (0.75" O.D., 0.03" thick), the
flux detector, 4 flux spectrum packages and radial, axial and
tangential flux wires (Fig. 2). The GETR flux density and spectra
are well known adjacent to the pressure vessel (1.5×10^{14} nv thermal).
The Phase 1 irradiation device was positioned approximately 15
inches from the pressure vessel (Fig. 3) to provide similar data
at the lower limit of useful flux density for silicon irradiations
(2×10^{11} nv thermal). Sixteen days of irradiation are required in
this location to produce 90 ohm-cm material.

After irradiation, the facility was allowed to decay in the
GETR storage canal for five days. The facility was disassembled
and the flux dosimeters transferred to the Analytical Chemistry
Laboratory for radioanalyses.

Two pairs of bare and CdOCu covered flux capsules were located
in the GETR pool at 12 inches and 17 inches from the pressure vessel
for neutron spectral determinations. The dosimeters consisted of
nickel, iron, aluminum-cobalt, titanium, and copper wires, ^{235}U,
silver, thorium and scandium salts, and ^{238}U, ^{237}Np, and ^{232}Th
powders. The salts and powders were encapsulated in quartz. Accom-
panying each dosimeter or pair or group of dosimeters was a 10 mil
1/8 inch diameter nickel disc which was used to normalize each
dosimeter to the reference positions.

Three sets of aluminum-cobalt, nickel, and iron flux wires were
encapsulated in aluminum tubes to obtain radial, vertical, and tangential flux profiles. Following irradiation the aluminum-cobalt,
nickel wires, and selected regions of the iron wires were cut into
segments and weighed. The radial wires were cut into 1/4 inch
lengths and the vertical and tangential wires were cut into 1/2
inch lengths.

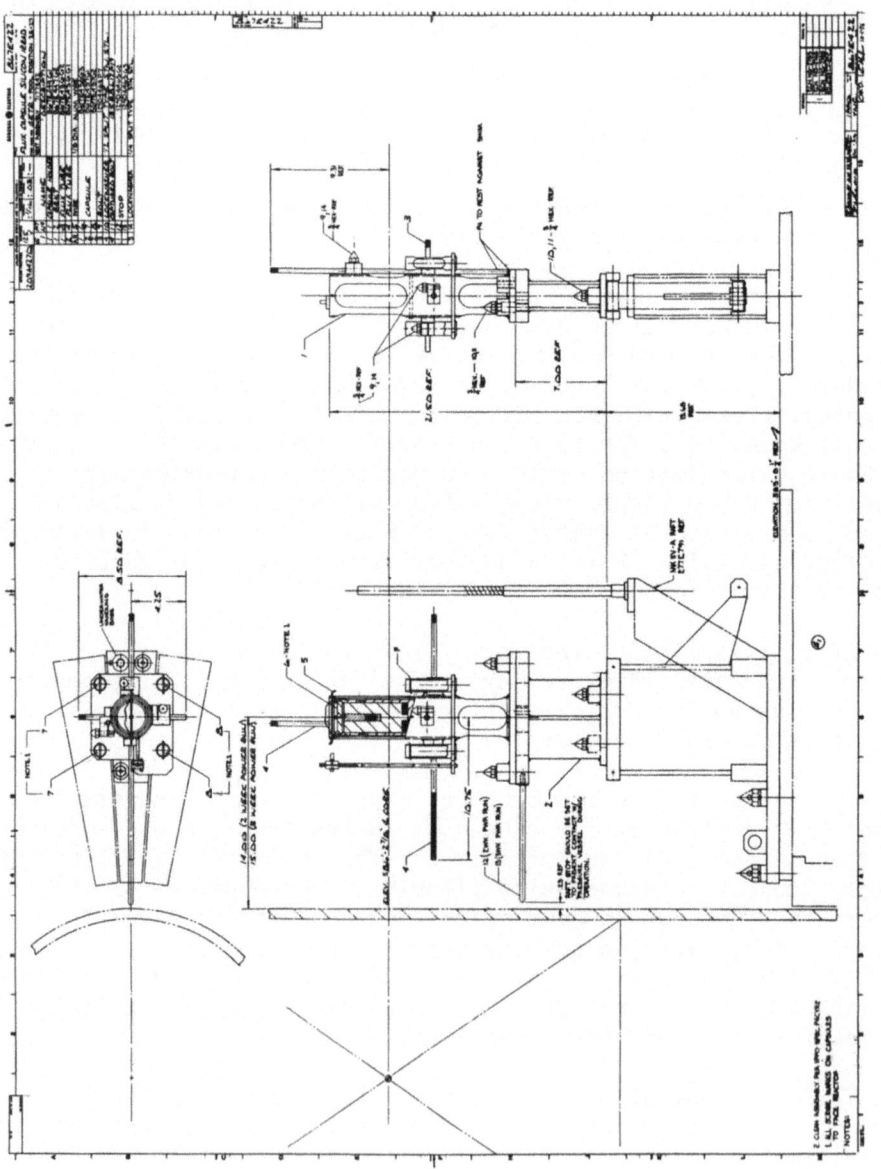

Fig. 1. Phase 1 – Silicon irradiation facility at GETR.

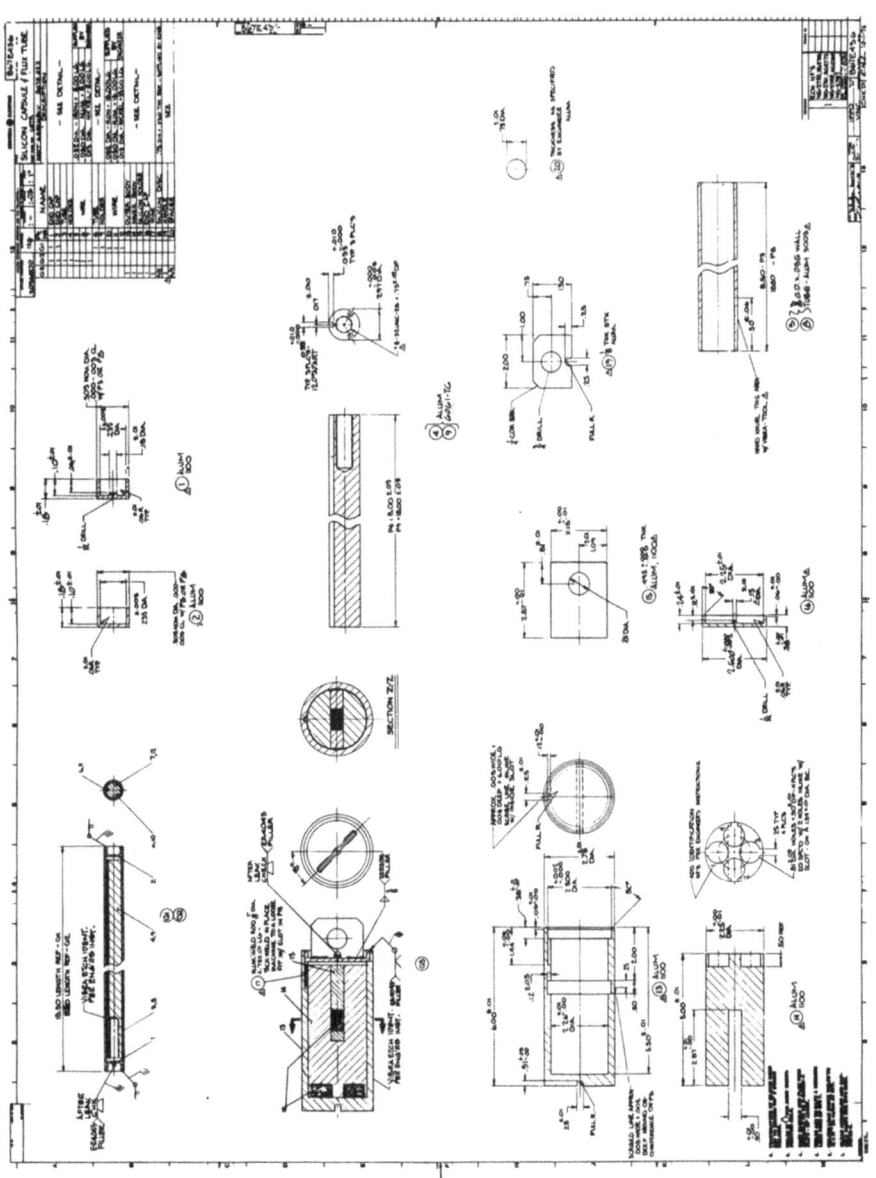

Fig. 2. Test packages contained in the Phase 1 facility.

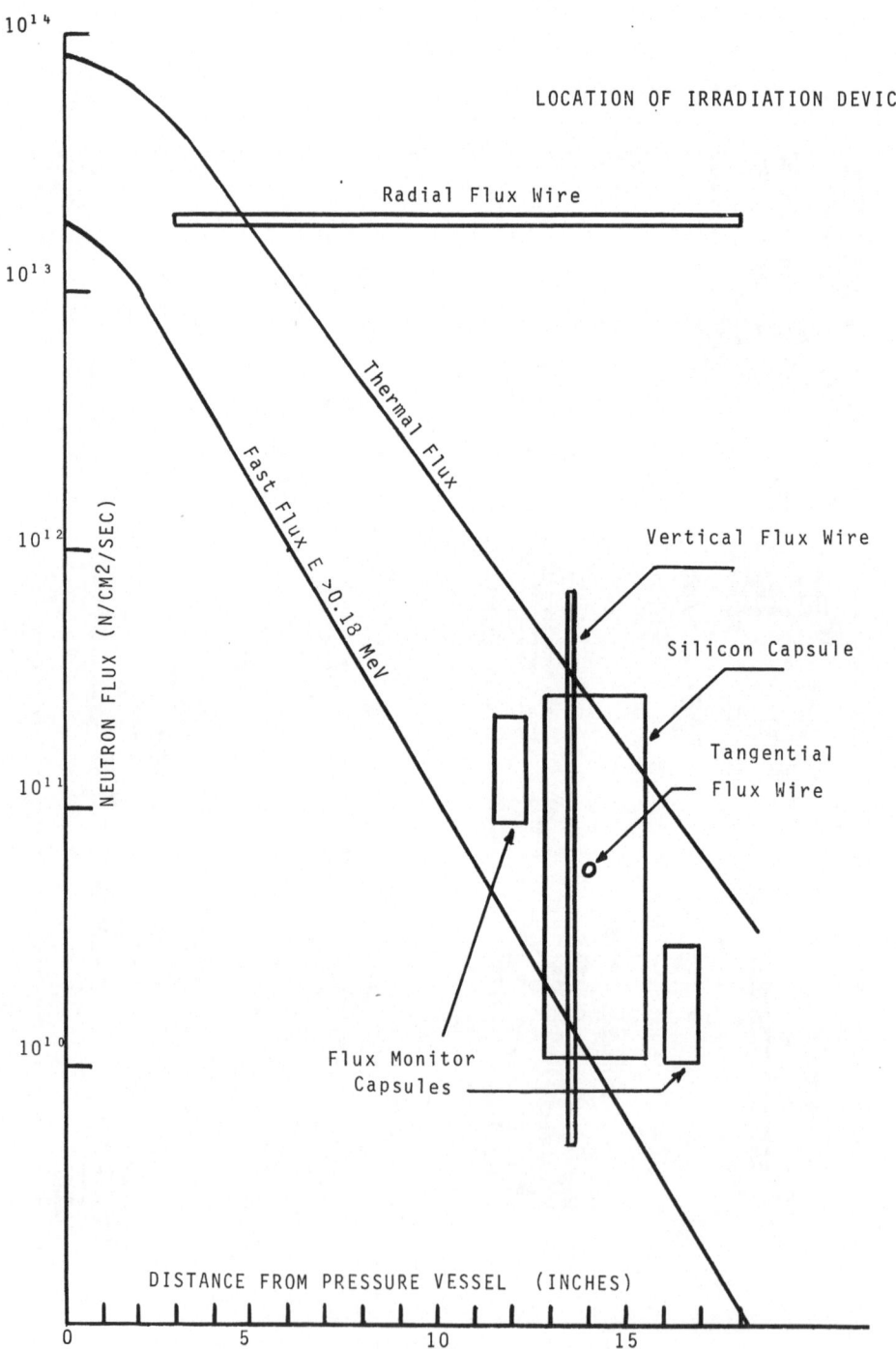

Fig. 3.   GETR 50 MW neutron flux pool position Z-6.

Following irradiation, capsule dosimeters, nickel discs, and gradient wires were analyzed for radioactivity content: reaction rates were subsequently calculated; neutron spectral shapes were determined at the reference locations; reaction rate profiles from the gradient wire analyses were determined; cross sections were calculated for the cobalt, nickel, and iron reactions; and thermal and fast flux density profiles were determined.

Sixteen of the eighteen discs were released unconditionally and sent to Corporate Research and Development for resistivity measurements. Two were retained at VNC as their radioactivity levels were too high to release due to surface contamination.

The average resistivity of 76 ohm-cm was below the target of 90 ohm-cm but was well within the accuracy expected for the initial irradiation located approximately 15 inches from the pressure vessel. The variation in resistivity from the discs closest to the core (67 ohm-cm) to the discs farthest from the core (79 ohm-cm) is only 18% whereas the variation predicted from nuclear physics calculation would approach 50% or greater for the separation distance of an inch or more without target rotation. These results indicate that large diameter discs or ingots if rotated during irradiation, would have almost flat radial resistivity profiles. It was previously thought that a non-absorbing moderator such as graphite would have to be used in large quantities to flatten the normally steep exponential flux gradient in the pool to attain flat radial resistivity profiles on large diameter ingots. While actual testing of large diameter ingots or discs is still required to verify this phenomenum, it appears that a large capacity silicon irradiation facility would not require special moderation.

Current readings (proportional to neutron flux) were obtained periodically throughout the irradiation and the integral of these readings with respect to time is proportional to the average silicon resistivity of 76 ohm-cm. Using this information and correcting for geometrical differences between the Phase 1 and Phase 2 facilities, irradiations in the Phase 2 facility can be controlled to reasonably tight average resistivity values by terminating the irradiation once the desired integrated detector reading is reached. Calibration of the flux detector will be continued in the Phase 2 activity.

Analysis of the flux spectrum and flux wire data indicated the ratio of thermal flux to fast flux is 10 to 1 or greater in the region of interest.

3. PHASE 2 - JOINT DEVELOPMENT OF PILOT SCALE CONTROLLED IRRADIATIONS

Phase 2 will be a joint development program between General Electric and other commercial companies and includes a precision and reproducible irradiation facility for limited production of NTD silicon. Development Program participants have the opportunity to irradiate a limited number of silicon samples of up to 3 inches in diameter and evaluate the results of the irradiation by directly measuring the resistivity of wafers or the electrical properties of devices manufactured from the NTD silicon.

The routine production of NTD silicon at specified resistivities and ± 1% radial variation can only be achieved in a well characterized and controlled irradiation environment. Phase 2 will achieve this for a relatively limited production volume. Control will be exercised by incorporating the calibrated neutron detector (integrated signal is proportional to nvt which is directly proportional to phosphorus concentration and hence resistivity) and the neutron spectra and shapes from Phase 1, with two additional degrees of control, rotation, and irradiation time. The Phase 2 irradiation facility will rotate about the vertical axis at a controlled angular velocity (1 rpm) and the silicon sample will be removable from the neutron flux upon demand. The irradiation facility will be located relatively distant from the reactor core to allow the irradiation to take place in tens of hours, rather than a few hours or minutes, and will be in a well thermalized neutron flux. Incorporation of a calibrated neutron detector and relatively long irradiation times reduces the error in absolute nvt to $\ll 5\%$.

Irradiation parameters

Thermal Neutron Flux 3 to 4×10^{12} nv (up to 1.5×10^{14} nv available)
Thermal to Fast (> 1 Mev) ratio > 10:1
Irradiation Time for 200 ohm-cm $\sim$ 10 hours
                          100 ohm-cm $\sim$ 20 hours
                           50 ohm-cm $\sim$ 40 hours

Resistivity variation

± 5 on average resistivity
± 1 to 3% radial variation, 3" O.D. sample
± 5% to axial variation, $\sim$ 6" length

4. PHASE 3 - SILICON PRODUCTION IRRADIATION FACILITY (SPIF)

Based on the results of Phase 2, and a business opportunity sufficient to justify the investment, General Electric would consider undertaking the design and fabrication of Silicon Production Irradiation Facility (SPIF). The present SPIF concept design includes a facility capable of handling hundreds of kilograms of silicon per month, probably as ingots, and of any diameter up to 82mm. The SPIF would provide for dry irradiation of the silicon and a charge-discharge mechanism which would permit continuous irradiation of silicon in a high volume production mode. The facility would incorporate primary and secondary calibrated neutron flux detectors and signal integration electronics to provide independent monitoring for silicon targets.

5. FUTURE

The encouraging results from Phase 1 indicate NTD silicon irradiations can be performed to very tight resistivity tolerances in the GETR. Assuming the resumption of GETR operation in mid-1978, Phase 2 irradiations would begin in late 1978.

FOOTNOTE

On October 27, 1977, the GETR was ordered shut down by the Nuclear Regulatory Commission due to the alleged presence of a new nearby earthquake fault. As of October 6, 1978, the GETR has not yet resumed operation. In the past year G. E. has undertaken a substantial program to investigate the geologic characteristics of the area and to install additional engineered safeguards to provide adequate protection to the GETR in the event of a seismic event associated with the hypothetical fault. It is G. E.'s position that based on the evidence to date, the fault does not exist and geologic data indicate the presence of a landslide. It is anticipated that the geologic work will be completed in late October and it is hoped that GETR operation can be resumed in early 1979.

AN AUTOMATED IRRADIATION FACILITY FOR NEUTRON DOPING OF LARGE

SILICON INGOTS

J. L. Bourdon and G. Restelli

Joint Research Center

Ispra, Italy

ABSTRACT

The tendency of the NTD silicon producers to increase their capacities and the extending range of devices in which NTD silicon is being used, calls for special reactor irradiation facilities. This is increasingly important at low resistivity values where the neutron dose requires a high neutron flux to keep within acceptable values the irradiation time. In addition an automated charging-discharging device for the silicon ingots is desirable to optimize the dead handling times with respect to the irradiation time.

The JRC heavy water moderated ESSOR reactor presents large irradiation volumes in large diameter experimental channels with neutron flux characteristics especially suited to the above requirements. The described facility has been conceived for installation in one of these channels; its main characteristics are indicated below.

The irradiation volume consists of a cylinder 50 cm long, 87 mm in diameter; the future extension of the diameter to 112 mm is feasible; the thermal neutron flux is equal to 2.7×10^{14}n cm^{-2}s^{-1} with a thermal to fast (> 100 keV) neutron flux ratio equal to about 400. The irradiation position in the channel is optimized by displacement during the reactor operation cycle so that the axial spread of the neutron flux is maintained within ± 4% over the total length.

The silicon ingots are irradiated in bored plastic material capsules immersed in a D_2O flux to ensure efficient cooling; a slow

rotation of the ingots is induced by a suitable shaping of the
capsule. The D_2O flux from the moderator carries the capsules from
a valving arrangement (SAS) to the irradiation position and keeps
them for the required irradiation time determined from a calibrated
three collectrons system with associated electronics where the
integrated neutron dose can be achieved with a precision of 3%.

At the end of the irradiation time, the capsule is discharged
by gravity into the SAS where the remaining D_2O is evaporated. The
capsules are introduced into the SAS and extracted by a suitable
system (TAU) which is manually loaded and unloaded with the capsules
also during the reactor operation; the TAU system holds up to 100
capsules, and advances them automatically to the correct position
for injection into the reactor.

Preliminary experimental production of NTD-Si in ESSOR has
shown the outstanding advantages of this reactor for such work.

1.   INTRODUCTION

The tendency of the NTD silicon producers to increase their
capacity and the extending range of devices in which NTD silicon
is being used, call for special reactor irradiation facilities.
This is increasingly important at low resistivity values where the
neutron dose requires a high neutron flux to keep within acceptable
values the irradiation time. In addition an automated loading-
unloading device for the silicon ingots is desirable to optimize
the dead handling times with respect to the irradiation time.

The JRC heavy water moderated ESSOR reactor was built to test
components such as fuel bundles for pressurized or boiling water
reactors, or complete channels of heavy water reactors.

The large irradiation volume with a good neutron flux homogeneity in large diameter experimental channels, in conjunction
with its neutron flux characteristics, make this reactor well
suited to the requirements for silicon irradiation.

In this article, a silicon irradiation facility which has been
conceived for installation in one channel of this reactor is briefly
described.

Attention has been especially devoted to obtain an automated
operation of the facility, and to optimize the characteristics of
the irradiation volume.

2. ESSOR REACTOR CHARACTERISTICS

The facility will be installed in one of the twelve experimental vertical channels (N°7). (See Figs. 1 and 2.) Since the operation of the facility is foreseen from the lower part of the channel, a maximum diameter of 120 mm is available (the channel diameter being 170 mm in the upper part). The channel is immersed in the moderator heavy water. The irradiation volume has been limited at the moment to 50 cm length by 7.7 cm diameter. A future extension of the diameter to 10.2 cm will not require major modifications.

The height of 50 cm, with respect to a core vertical dimension of 150 cm, has been selected in order to obtain a minimum axial spread of the neutron flux (± 4%).

A thermal neutron flux of $(2.7 \div 3) \times 10^{14} n\ cm^{-2}s^{-1}$ is available at the irradiation position with a thermal to fast (> 100 keV) neutron flux ratio larger than 400.

3. GENERAL DESCRIPTION OF THE FACILITY

The silicon crystals are loaded into a transport unit which can locate up to 100 ingots 77 mm in diameter by 500 mm length (Fig. 3). The ingots are loaded protected by a bored plastic capsule which defines the irradiation volume indicated above (77 mm x 50 mm). The capsules are then loaded into the reactor through a lock between the transport unit and the channel for transfer from the air to the heavy water circuit.

The capsule is conveyed up to the irradiation position by a heavy water flow in parallel with the moderator circuit and stopped by an adjustable shaft which is displaced during the reactor cycle to match at any time the region of maximum axial flux homogeneity. This device is supported together with three flux detectors (collectrons) by the upper plug of the channel.

The heavy water circulation assures efficient cooling of the silicon ingot during irradiation (the maximum crystal temperature should not exceed 70°C) and induces, by use of a suitable shape of the capsule, a slow rotation of the ingot for minimizing radial dispersion of the neutron fluence.

At the end of the irradiation, determined by the control system, the capsule is transferred into the transport unit where it stays for at least 4 days before being discharged thus providing for decay time between the irradiation and the extraction time. The main parts of the facility are described in the following sections.

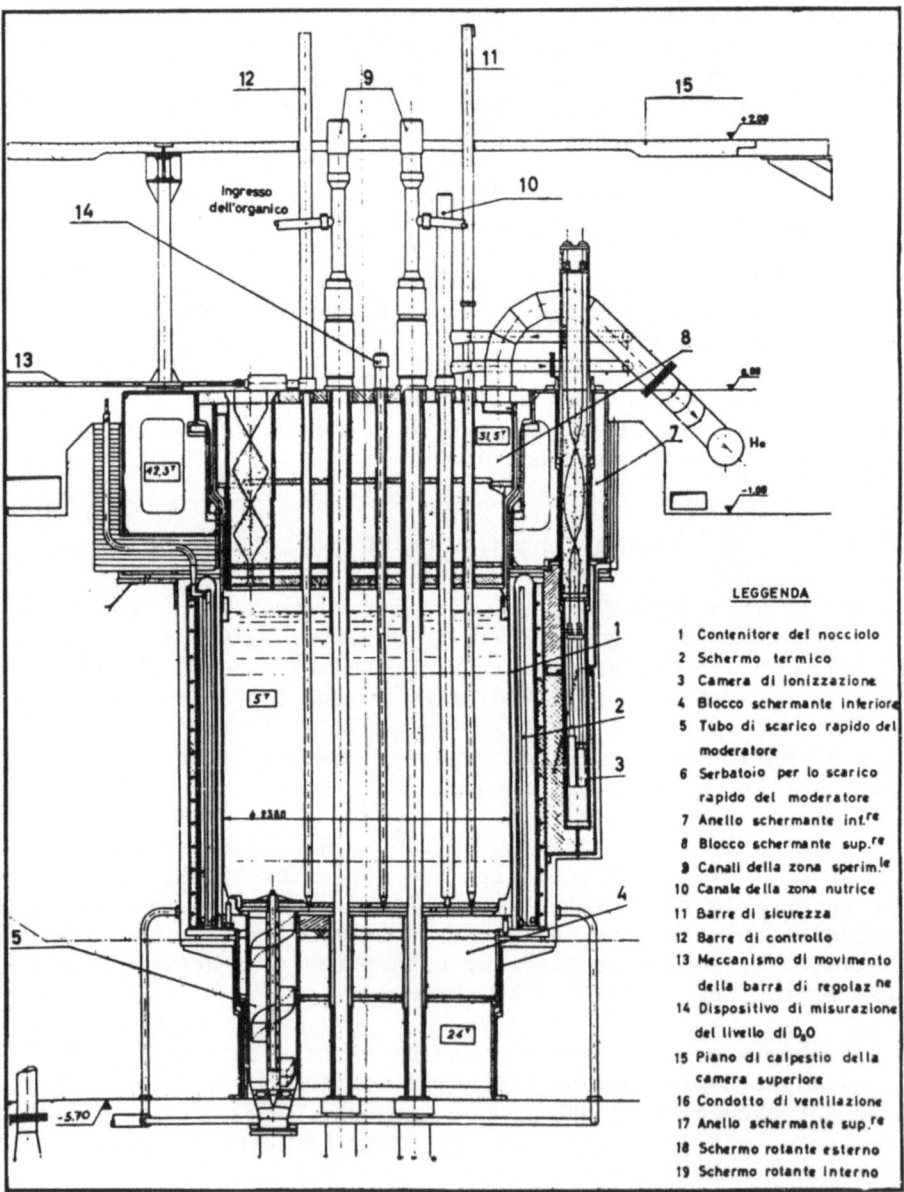

LEGGENDA

1 Contenitore del nocciolo
2 Schermo termico
3 Camera di Ionizzazione
4 Blocco schermante inferiore
5 Tubo di scarico rapido del
   moderatore
6 Serbatoio per lo scarico
   rapido del moderatore
7 Anello schermante int.re
8 Blocco schermante sup.re
9 Canali della zona sperim.le
10 Canale della zona nutrice
11 Barre di sicurezza
12 Barre di controllo
13 Meccanismo di movimento
   della barra di regolaz.ne
14 Dispositivo di misurazione
   del livello di D_2O
15 Piano di calpestio della
   camera superiore
16 Condotto di ventilazione
17 Anello schermante sup.re
18 Schermo rotante esterno
19 Schermo rotante interno

Fig. 1. ESSOR reactor - vertical section of the core.

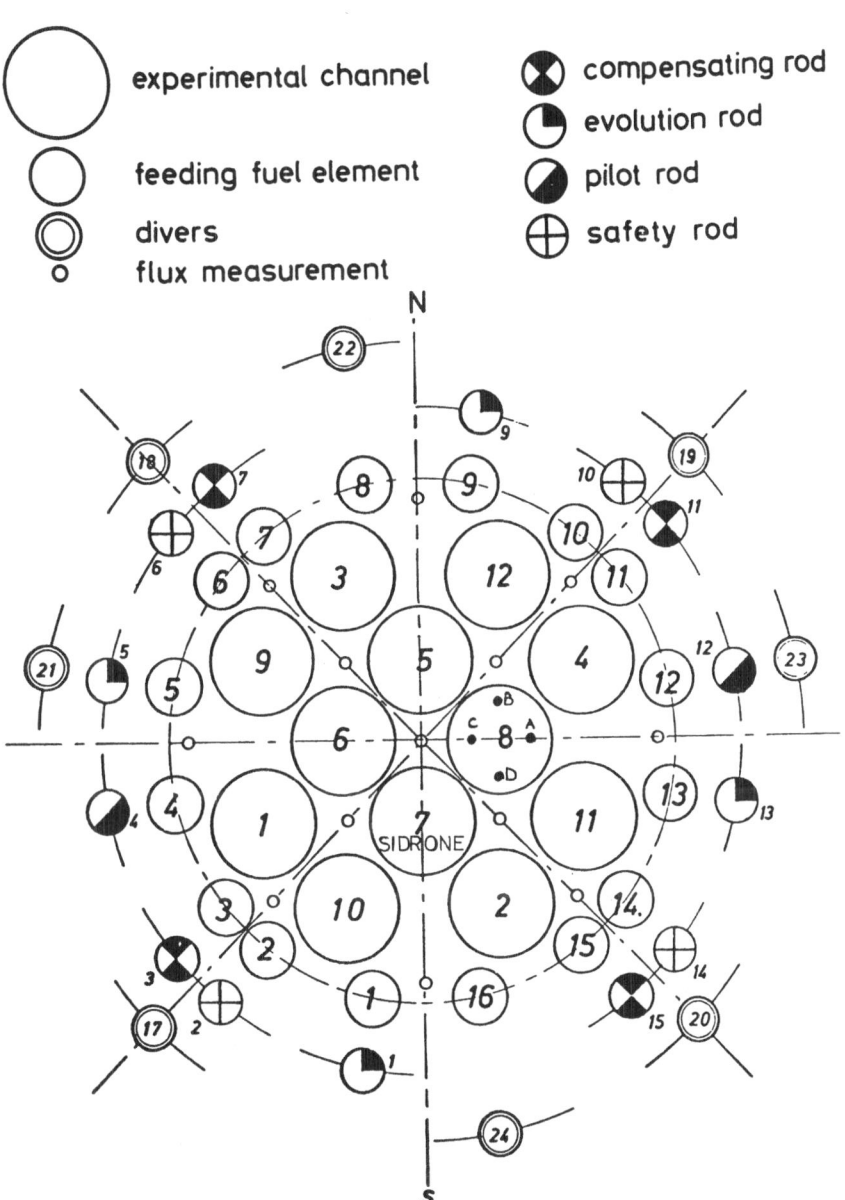

Fig. 2. ESSOR reactor - core map.

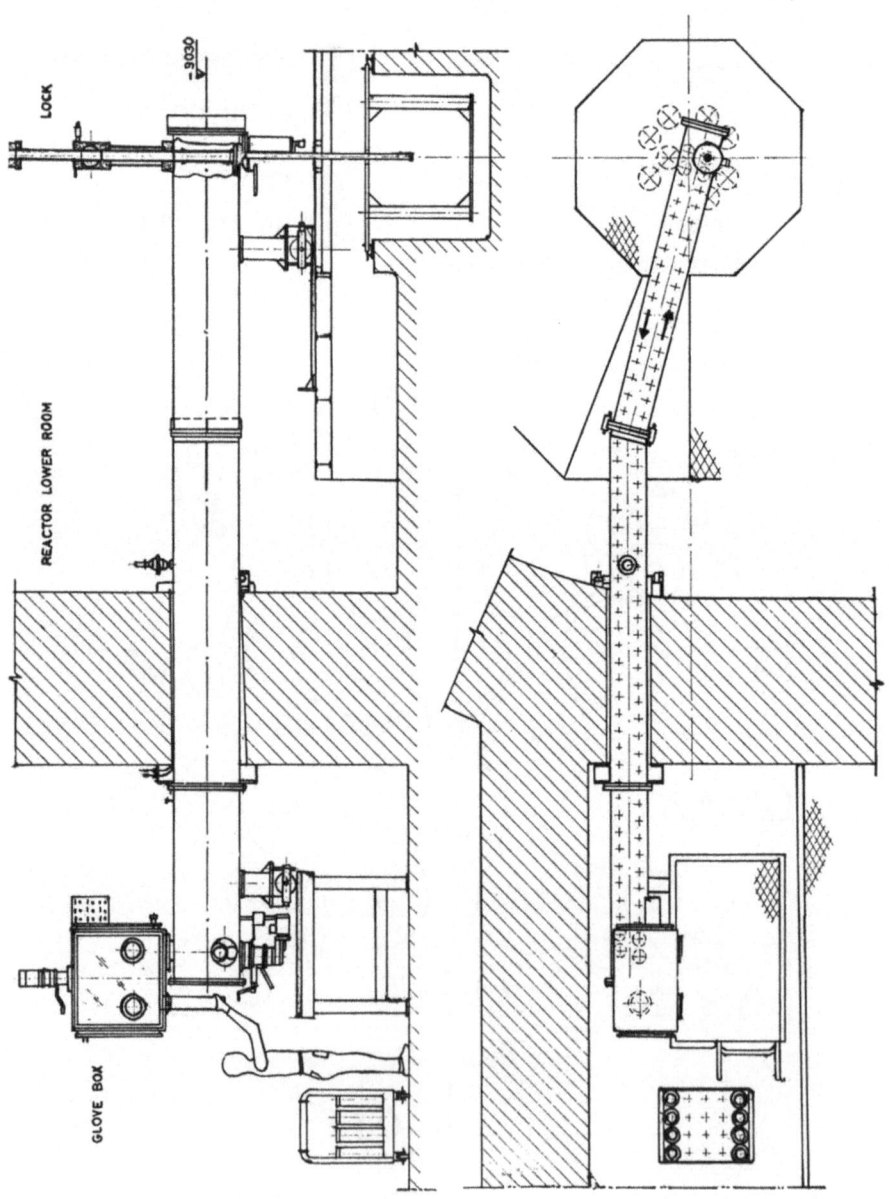

Fig. 3.   Section and layout of the irradiation facility.

4. GLOVE BOX

The presence of tritium coming from the heavy water in the facility requires the use of a glove box for the safe introduction and extraction of the capsule in the transport unit.

Each capsule is placed into a container used for the transport, and for hermetically-sealed transfer to and from the glove box. This container is settled by hand under the glove box.

An air operated handling mechanism transfers the capsule from the container to the transport unit, or vice versa, through an aperture, normally closed by a cover.

5. TRANSPORT AUTOMATIC UNIT (TAU)

The transport unit consists of a rectangular tube in stainless steel, of 330 x 610 mm section, 7.5 m long, which contains two rails with end curves, thus realizing a closed circuit. One hundred small wagons, connected together by an endless chain, slide on those rails; the chain is moved by a geared motor. Each wagon receives one capsule, and carries a device which allows unequivocal identification by means of proximity detectors.

6. LOCK

The lock (Fig. 4) consists of two ball valves and an inter-space that can receive one capsule, in vertical correspondance with the channel. An air cylinder underneath this position is used to insert the capsule into the lock, the first valve of which is open. Meanwhile as the cylinder piston goes down, the capsule is retained by a shaft introduced horizontally. Then the first valve is closed, the second one is open, and the heavy water flow conveys the capsule to the irradiation position. The heavy water returns directly to the reactor vessel at the upper part of the channel.

At the end of the irradiation time, the capsule is discharged by gravity into the lock where it is dried by a hot helium flow from the reactor vessel circuit. The introduction and extraction velocities of the capsule induce a systematic difference in the irradiation time between top and bottom of the capsule of the order of 10 seconds.

After returning to the transport unit, the TAU chain is moved one pitch, and the cycle is repeated for the next capsule. All of these operations are automated.

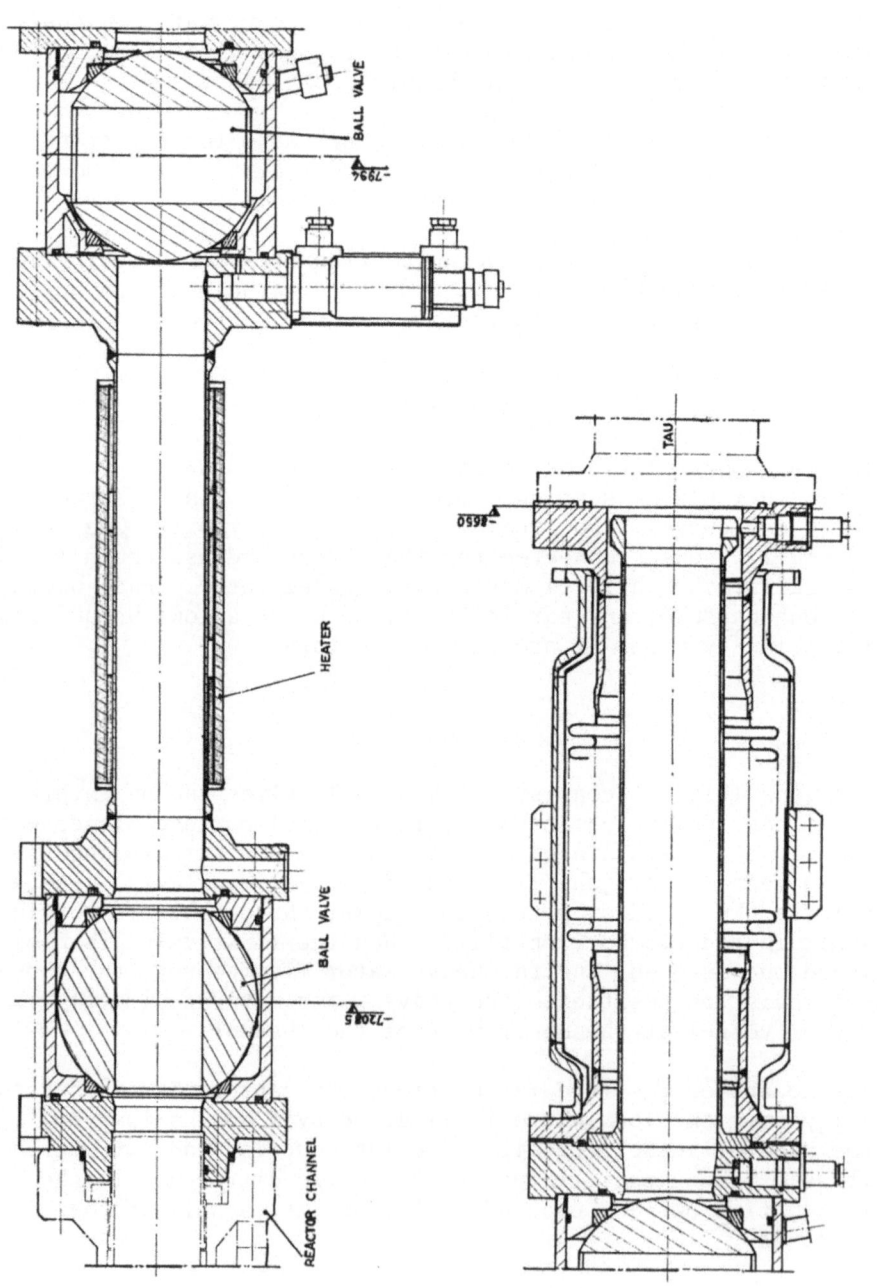

Fig. 4. Lock system.

7.   CHANNEL AND INSTRUMENTED UPPER PLUG

The channel consists of an aluminum tube of 95/87 mm diameter
(Fig. 5). This tube is intended to be replaced by a Zircaloy tube
of 116/112 mm diameter for future irradiation of 4 inch silicon
rods. The level of the irradiation position can be adjusted to
match the flat maximum of the neutron axial flux curve at any time
of the reactor cycle, so that the axial spread of the neutron flux
is maintained within ± 4% over the total irradiation length. For
this purpose the capsules are stopped at their irradiation position
by an adjustable shaft, supported by the upper plug. This plug
completes the shielding of the channel and supports the adjustable
shaft and collectrons, a system which is vertically positioned by
a geared motor.

8.   COLLECTRONS

For the control of the irradiation time, the neutron flux is
evaluated by three vanadium self-powered neutron detectors
(collectrons) placed around the irradiation position. The detectors are 500 mm long, clad in stainless steel, and exhibit a
consumption of 0.1% per month of operation.

A suitable calibration using silicon crystals is an essential
prerequisite for the operation of the facility and assures a good
precision for several months in the subsequent irradiations. The
current from the collectrons is suitably processed electronically
to give a signal for the control of the irradiation time.

Previous experiments on silicon ingot irradiation have shown
that a precision of 2% can be reached in the neutron fluence
evaluation (O. Simoni, 1975, unpublished).

9.   CAPSULES

The plastic material for the capsule (Fig. 6) must be chosen
as a function of its qualities of mechanical behavior, radiation
resistance, heat resistance and low radioactivation. On the basis
of the three first requirements, some commercial plastics have been
selected and samples of each irradiated in the HFR reactor of
Petten (Nederland) or in ESSOR, in order to choose the best material
(Fig. 7). None of the tested materials is completely satisfactory
so that we have foreseen the use of Noryl 731 for the preparation
of capsules to be used for short irradiation times, and polystyrene
or PPO for long irradiation times.

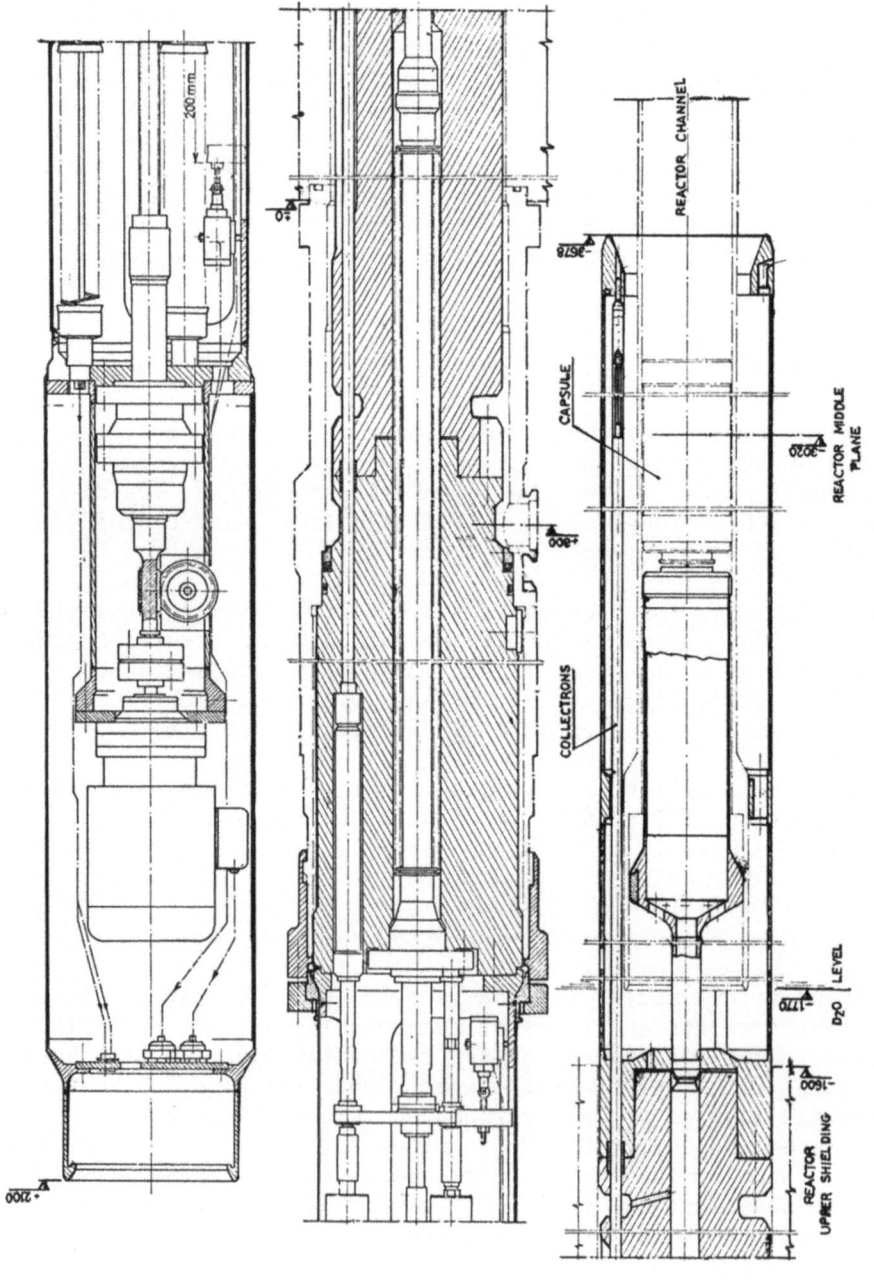

Fig. 5.   Instrumented upper plug.

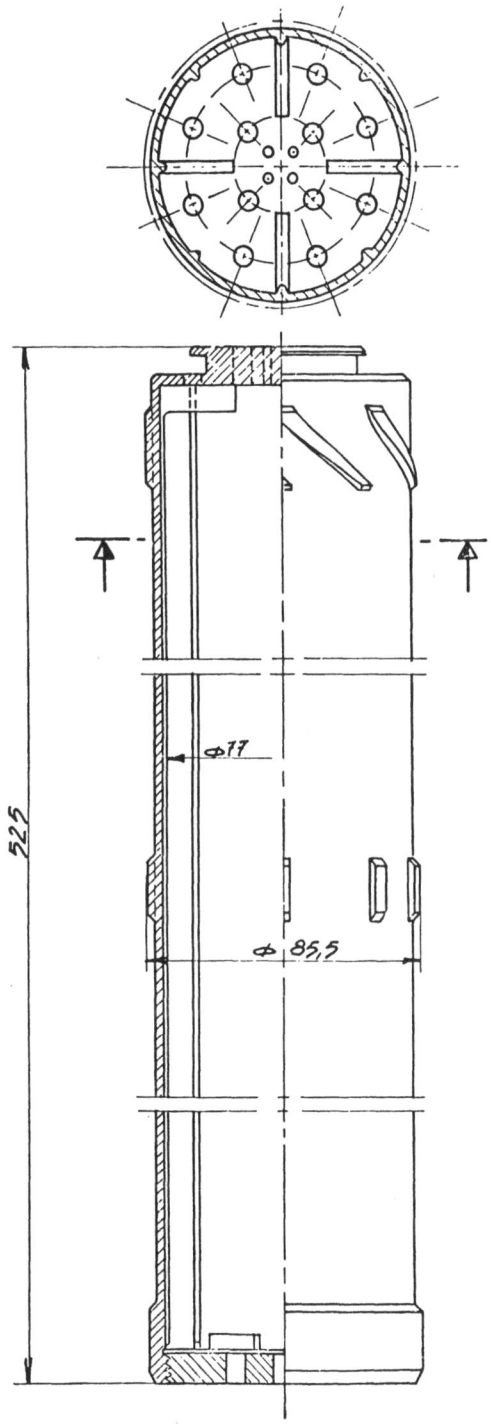

Fig. 6.   Irradiation capsule.

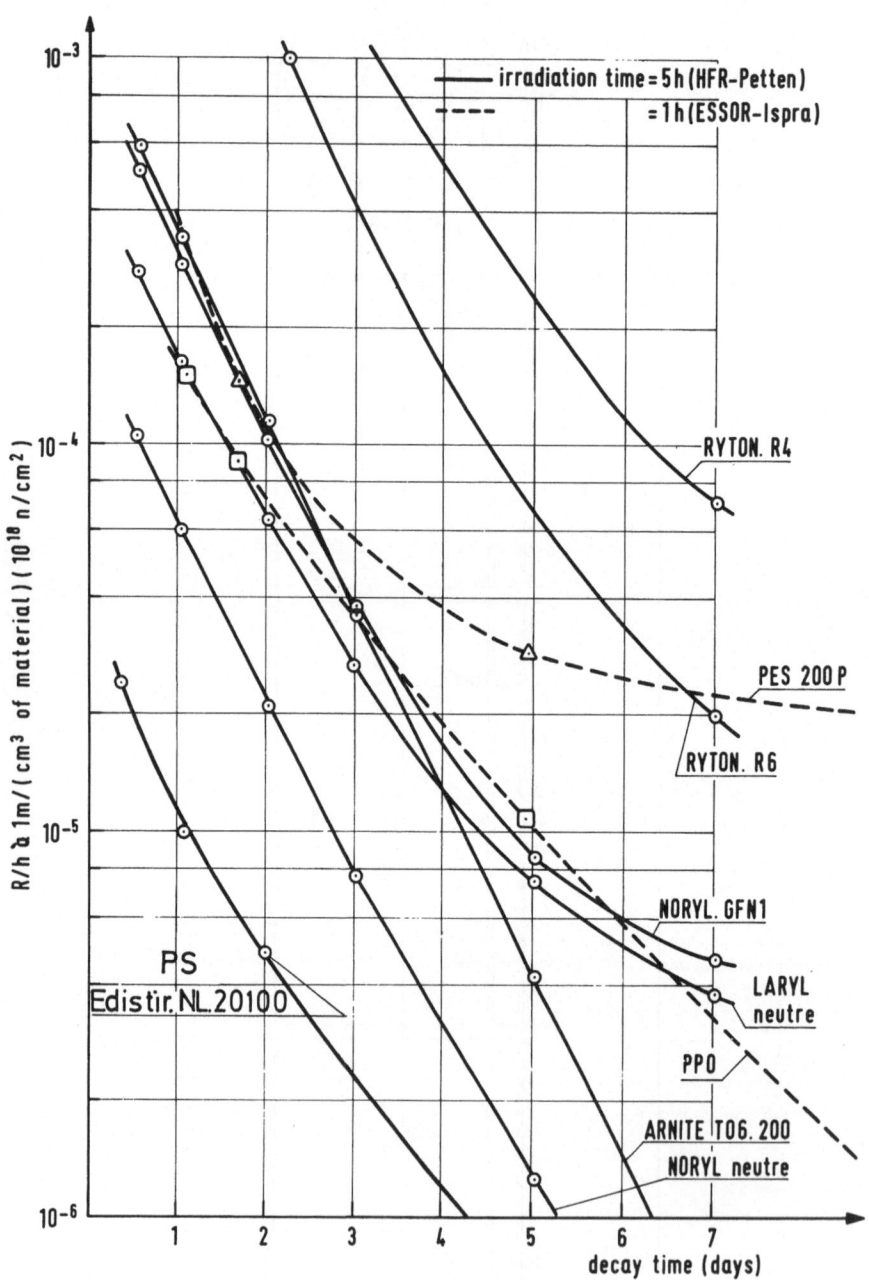

Fig. 7. Radioactivation of some plastic materials.

10. DISCUSSION

It is accepted that the most likely substantial market for NTD silicon is that presently covered by conventionally doped n-type float zone material of resistivity greater than 10 ohm-cm. This part of the market in terms of metric tons per year represents a considerable amount which poses to the reactor irradiation facilities severe requirements which should obviously be those of a production unit. This is especially true for the lower resistivity range where a reduction of the irradiation time (and cost) necessitates the availability of high thermal neutron flux densities. It must be remembered that also the time for radioactivity cooling (formation of ^{32}P) is inversely proportional to the neutron flux density. Starting from these premises, it was decided to design the facility described taking into account the advantages offered by the ESSOR reactor. These advantages are a high thermal neutron flux density, very uniform and easily controllable over a large irradiation volume, a good thermal to fast neutron flux ratio in conjunction with the possibility to irradiate the capsule immersed in D_2O which assures an efficient cooling of the Si crystal.

The high thermal to fast neutron flux ratio assures fewer lattice defects as demonstrated by the fact that a thermal annealing of 5 min. at 750°-800° has been found sufficient to achieve complete recovery of the final resistivity (Fig. 8).

Preliminary tests have demonstrated an extremely reduced surface contamination of the Si ingot after irradiation so that a simple washing of the crystal removes any surface activity to below the exempt limit.

The aim of reaching a maximum exploitation of the reactor operating time using extended automation of the facility has imposed the choice of a plastic capsule. In fact, after the residence time of the irradiated silicon into the transport unit (4-6 days) only the activity due to the capsule represents a problem for an immediate and safe handling of the silicon ingot. The use of a plastic material for the capsule appears a good choice even if the commercially available plastics represent only an acceptable compromise (see par. 9).

An approximate evaluation of the potentiality of this facility in terms of N irradiations per year can be obtained from the following formula:

$$N = \frac{H}{(3600 \times 2.7 \times 10^{14})^{-1} \Phi + 1}$$

where H is the reactor operating time (hours per year) and Φ is the irradiation neutron dose requested. The correction (+ 1 hour)

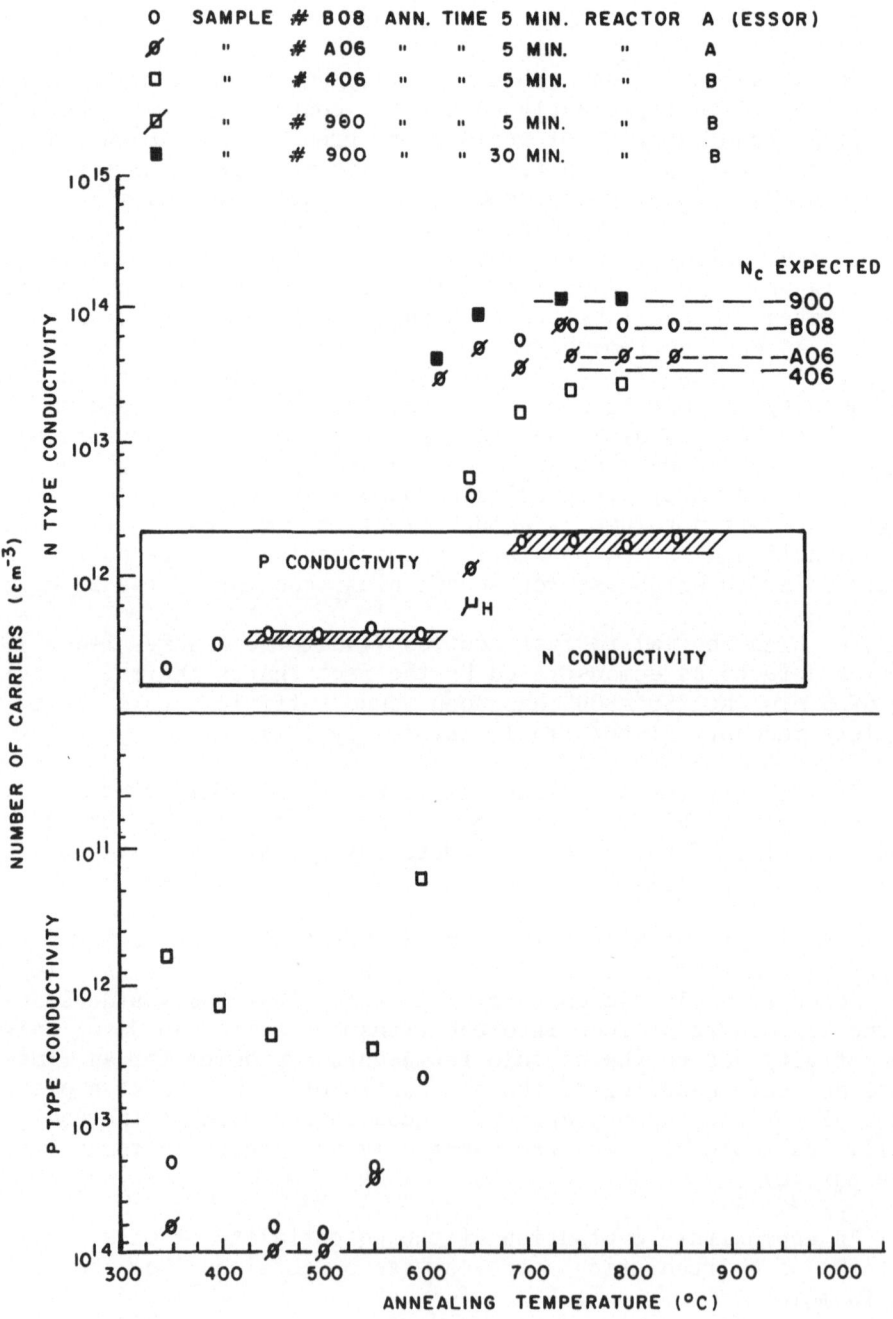

Fig. 8. Number of carriers (and Hall mobility) versus annealing
        temperature (isochronal anneals).

accounts for the time needed to dry the capsule from D_2O before discharging it to the transport unit and other operations of loading and unloading the transport unit.

ACKNOWLEDGEMENTS

The authors would like to thank their colleagues for their cooperation in the design and construction of this facility.

HIGH PRECISION IRRADIATION TECHNIQUES FOR NTD SILICON AT THE UNIVERSITY OF MISSOURI RESEARCH REACTOR

S. L. Gunn, J. M. Meese*, and D. M. Alger**

University of Missouri Research Reactor Facility

Columbia, MO 65211

ABSTRACT

The design factors which governed the construction of present NTD irradiation facilities at MURR were efficient utilization of reflector space, good doping accuracy, good doping uniformity, and potential for growth. The present MURR flux integration system will be described. This system consists of 1.5 mm diameter, self-powered neutron detectors and high precision analogue current integrators. These are utilized at each sample position to achieve an overall fluence accuracy of better than ± 1%. The integration system is used to determine the 50% of total fluence point at which samples are flipped end-for-end. This technique, as opposed to flux flattening or sample spiraling, provides a very efficient utilization of reflector space without compromising axial uniformity. Axial uniformity obtainable in any position is better than ± 4% of target over sample lengths of 250 mm. Sample rotators are used to obtain radial uniformity of better than ± 1% up to 86 mm diameter. Present irradiation positions consist of two 750 mm long x 85 mm diameter, two 750 mm long x 60 mm diameter, seven 750 mm long x 80 mm diameter, and a high precision variable flux facility. The present MURR capacity is of the order of 15-20 ppb-tonnes per year.

1. INTRODUCTION

We will describe in this paper the irradiation techniques used at MURR for the processing of commercial quantities of NTD-Si and also experimental facilities used for close compensation of detector

grade silicon. These techniques have evolved through experience
which has been gained from our initial irradiation tests in the
latter part of 1974 until the present. Although the techniques
described here may in many cases be transferred to other reactor
facilities, our procedures understandably reflect to a strong
degree, the influence of a particular reactor design on the methods
we have evolved. Other reactors will develop a different method-
ology because of differences in irradiation positions, core
configurations and flux. The common denominator for the procedures
adopted, however, will be the irradiation goals which are set by
the industry. We shall, therefore, discuss these goals briefly.
Because the requirements of the semiconductor industry are likely
to change with time, volume of NTD sales and price structures,
a certain amount of flexibility in the design of irradiation
facilities is highly desirable and has been incorporated into our
design philosophy. It is anticipated that our irradiation procedures will continue to evolve to meet the demands of this rapidly
growing industry.

2. SYSTEM DESIGN FACTORS

Several specific factors have been considered in the design
of NTD irradiation facilities at MURR and are listed below.

1. Radial and axial uniformity
2. Doping accuracy
3. Sample sizes and volumes
4. Efficient utilization of irradiation space
5. Holding time, sample storage and handling
6. Activity monitoring
7. Sample cleaning
8. Sample inventory accounting
9. Feedback on irradiation procedures
10. High yield
11. Potential for system growth
12. Prevention of breakage
13. Reduction of cost

The irradiation techniques and handling procedures which have
evolved are a function of the goals we have set for the factors
above.

The achievement of the highest radial uniformity attainable
dictates that the samples be rotated. Once this is done, radial
uniformity is essentially independent of reactor type and is better
than ± 1% for ingots less than 83 mm diameter. If resultant radial
uniformities in NTD samples are much worse than 1%, this is caused
by too small a doping ratio[1] or extreme non-uniformities in the
starting material.[2]

Axial uniformity is very much a function of the flux profiles characteristic of specific reactor types. In a small core, high flux density reactor such as MURR, the axial uniformity is the most difficult parameter to control. We have, therefore, developed a unique approach to this problem which utilizes our high flux density to the fullest but maintains a uniformity which is competitive with doping accuracies presently being marketed. Our design goal of ± 7% has usually been exceeded experimentally but has required that our fluence accuracy be held to very tight tolerances.

Fluence accuracy can potentially be exceedingly good in the NTD process if careful flux monitoring is utilized. In our early experimental test runs in 1975, we confirmed that control blade motion in our high flux density reactor produced flux variations which made it impractical to control sample fluence by monitoring irradiation time. A further requirement of a research program to produce exact compensation of residual boron in IR detector material demanded an even greater fluence accuracy of ± 1/2%. We, therefore, developed a flux monitoring system which could operate equally well at the high flux levels used for the commercial NTD production and also at the much lower flux levels required for the high precision work.

Consideration of items 5, 6, 7, and 8 above are discussed elsewhere in this conference.[3-4] It is important that feedback of irradiation results be provided by the user so that the fullest utilization of the flux monitoring system is realized and that the number of samples falling within the tolerance requirements of the user is maximized. Sample yields will be discussed briefly below. It should be mentioned that an active semiconductor NTD research program at the reactor facility has been extremely useful in providing immediate feedback of experimental data. Results of this program have been described elsewhere.[5]

Because of the varying requirements of a number of users and their customers, it is desirable to maintain a certain amount of flexibility in the design of irradiation facilities. The removable wedge irradiation facility concept at MURR has provided this flexibility and provides a method of future update.

3. REACTOR DESCRIPTION

The University of Missouri Research Reactor is a pressurized light water reactor operated at 10 MW for 150 hours per week. A cross-sectional view of the reactor core, reflector and swimming pool irradiation positions is shown in Fig. 1. The core is comprised of eight pie-shaped enriched ^{235}U fuel elements surrounding

the flux trap.   The flux trap is available for sample irradiations
and provides thermal flux from 6 x 10^{14} to 1 x 10^{14}n/cm^2sec with
a Cd ratio on Co of about 8:1.   The MURR reactor has an inner
neutron reflector made of beryllium and an outer reflector made of
graphite.   Reactor control is achieved by the insertion of neutron

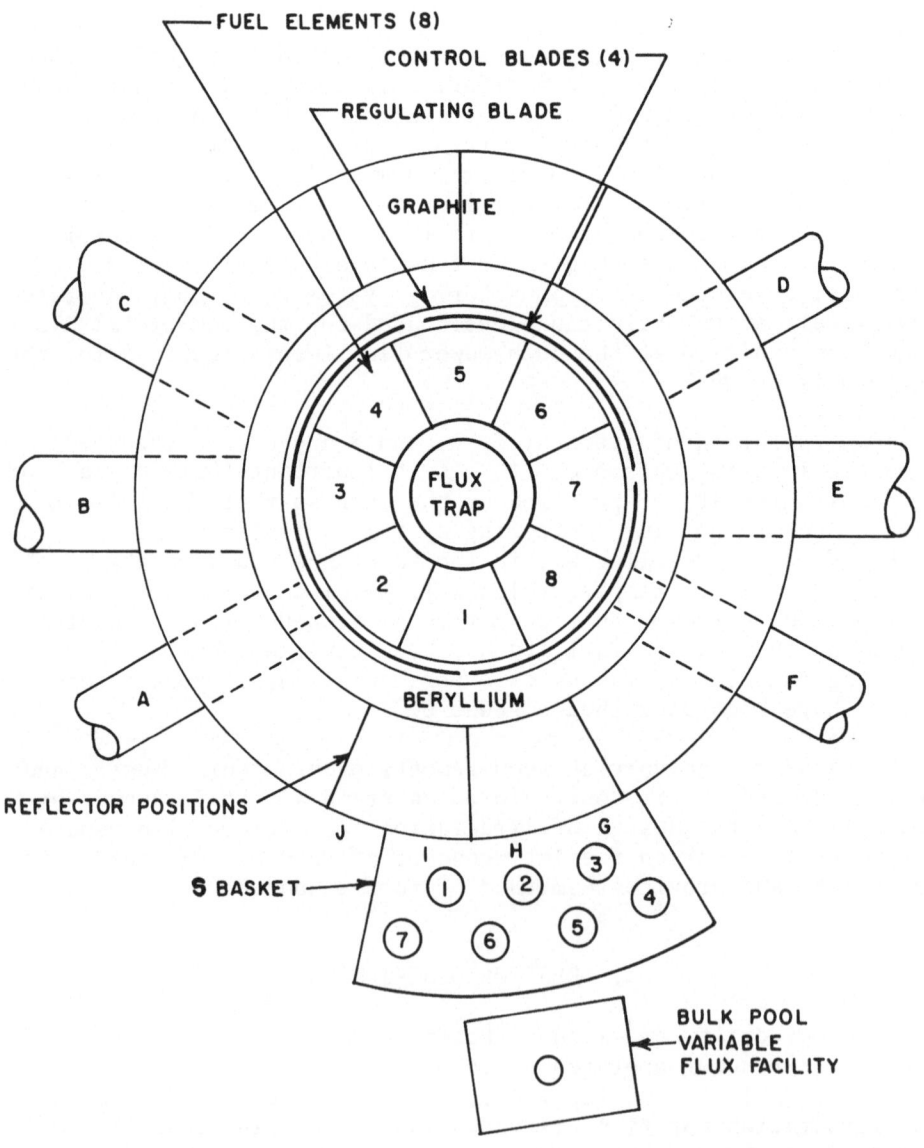

Fig. 1.   Cross-sectional view of the University of Missouri Research
          Reactor (MURR).

absorbing control blades between the beryllium reflector and the core. Sample irradiations are performed in irradiation holes in the graphite reflector, where the thermal flux ranges from 1×10^{14} n/cm^{2}sec to 1×10^{13}n/cm^{2}sec at a Cd ratio of approximately 10:1 to 20:1, or in the bulk pool facilities where the flux varies from 1×10^{13}n/cm^{2}sec down to 1×10^5n/cm^{2}sec with a Cd ratio of 12:1 to 30:1.

At the present time, all commercial silicon production is performed in one of the graphite wedges or one of the two bulk facilities, the S-basket and the variable flux facility. The variable flux facility is also used for Research and Development.

4. SILICON IRRADIATION WEDGE DESIGN

Figure 2 is a top view of the silicon irradiation facility wedge which is located in the graphite reflector. Rotating collars which are geared together and driven by a single drive shaft are under construction. At present, each irradiation hole is rotated by separate rotation shafts. Rotation speed is 1/9 RPM. Radial uniformity of ± 1% is obtained for diameters up to 83 mm.

The numbers in Fig. 2 in each cylindrical sample tube indicate the maximum sample-can diameter which can be irradiated in that position. The 32 mm facility is not usually used for silicon production. Each of the four silicon positions is designed to handle three 250 mm long, sample cans per position.

Figure 3 is a photograph of the silicon wedge irradiation facility taken before insertion into the reactor. The cables that run into the conduits at the left of the wedge are signal cables from the neutron detectors which are installed adjacent to each irradiation position. The largest sample tubes which accommodate the 83 mm sample holders are 100 mm in diameter. The total height of the wedge is 1 m.

A view of the wedge installed in the graphite reflector is shown in Fig. 4. It should be mentioned that the present wedge can be removed and rebuilt to accommodate other sample diameters which might be required at a later date. There are additional wedge positions adjacent to the present wedge and also at the rear of the reactor (Fig. 2) in which additional irradiation facilities of similar capacity can be installed. Also seen in Fig. 4 are the S-basket, the variable flux facility and two drive shafts used to rotate the silicon. The tops of the sample holder tubes can just be seen in Fig. 4 at the bottom of the drive shafts. These rotation tubes are 900 mm long, have a wall thickness of 2 mm and a diameter which depends on the irradiation position.

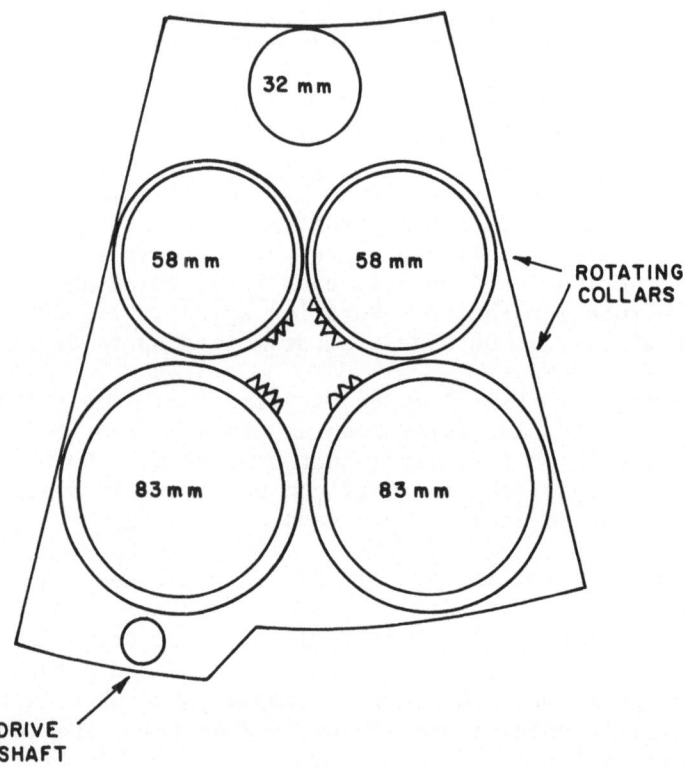

Fig. 2.   The silicon irradiation wedge located in the graphite
          reflector.

Fig. 3. Photograph of silicon irradiation wedge and detector conduits.

Fig. 4. View of reactor pool which shows wedge positioned in graphite reflector, the S-basket and the variable flux facility

5. IRRADIATION TECHNIQUES

Figure 5 shows our aluminum irradiation cans used for silicon
irradiation at MURR. All silicon is encapsulated in watertight ex-
truded aluminum cans which are 250 mm tall and have a wall thickness
of 0.75 mm. The can diameters are different for different irradiation positions. The purpose of these aluminum cans is twofold.
First, the cans protect the silicon from any induced activity
which might be collected from the pool water. Secondly, the cans
protect the silicon from breakage and also prevent silicon chips
from circulating throughout the pool coolant system.

The total neutron dose (fluence) is controlled by a real time
neutron monitoring system represented schematically at the top of
Fig. 6. This system consists of rhodium wire self-powered detectors[6] and analogue current integrators[7] with presets. The neutron
detectors utilize the β-decay of ^{104}Rh to produce a current which
is proportional to the neutron flux. The sensitivity of the
detectors is greatest at the absorption resonance near 1 eV shown
in Fig. 6. This resonance is located near the Cd absorption edge
and would, therefore, not be suitable for the measurement of
thermal neutron flux for the irradiation of some materials, however,
the resonance is in about the center of the 1/v absorption range in
silicon.[8]

Fig. 5. Photo of aluminum irradiation cans.

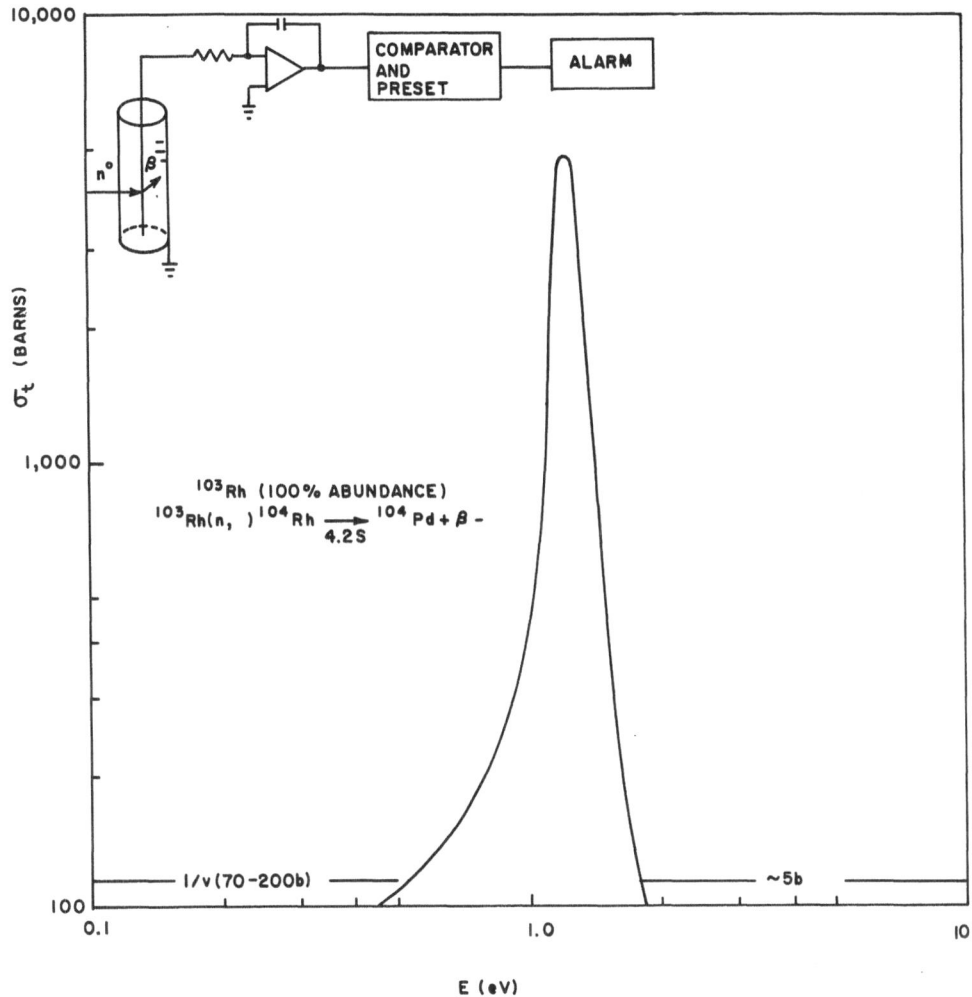

Fig. 6. Neutron absorption cross-section for Rhodium wire detec-
         tor. The insert at the top represents the detector
         current integration system schematically.

Below this resonance, the detector responds with a 1/v sensitivity while the absorption cross-section of the detector falls
to about 5 barn somewhat above the resonance. The current
integrator used to measure the detector current (which is proportional to neutron flux) is manufactured by Brookhaven Instruments

Corporation.[8] This current integrator has current scales ranging
from the mA to the nA range, an analogue current meter and a
digital indicator of integrated current units which is compared
with a digital preset. When the integrator counts reach the preset
value, an alarm is sounded to signal a sample changing event to
the reactor operators who then perform the scheduled event. Sample
handling events are discussed elsewhere in this volume.[4]

Figure 7 is a photograph of a bundle of three self-powered
detectors which are used in a single wedge position for top,
center, and bottom irradiation heights in that position. The
active length of each detector is 200 mm while the diameter is
1.5 mm. These detectors are quite flexible and have been spiralled
around the sample tube in the high precision variable flux facility.
A bank of current integrators is shown in Fig. 8. A single integrator is dedicated to each detector. The reliability of the B.I.C.
Model 300c integrators has been well tested in the accelerator field
and is legendary. Our experience has been that this instrument is
extremely immune to misuse. We have experienced no instrument
failures for over two years in our own integrator array. During
this period, the integrated current calibration accuracy has always
been better than ± 1/2% which represents the stability of the
constant current source we use to test our integrators. The manufacturer's specified calibration accuracy is ± 0.02%.[7]

Using the current integrators and self-powered detectors have
allowed us to compensate p-type silicon wafers to resistivities
over 100,000 Ω-cm consistently in the high precision variable flux
facility which has a flux of 5 x 10^{11} n/cm^2 sec. The current from
the detector at this flux range is about 2 x 10^{-8} A. Figure 9
shows the resistivity change with fluence for a p-type sample with
an initial resistivity of 1000 Ω-cm. The upper scale in the figure
is in units of $(1 - \phi/\phi_c)$ where ϕ_c is the target fluence necessary
to produce exact compensation or the fluence required to make the

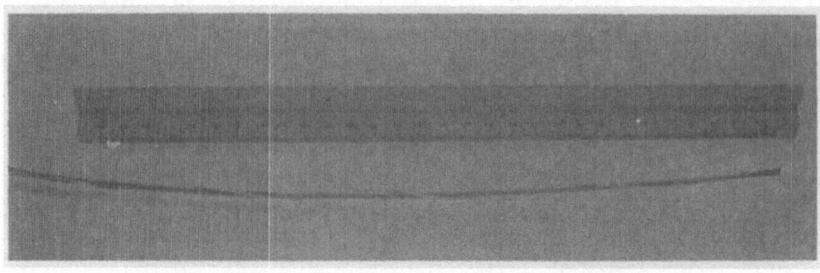

Fig. 7. Bundle of three self-powered neutron detectors.

Fig. 8. Current integrator array.

sample intrinsic. Therefore, 100 $(1 - \phi/\phi_c)$ represents the per-
centage the actual fluence ϕ differs from that fluence required
for exact compensation. It can be shown analytically, that
fluences within ± 0.9% of the target fluence will produce resistivities greater than 100,000 Ω-cm. Therefore, we have been able
to test the fluence accuracy of our integration system to this
accuracy. The ability to achieve these high resistivities is
very dependent on the accuracy with which the starting material
resistivity is measured. To achieve NTD compensated resistivities
higher than 100,000 Ω-cm, the starting material resistivity must
be measured to a similar accuracy. This is, of course, impossible
to achieve on an ingot basis because of variation in the starting
material.

The effects of initial resistivity measurement accuracy on
the final NTD doping accuracy are illustrated in Fig. 10. The
mean target error for a 23 sample commercial lot is shown in this
plot. The target resistivity is 43 Ω-cm. The initial resistivities
of these ingots were characterized using 4 point probe techniques
with a 3σ uncertainty of about ± 3%. From the distribution curve
obtained in Fig. 10, it is clear that effects of the initial resistivity characterization determine almost entirely the distribution
in the error of missing the target again suggesting that our inte-

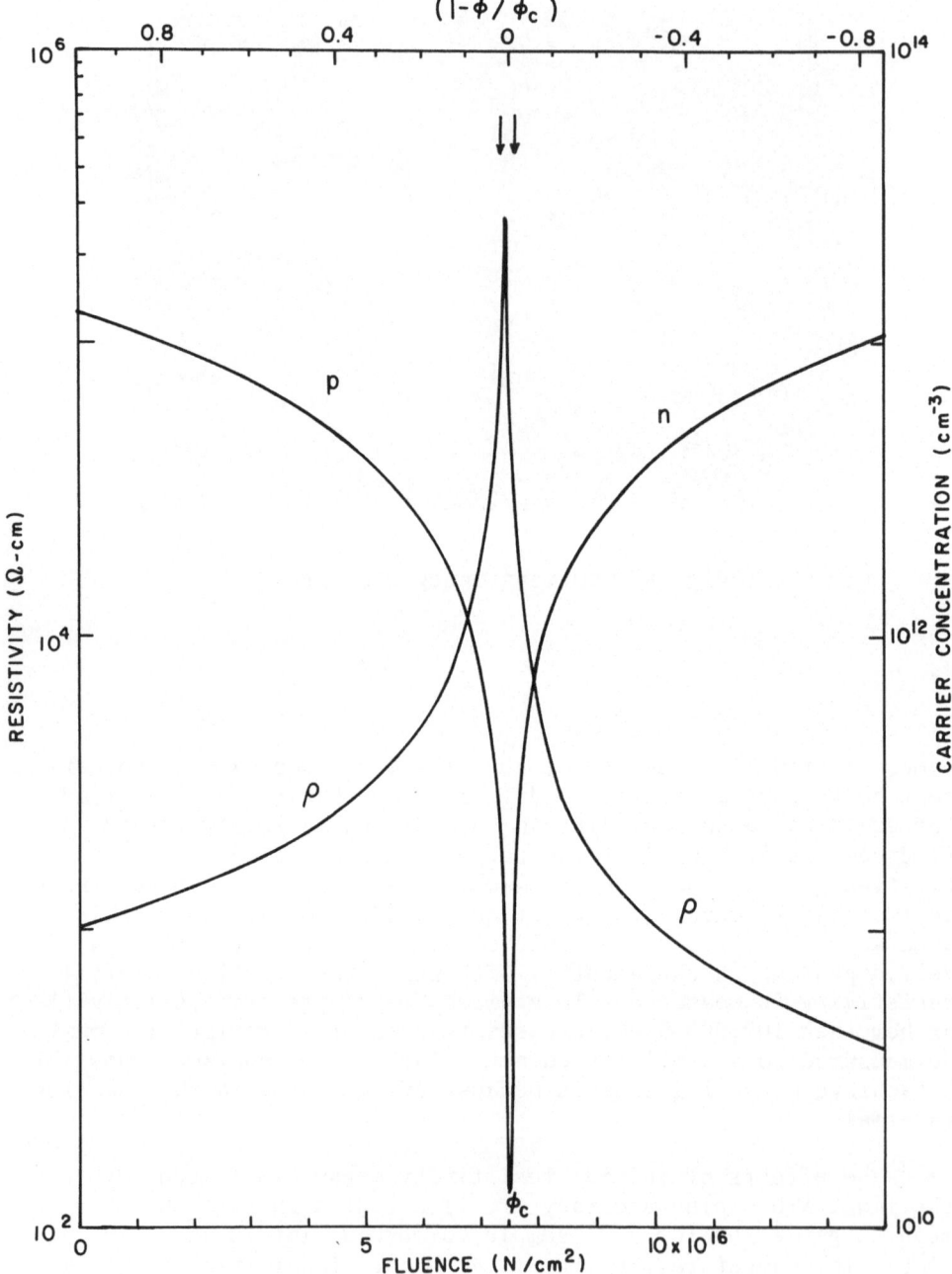

Fig. 9. Resistivity and carrier concentration vs. fluence for a
p-type sample with an initial resistivity of 1000 Ω-cm.

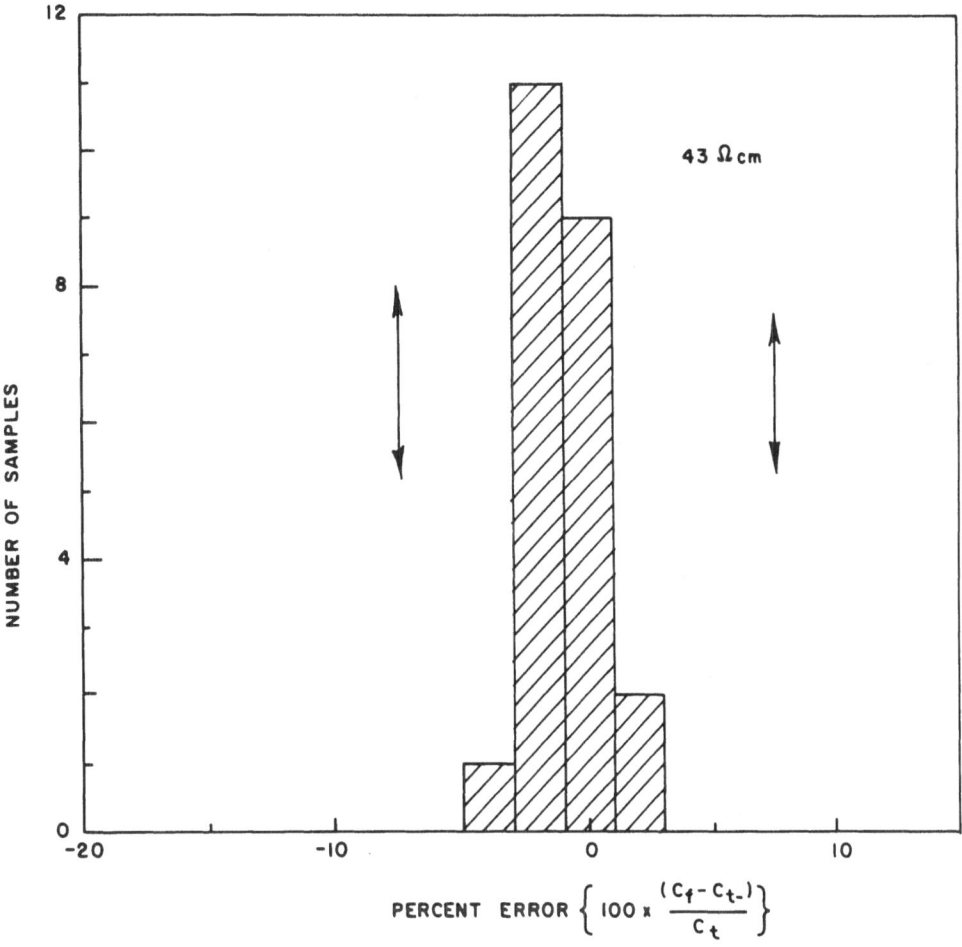

Fig. 10.   Mean target error plot for a 23 sample lot with a target
           resistivity of 43 Ω-cm.

grator system is considerably more accurate than the industrial
standards for resistivity accuracy.[9]

Figure 11 is a plot of the target error distribution for 1000
ingots irradiated at MURR. These data include all positions and
target resistivities from 25 to 150 Ω-cm as well as all calibrations which were made during this period to determine ppb/integrator
count. These data, therefore, represent a worst case situation.
The increased error is primarily due to three things, the difficulty

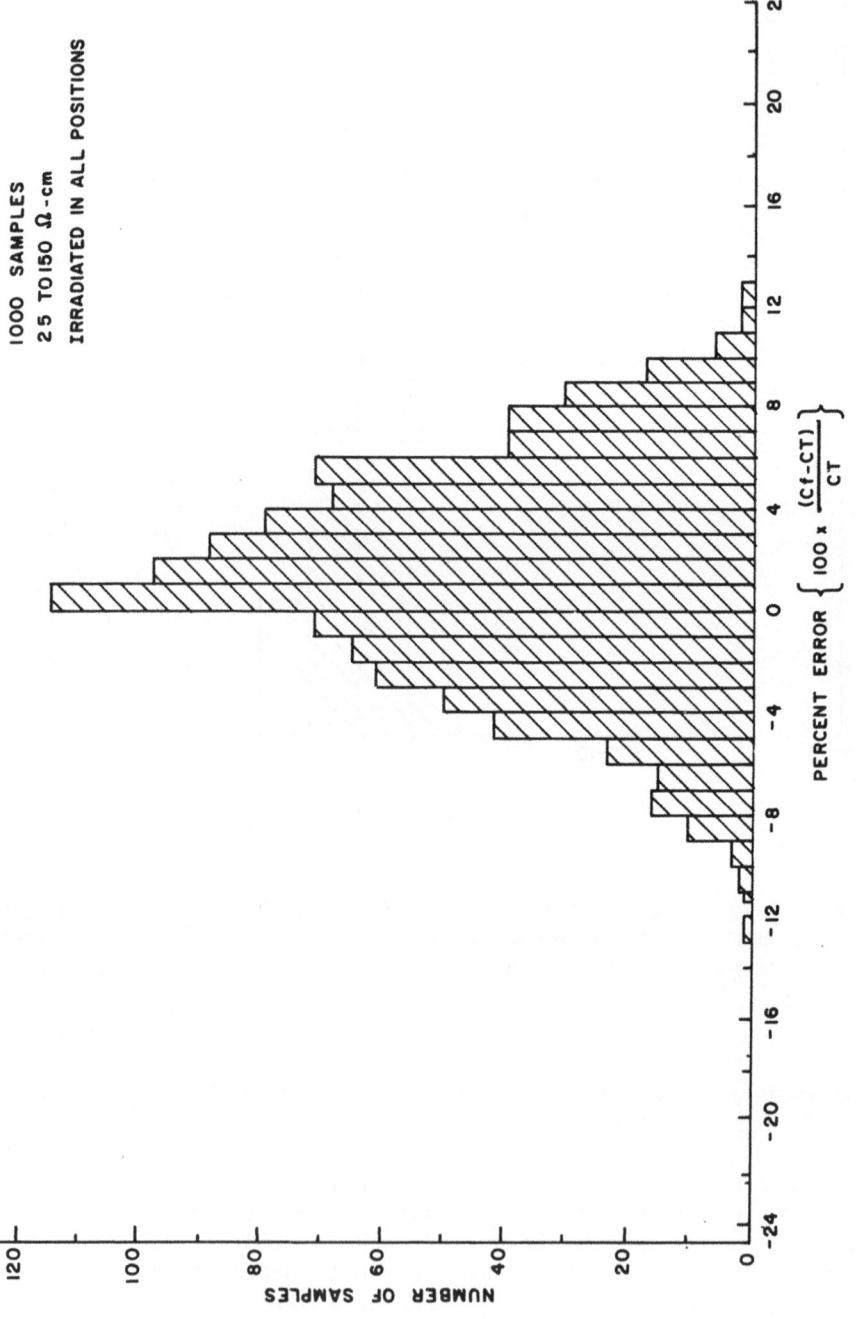

Fig. 11. Mean target error for the last 1000 samples irradiated at MURR in all positions and including calibration runs.

in hitting high resistivity targets, systematic errors in zeroing
in on new calibration factors and mischaracterization of the starting material. Still in all, 98% of the samples fall within ± 10%
of the doping target. This result exceeds conventional doping by
a comfortable margin.

As mentioned previously, the axial profile in MURR has ·a steep
gradient because of the reactor's high flux density and compact
size. It is interesting to note that the compact size of MURR
produces flux excursions as a function of control blade height
which forced the development of our integrator system. The use
of this integrator system, has, however, led to a solution of the
axial profile problem using a method which utilized reactor space
and flux to its fullest.

Several approaches to axial uniformity were rejected as impractical. The first approach involves flux profile modification
using neutron absorbers. An inspection of the MURR flux profile
in Fig. 12 shows that absorbing neutrons selectively to a total
uniform flux of $1 \times 10^{13} n/cm^2 sec$ would reduce irradiation capacity
by $(25"/30") \times (1 \times 10^{13}/2 \times 10^{13}) \times (1 \times 10^{13}/2.3 \times 10^{13}) = 0.18$,
or about 20% of the flux capacity available. This would, there-
fore, represent a considerably increased cost for NTD silicon irradiations.

A somewhat better approach would be to oscillate small samples
vertically through the whole 30" height. If the sample length were
10", this would reduce irradiation capacity by 2/3. Further disadvantages would be the expense of an oscillating mechanism and
the inability to determine fluence accurately.

Our integrator system has give us a third solution which
maintains irradiation capacity at 100%. By flipping the sample
at precisely 50% of the total fluence, the sample then sees a
mirror image of the flux profile as shown in Fig. 12 for the three
250 mm vertical zones. The center curve between these mirror
profiles is the average fluence as seen by the sample for the
entire irradiation. This system depends, of course, on the ability
to detect the 50% of total fluence point with high accuracy which
is possible because of the integration system. It can be seen in
Fig. 12 that this technique works best in the linear portions of
the flux profile in the top and bottom zones where the axial flux
profile is estimated to be uniform to better than ± 1%. The center
zone has a uniformity of ± 5% which is suitable for most commercial
applications.

Representative two point probe axial profiles are shown in
Fig. 13. For these three bars, the maximum deviations from mean
range from 1.4% to -5.75%. These deviations agree well with
estimates obtained from the flux profile shown in Fig. 12.

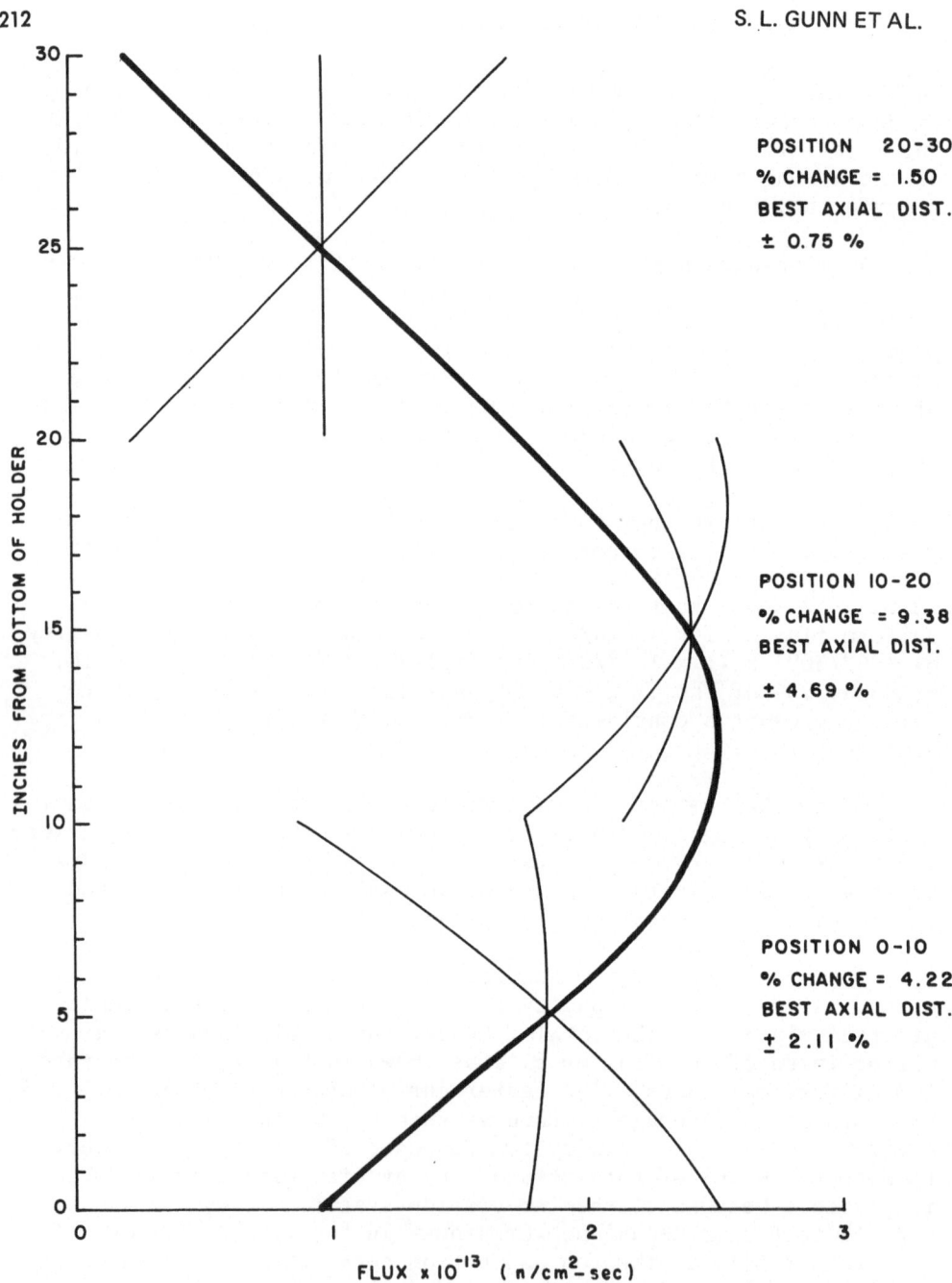

POSITION 20-30
% CHANGE = 1.50
BEST AXIAL DIST.
± 0.75 %

POSITION 10-20
% CHANGE = 9.38
BEST AXIAL DIST.
± 4.69 %

POSITION 0-10
% CHANGE = 4.22
BEST AXIAL DIST.
± 2.11 %

INCHES FROM BOTTOM OF HOLDER

FLUX x 10^{-13} (n/cm^2-sec)

Fig. 12. Typical silicon wedge hole flux profile represented by the
solid heavy line. The light lines represent the flux
profile inverted as seen by a sample which has been rotated end-for-end. The average of these two profiles is
shown as the center vertical line for each of the three
height positions.

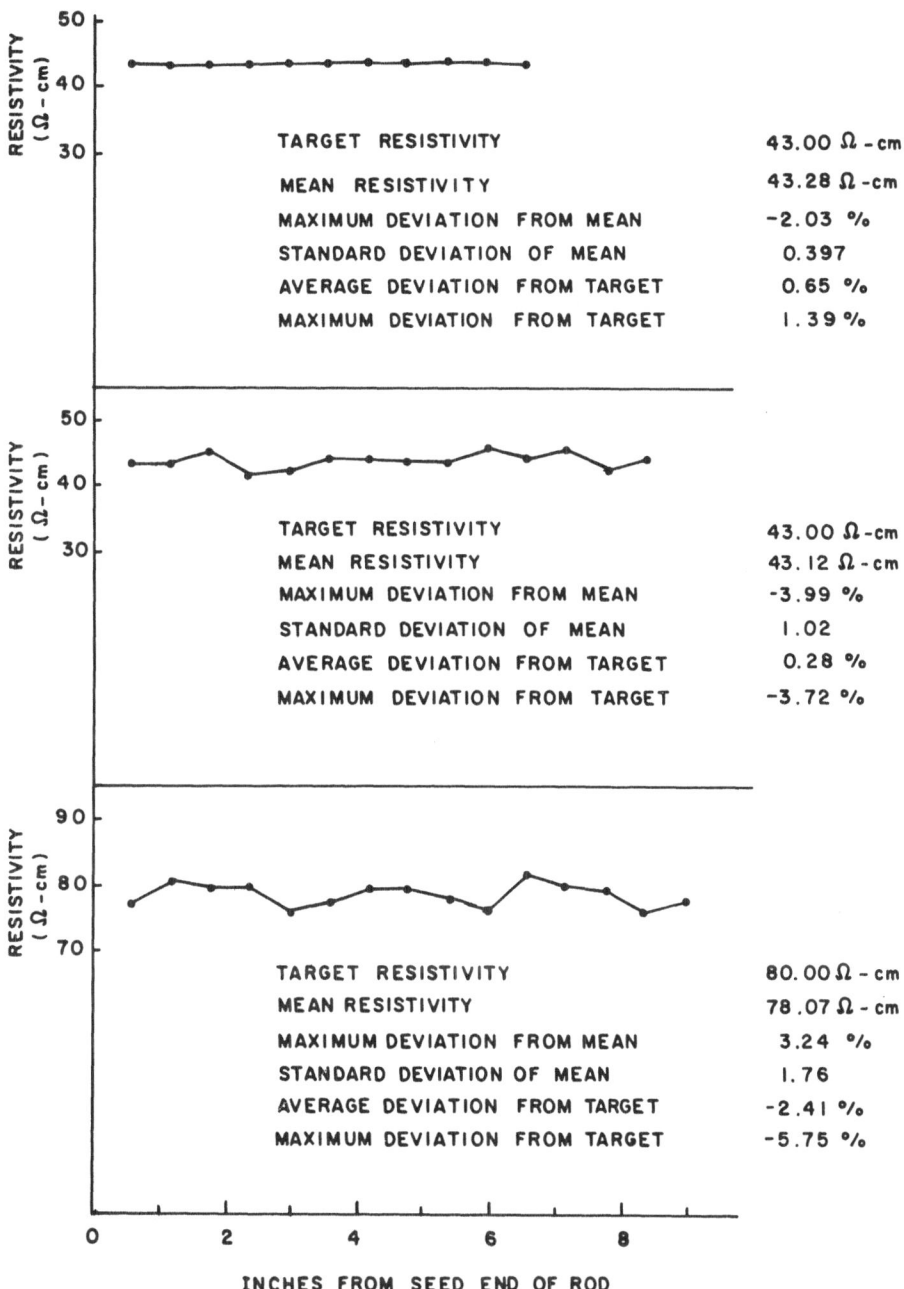

Fig. 13.   Two point resistivity axial profiles for three representative rods selected at random.

6.   IRRADIATION CAPACITY

As a final topic, we should discuss irradiation capacity. Since the time an ingot remains in the reactor is a function of final resistivity as well as initial resistivity, a reasonable unit of capacity is ppb-gm where 1 ppb = 5 x 10^{13}P/cm^3 atoms added to the original ingot. Also, if the starting material is reasonably pure, then a 50 Ω-cm final resistivity corresponds to approximately 2 ppb added.

The capacity of the present silicon wedge at MURR is 30 ppb-metric tons per year where a metric ton is 1000 kg. This translates to 15 metric tons of 50 Ω-cm material per year per wedge. With little modification, an additional three wedges could be added to bring MURR's capacity to 60 tons of 50 Ω-cm.

Although the capacity is certainly sufficient to handle the envisioned float zone business, it is clear that irradiation of Czochralski to 5 Ω-cm for the LSI device industry would push the capabilities of any single reactor.

7. SUMMARY

We have outlined the design philosophy which governed the present configuration of irradiation facilities at MURR. The particular characteristics of the MURR reactor have suggested solutions once specific irradiation goals were set. We have tried to maintain a certain amount of flexibility in the facility design to accommodate future expansion.

REFERENCES

1)   H. M. Janus, IEEE Trans. Electron. Devices ED-23, 797 (1976).
2)   J. M. Meese, this conference.
3)   B. D. Stone, D. B. Hines, S. L. Gunn and D. McKown, this conference.
4)   R. Berliner and S. Wood, this conference.
5)   J. M. Meese, Silicon Detector Compensation by Nuclear Transmutation, Final Technical Report AFML-TR-77-178, Air Force Materials Laboratory, (1978).
6)   Reuter Stokes, 18532 S. Miles Parkway, Cleveland, Ohio, 44128.
7)   Brookhaven Instrument Corporation, Box 3136, Austin, TX.
8)   D. E. Cullen and P. J. Hlavac, ENDF/B Cross Sections, Brookhaven National Laboratory, Upton, NY (1972).
9)   A.S.T.M. Standard F84-73, 1916 Race Street, Phil., PA.
*   Also Dept. of Physics, University of Missouri-Columbia.
**   Also, Dept. of Nuclear Engineering, U. of Missouri-Columbia.

A COMPUTER CONTROLLED IRRADIATION SYSTEM FOR THE UNIVERSITY OF MISSOURI RESEARCH REACTOR

R. Berliner

University of Missouri Research Reactor, Columbia, MO

S. Wood

Brookhaven Instrument Corporation, Austin, TX

ABSTRACT

A computer controlled irradiation monitoring system has been designed and is being installed at the University of Missouri Research Reactor. The system utilizes rhodium wire self-powered neutron detectors and a multiplexed data input system with BIC current integrators to monitor the specimen flux. A PDP 11/03 computer is used to maintain a memory image of the current status of all the in-pool specimens and to alert the reactor operators to required specimen handling procedures. The PDP 11/03 computer is in periodic communication with a PDP 11/40 which maintains in its disk files a record of previous irradiations and the schedule for future irradiations. Separate programs which are run on the PDP 11/40 provide for specimen logging, calculation of irradiation parameters, scheduling of irradiations and continuous recalibration of the detectors.

The system has been designed to be resistant to failure of either of its computers and to provide protection against power interruptions.

The system hardware will be described and the system software block diagrams will be discussed.

1. INTRODUCTION

Transmutation doped Si has several obvious advantages in the production of n-type substrates for electronic devices. These advantages--dopant uniformity and dopant concentration control-- have led to a rapid increase in the volume of NTD-Si production. The quantity of Si ingots commensurate with world production pre- dictions for NTD-Si of 20-50 tons/year by 1980[1] make it imperative that reactor facilities automate to the extent practical, the handling, irradiation monitoring and record keeping that accompany the NTD-irradiation processing.

The current Si production at the University of Missouri Research Reactor is in excess of 1000 cans/year and imposes a serious paperwork burden on the Reactor Services staff. In addi- tion, the increased irradiation volume and capability represented by recently installed facilities make it impractical to continue procuring n-flux monitor current integrators as each new irra- diation position is put on line. For these reasons, a computer controlled irradiation monitoring system has been designed and is being installed at the Reactor Facility. This system was designed for the irradiation monitoring, control and accounting associated with NTD-Si production but it is sufficiently general so that it will be extended to control all in-pool irradiations. In the material that follows, the irradiation monitoring and specimen accounting system will be described and the hardware and software design of our computerized system will be outlined.

2. IRRADIATION MONITORING

2.1. Existing Irradiation System

The University of Missouri Research Reactor is a 10 MW open pool, light water moderated flux trap reactor. A schematic cross sectional drawing of the core and the main irradiation positions is shown in Fig. 1. In this figure the relationship between the flux trap, fuel elements, control blades and the Be and C reflectors can be clearly seen. Irradiation positions for Si, until very recently, consisted of three separate facilities. First, H_2 and I_2 denote the holes located in the C reflector. These each had a capacity of three 10-inch long cans 2-inches dia. at a flux level of approximately $3 \times 10^{13} n/cm^2$. Second, the S-basket denotes an irradiation facility placed just beyond the C reflector with seven 3-inch dia. holes, each capable of accomodating three 10-inch long cans at a flux of approximately $5 \times 10^{12} n/cm^2 sec$. Finally, there is the variable flux facility which is used for very high precision doping and is a 2-inch dia. 20-inch long position at a maximum flux of $5 \times 10^{11} n/cm^2 sec$.

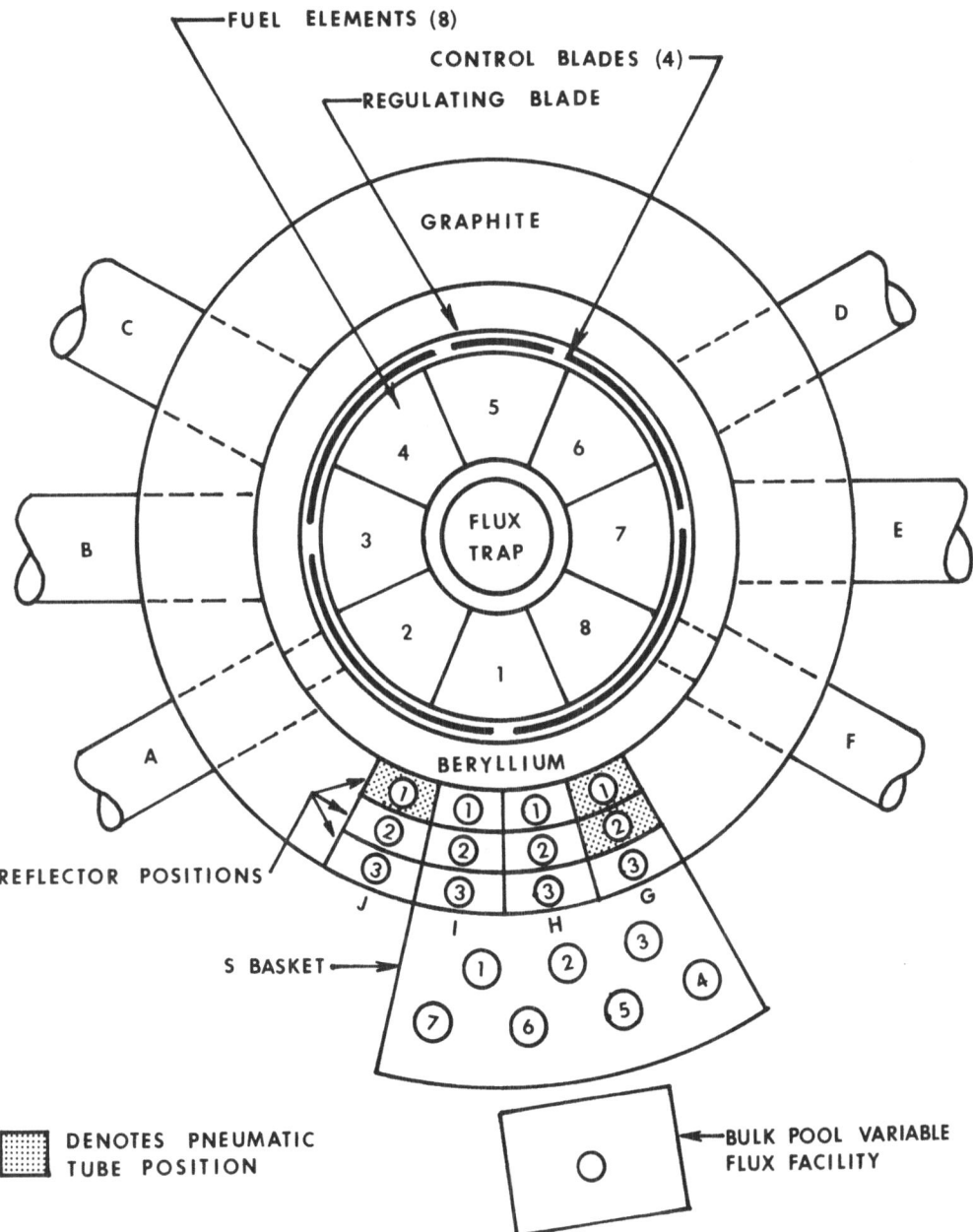

Fig. 1. Cross-sectional view of the University of Missouri Research
Reactor core and irradiation volume.

Recently installed irradiation facilities replace the old C-reflector positions with a new wedge which is shown in Fig. 2. This facility has a capability of six 10-inch long 2-inch dia. ingot positions and six 10-inch long 3-inch dia. positions.

Accurate control of the n-fluence and characterization of the starting material are necessary for precision tailoring of the electrical characteristics of NTD-Si. Very early irradiations at MURR were accomplished by control of the time an ingot was exposed to the neutron flux. Because of flux variations due to factors such as control blade height, this technique did not produce reliable results.[2] Neutron fluences are now monitored by Rh self-powered neutron detectors. The use of these detectors arose out of detector tests that were performed by Rosemont in cooperation with MURR. The detectors are 4- to 8-inches long, 1/16-inch dia. and have at their center an Rh wire. The isotope ^{103}Rh has a

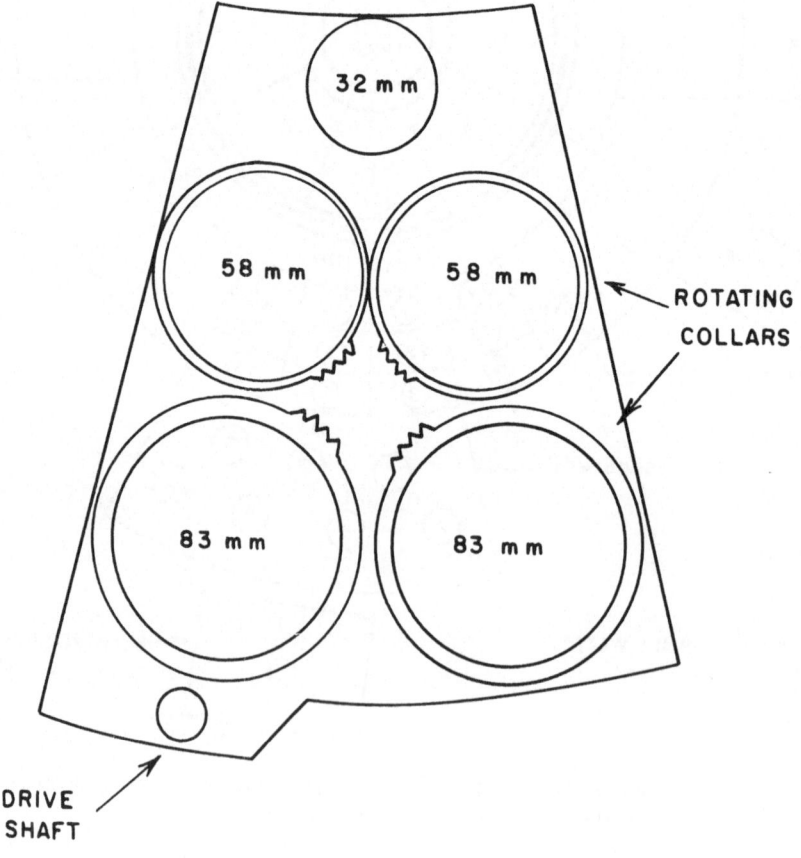

Fig. 2. Silicon irradiation positions.

resonance near 1 eV, $^{103}Rh(n,\gamma)^{104}Rh \rightarrow ^{104}Pd + \beta^-$. This reaction results in a net positive charge on the wire. The detectors have an absolute calibration factor of approximately 1 μa/10^{14}n/cm²sec and they are placed adjacent to the ingot irradiation positions as shown in Fig. 3.

The current integrators[3] used for monitoring detector currents are well-known to ion-implantation workers. They are often used to measure ion-implant does for device production and research.

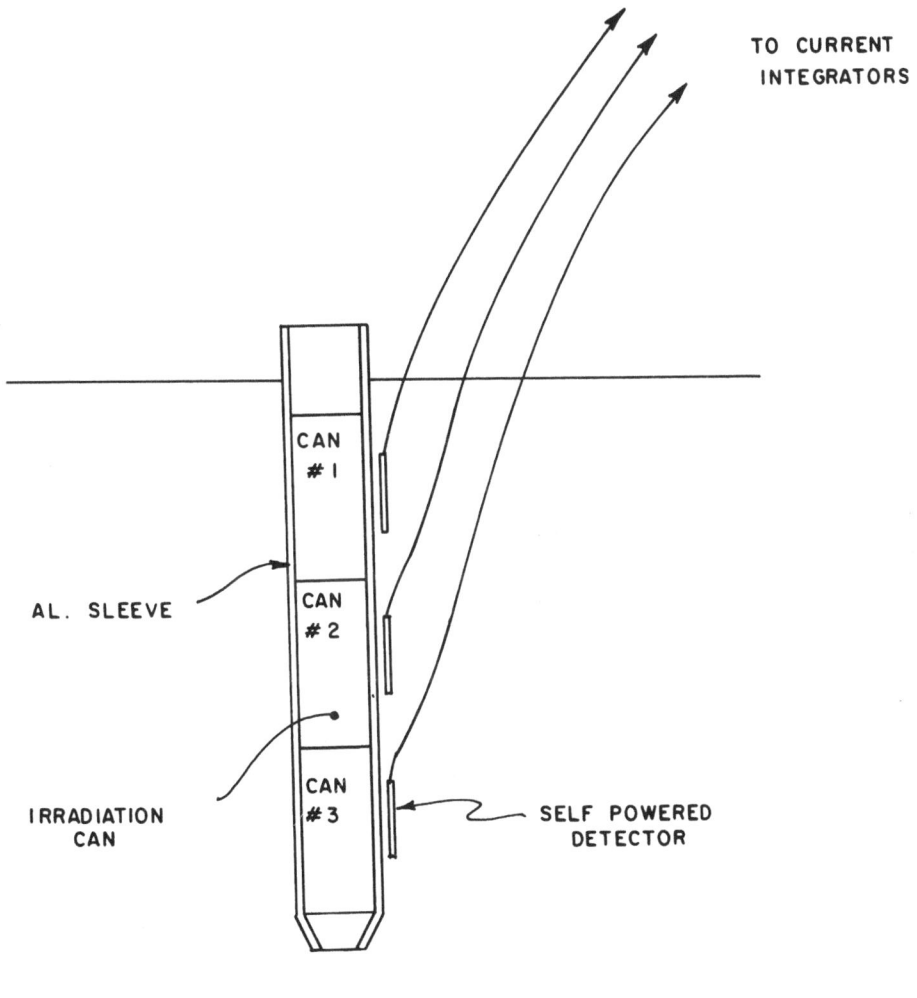

Fig. 3   Arrangement of Si encapsulation cans and self-powered neutron detectors.

In order to obtain high axial and radial doping uniformity, the ingots are rotated during irradiation and turned end-for-end at their irradiation midpoint. These procedures have resulted in highly accurate and reproducible doping results.[4]

2.2.   Automated System Hardware

Functionally, the computerized system fulfills two requirements. First, it economically provides a method of monitoring the n-flux at a large number of irradiation positions and second, it considerably reduces the volume of hand generated records that are required for the processing of each Si ingot.

A block diagram of the hardware configuration of the flux monitor system is shown in Fig. 4. The Rh self-powered neutron detectors are connected to current integrators and PDP 11/03 through a controller/multiplexer. The structure of the multiplexer is shown in Fig. 5. The system is presently configured with one expander which accomodates up to 20 detectors. Three additional expanders will permit the system to handle up to 80 detectors. The multiplexer is connected to between 1 and 9 modified BIC current integrators. Each integrator is set to a different current range. In order to sample the current from a detector, the computer places the detector address (a number between 1 and 80) and the appropriate current range on the 11 input lines of the multiplexer and strobes the read-start line. The detector signal is routed to

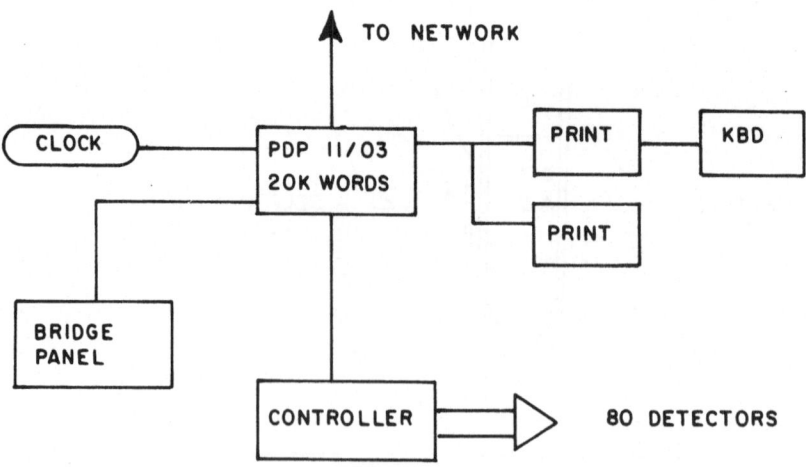

Fig. 4. Automated irradiation monitor system block diagram.

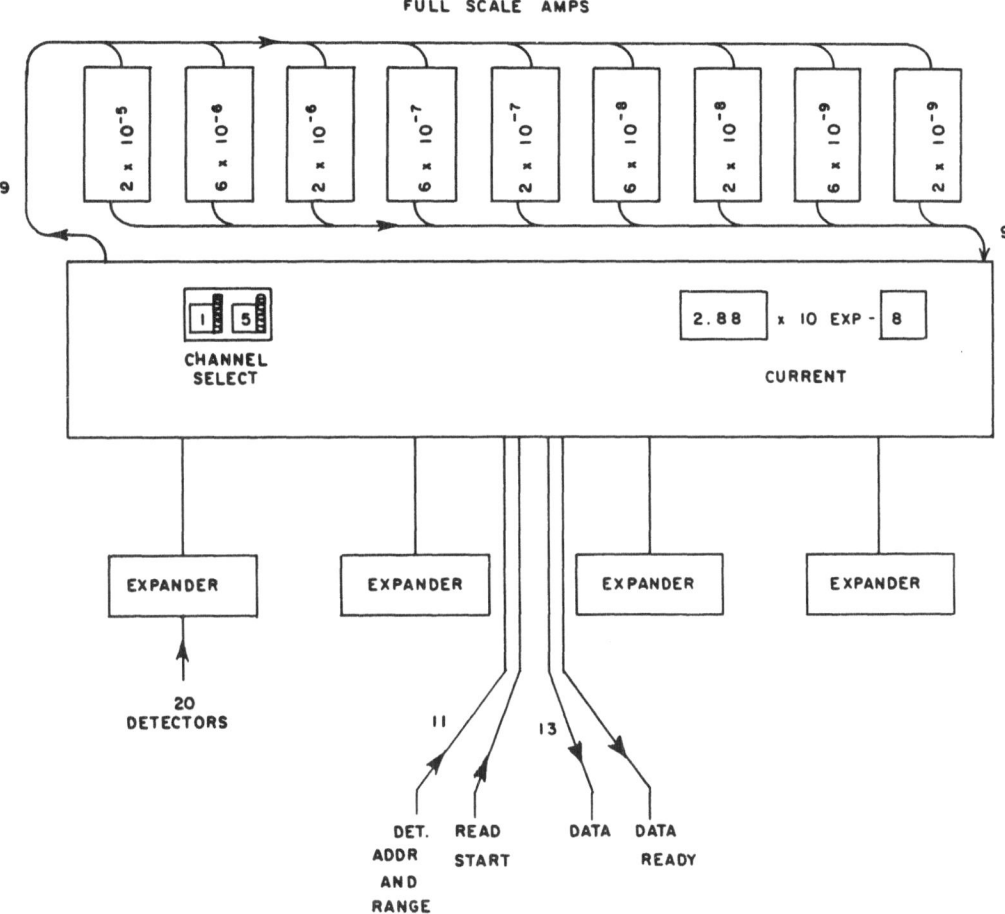

Fig. 5. Current integrator multiplexer system.

the appropriate integrator and the BCD encoded current on the integrator from the detector is available 200 ms later on the multiplexer data output lines (except for the 2 x 10^{-9} A range where 1.5 sec is required).

The PDP 11/03 is connected to two printers, the bridge panel switch board and a real time clock. One of the printers is used to inform the operators of required specimen evolutions, detector malfunctions, error conditions, and pending specimen change operations. The other printer maintains a complete hardcopy record of all specimen transactions.

The computer is informed of specimen evolutions by the status
of the bridge switch panel. As specimens are inserted into or
withdrawn from their irradiation positions, the appropriate
switches on the bridge panel are set by the operators. The system
software examines the condition of these switches and uses their
states to cross-check for errors in specimen change operations.

The PDP 11/03 computer forms part of the MURR beam port floor
computer network. The structure of the MURR computer network is
shown in Fig. 6. The system consists of a host PDP 11/40 with
32 k words of memory, 10 M bytes storage on two hard disks and a
dual floppy disk system. The network supports 6 PDP 11/03 computers under the DEC operating system RT11-Remote 11. Each
computer controls a single experiment and each can operate independently of the host PDP 11/40. This system has several advantages
over the use of a single central computer system. First, since all
of the PDP 11/03 computers are essentially identical, failure of

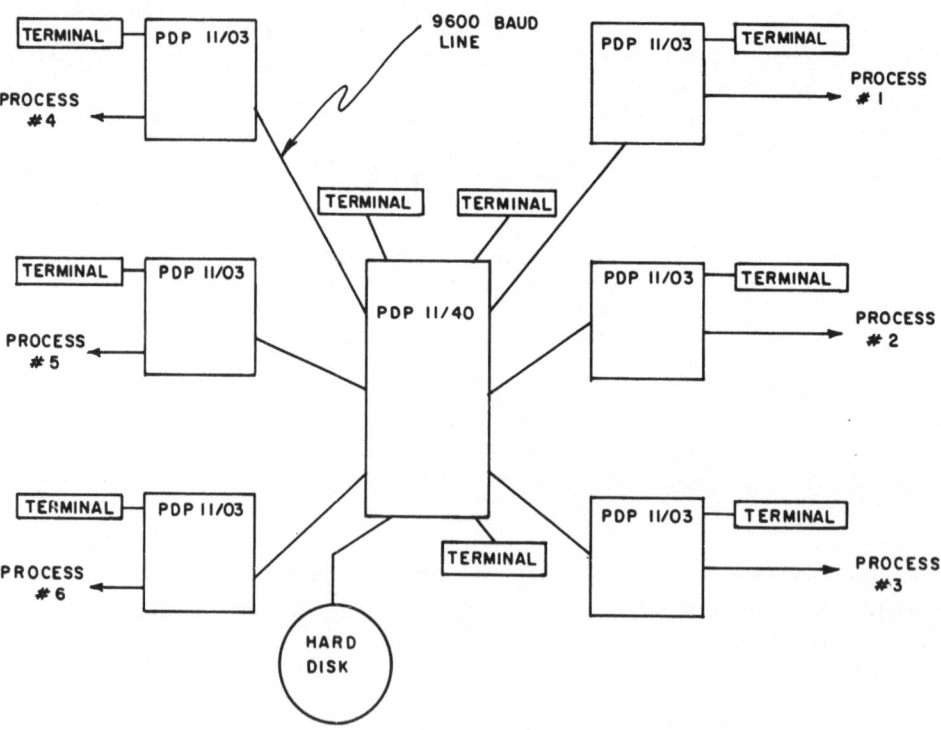

Fig. 6. MURR computer network.

any network element will only shut down one experiment (or process controller) and high priority applications can be supported by simply exchanging computers or parts with low priority applications. Second, considerable cost savings are realized by having the satellite computers share the expensive disk memory units.

In the RT11-Remote 11 operating system environment, PDP 11/03 applications programs and operating systems are stored on the PDP 11/40 system disks. Each PDP 11/03 can access the disks to obtain these programs or to record data. In case of PDP 11/40 or disk failure, the satellite systems can be loaded with paper tape.

2.3. Si Irradiation Management

We have chosen to implement the automation system software in a manner which mimics the NTD-Si accounting procedures that are currently used at the facility. This has the advantage that changeover retraining problems are minimized.

Silicon ingots arrive at MURR in quantities denoted "Lots" which can each consist of up to 30 bars. A Lot is composed of ingots that have nominally the same neutron fluence requirements. Because the ingots often do not fill the entire encapsulation can, Si spacers are used to position the bar in the middle of the irradiation volume. The bars are sealed into welded Al cans and labeled with an identification number. All of a Lot is scheduled into a single irradiation position depending on its availability, the Lot required delivery date, and the doped ingot target resistivity. For instance, a Lot which has a light irradiation requirement is not scheduled into a high flux position because the irradiation times become inconveniently short.

The desired integrated detector current is calculated for each ingot using the known calibration factor for the detector corresponding to the irradiation position. The cans comprising a Lot are then loaded sequentially into their irradiation position and turned end for end at 1/2 the desired total fluence. A record of fluence and total irradiation time is kept for each bar. After the irradiation, these records are compared to the fluence per ppb-P added and the fluence per unit time in order to provide a continuous recalibration of the detectors which have a nuclear burn-up of approximately 1%/Mo. This process is summarized in Fig. 7.

The system software has been divided into two parts. The PDP 11/40 executes the programs which provide for log-in of arriving Si Lots, irradiation scheduling, spacer calculations, ID# assignments, flux calculations, and record keeping. The PDP 11/03 software controls the actual irradiation monitor functions and

IRRADIATION MANAGEMENT

RECEIVE LOT

↓

ASSIGN CAN #'S
CALCULATE SPACERS
WELD CANS

↓

ASSIGN TO HOLE
SCHEDULE

↓

IRRADIATE
TURN AT $\phi/2$

↓

WAIT FOR DECAY
PROCESS FOR SHIPPING

↓

COMPARE RESULTS
ϕ/PPB
ϕ/UNIT TIME

↓

UPDATE DETECTORS

Fig. 7.   Flow chart for Si irradiation processing.

updates the PDP 11/40 disk files as the irradiation proceeds.
These relationships are summarized in Fig. 8. Both computers
access the main data base which is maintained on the network disks.

The data base on the system disk consists of two files--the
Lot File and the Schedule File--and a table, the Position Table.
One element of the Lot File is shown in Fig. 9. The upper portion
of this element contains information which is common to all of
the ingots in the Lot. The lower portion of the element consists
of information specific to the individual bars in the Lot. Briefly,
the Lot No., the bar diameter, date, target resistivity and date
due are entered into the system when a Lot is received at the
facility. Each bar has a number and is assigned a can No. (in-
scribed on the Al encapsulation can) and an MU ID No. The length
and weight of each bar is also recorded as is the net ppb-P to be
added. Once this information is added to the Lot File the scheduling program can be run. This program assigns the Lot automatically
to an irradiation position and selects a priority to permit comple-
tion of the irradiation in time to satisfy the required due date.
The start and stop dates are estimated for the Lot and for each
bar, and the status flag information is set to indicate the status

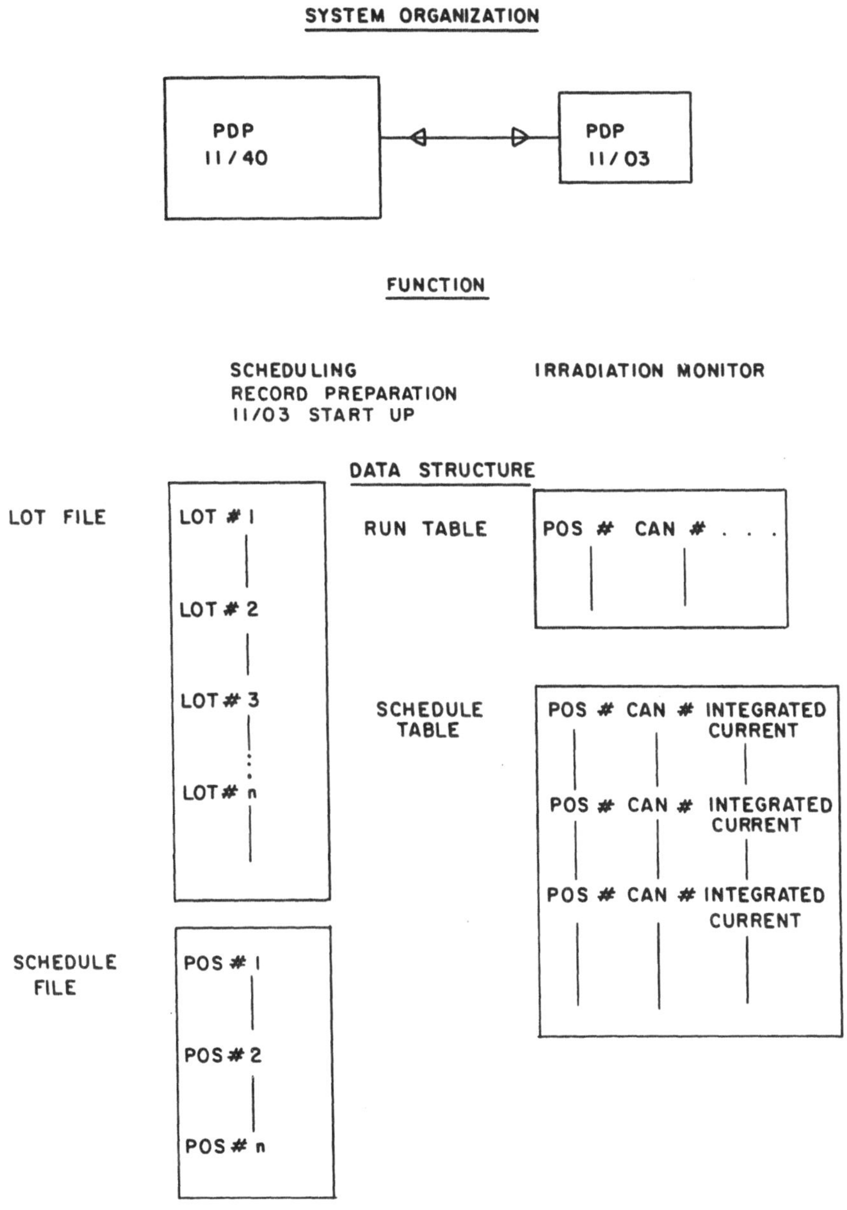
SYSTEM ORGANIZATION
PDP 11/40
PDP 11/03
FUNCTION
SCHEDULING
RECORD PREPARATION
11/03 START UP
IRRADIATION MONITOR
DATA STRUCTURE
LOT FILE
LOT # 1
LOT # 2
LOT # 3
LOT # n
SCHEDULE FILE
POS # 1
POS # 2
POS # n
RUN TABLE
POS # CAN # . . .
SCHEDULE TABLE
POS # CAN # INTEGRATED CURRENT
POS # CAN # INTEGRATED CURRENT
POS # CAN # INTEGRATED CURRENT

Fig. 8. Hardware/software partition of the computer activities for the network host computer and the irradiation monitor microcomputer.

PDP 11/40

LOT FILE

LOT #_____POS #_____PRI #_____*START_____*STOP_____

DIA_____TARGET RESISTIVITY_____CAN SIZE_____DATE_____

REQD. DECAY TIME_____DATE DUE_____

POS #_____PRI #_____STATUS FLAGS_____

BAR #   CAN #   MUID #   LENGTH   WT.   NET PPB   COUNTS   *START   *STOP

Fig. 9.   Representation of one element of the Lot File maintained
          in the host computer disk files.

of the Lot. At this time, the Schedule File is updated. An
element of the Schedule File is shown in Fig. 10. The File is
organized by irradiation position. Each element consists of a
list of Lots scheduled into a single irradiation position. The
estimated start and stop times for each lot are listed as are the
priority, Lot No., date due, and status flags. As in the Lot File,
the status flag information indicates first whether the Lot is
currently being irradiated (if not, the start/stop times are
estimated) and if so, the can number currently being irradiated
and its position in its irradiation schedule. Finally, the third

PDP 11/40

SCHEDULE FILE

POS #_____

PRIORITY   LOT #   *START   *STOP   DATE DUE   STATUS FLAGS

Fig. 10.   Representation of one element in the computer Schedule
           File.

element of the data base is the Position Table which contains the
position numbers, detector numbers and attributes of all the irradiation positions. These include the hole diameter, detector
current/ppb-P, current/unit time, detector number, and current
integrator scale appropriate to the irradiation position. By
altering this table, it is possible to assign any detector to an
irradiation position, and thus, continue operation in a position
with a faulty detector by temporarily using another detector after
properly adjusting its calibration factors.

Communication with this accounting data base on the system
disks is accomplished with a program on the PDP 11/40 which admits
the use of commands to log in a Lot, print that portion of the Lot
File corresponding to any Lot, edit the information in the Lot File
element, and to hand schedule (or unschedule) any Lot. In addition,
the Lot File can be searched for elements that satisfy certain
criteria such as scheduled/unscheduled, irradiated/unirradiated/in-
process and so on.

The PDP 11/03 software has three functions. First, it
provides the quasi-continuous monitoring of each detector by manip-
ulation of the current multiplexer. Second, the system provides
the reactor operators with a set of messages corresponding to upcoming events connected with the ingot irradiations. Any manual
operation such as the introduction of a specimen into an irradiation position, withdrawal of a specimen or mid-fluence end-for-
end rotation is called an event. Finally, the PDP 11/03 periodically transmits to the system disk files updated information on
the irradiation status of the in-pool specimens.

Because of the construction of the irradiation positions, an
event in any position requires that three specimens can be removed
from their irradiation hole. This information is communicated to
the PDP 11/03 by the bridge switch panel shown schematically in
Fig. 4. This panel has a switch corresponding to each irradiation
hole. When a switch is thrown, the computer ignores the detector
readings corresponding to those positions. Much of the PDP 11/03
software structure revolves around checking the status of the
bridge panel switches and comparing their status to the proper configuration as indicated by the irradiation records. The main PDP
11/03 activity consists of execution of a loop which sequentially
samples the current readings from all of the active detectors. As
each detector is sampled, the flux on each specimen is compared to
its preset values, the status of the bridge switch panel is checked
and the clock is read. Deviations of the detector current from its
normal values will result in a message alerting the operators to a
detector failure. Upcoming sample evolution events also result in
appropriate messages being transmitted to the operators. The PDP
11/03 transmits an updated record of the in-pool specimen irradiation status to the network disks at least every 5 minutes. This

record, coupled with a battery back-up on the PDP 11/03 memory, makes the system very resistant to catastrophic data loss due to power failures. In the event that a PDP 11/03 failure results in a short term computer shutdown, the irradiation status can be quickly recovered from the data stored on the network system disks. Finally, failure of the network PDP 11/40 or the system disks will not affect the operation of the PDP 11/03 except that as new Lots enter the irradiation phase of processing the information relevant to their irradiation control will have to be entered into the PDP 11/03 by hand.

3. SUMMARY

The computer automated irradiation monitoring system we have described provides a cost-effective solution to the problem of handling large volumes of Si for NTD processing. The incremental cost of the system is on the order of $20k and it can be easily adapted to computer environments existing at other installations. The system has been explicitly designed to minimize data loss from computer failure and requires the simultaneous shutdown of two computers to produce significant delays in irradiation processing.

REFERENCES

1)   Hans Janus. This conference.
2)   S. L. Gunn, University of Missouri. Private communication.
3)   Brookhaven Instruments Corporation, 11124 Jollyville Road, Austin, Texas.
4)   S. L. Gunn, J. M. Meese, and D. M. Alger. This conference.

ATOMIC DISPLACEMENT EFFECTS IN NEUTRON TRANSMUTATION DOPING*

H. J. Stein

Sandia Laboratories**

Albuquerque, NM 87185

ABSTRACT

The production of defects in silicon by neutron irradiation and the effects of subsequent thermal annealing are reviewed for their impact on neutron transmutation doping. Recoiling atoms from fast neutron interactions are the dominant contribution to the energy available for producing atomic displacements. Energy deposition considerations indicate that defect clusters are created but amorphous zone formation is highly unlikely. Extensive defect reordering occurs at room temperature with resultant vacancy-vacancy, vacancy-impurity, interstitial-insterstitial, and interstitial-impurity interactions. Many irradiation-produced defects have been characterized and identified for annealing temperatures < 500°C, and the role of impurities is pervasive. For temperatures near 500°C defect evolution processes yield vacancy and interstitial loops with impurities still participating. Additional work is needed to more fully characterize and identify residual lattice imperfections which remain in irradiated silicon after annealing to temperatures > 500°C, and to understand the interaction processes in their formation. Although the concentrations of residual defects are relatively low, these residual defects are of primary importance for neutron transmutation doping and also for ion implantation.

1. INTRODUCTION

The introduction of particle irradiation into the processing of semiconductor materials and devices creates a new need for additional understanding of atomic-displacement-produced defects in semiconductors. Irradiation-produced-defect studies have been supported over a long period of time by space and military agencies. That support, however, peaked in the late 1960's partly because of the success of the studies, and partly because ionization-associated irradiation effects became a more urgent problem in space and military hardware. The renewed interest in displacement-produced defects, which has been stimulated by neutron transmutation doping (NTD) as well as by ion implantation, is primarily concerned with residual defects remaining after material and device processing at temperatures > 500°C. There might, at first, be some question about the applicability to NTD of results from previous studies directed largely at defects produced at device operating temperatures because most of those defects anneal out at temperatures < 500°C.[1-2] But the situation is viewed as one of evolution so that residual defects remaining at temperatures > 500°C are products of primary and secondary defects through interactions with each other and with impurities and imperfections in the host crystals. Therefore, the previously obtained information is applicable, and new information from studies in support of NTD should add to the existing knowledge of displacement-produced defects.

This paper is intended to highlight basic ideas on neutron displacement effects and is not intended to be an exhaustive review. The discussion is centered on silicon because more detailed information is available on irradiation-produced defects in silicon than for any other material.

Based upon energy deposition, it is shown that fast neutrons dominate the displacement-damage effects for typical thermal-to-fast neutron ratios used in NTD processing of silicon. The clustered nature of neutron damage, and the possibility of overlapping clusters are considered. Defects identified in neutron damage are briefly reviewed and results from fast neutron damage are combined with results from ion implantation studies to discuss the defects remaining in NTD silicon at temperatures > 500°C.

2.   NEUTRON-PRODUCED ATOMIC DISPLACEMENTS

2.1.   Energy Deposition into Displacement Processes

Recoil atoms from gamma ray or particle emissions after thermal neutron capture, and recoil atoms from elastically and inelastically scattered fast neutrons, produce atomic displacements in solids.[3] The relative importance of the displacement damage produced by thermal and fast neutrons can be estimated by using isotope concentrations, capture or scattering cross-sections, and recoil energies. Such an estimate has been made for silicon and the results are presented in Table 1. The first three columns list

Table 1.   Energy available for producing displacement damage in silicon by neutron transmutation doping.

THERMAL NEUTRON CAPTURE RECOIL
(Energy into displacements, $\nu = 780$ eV)

Silicon Isotope	N Concentration (10^{22} cm^{-3})	σ Cross-Section (10^{-24} cm^2)	$N\sigma$ (cm^{-1})	$\nu\, N\sigma$ (eV/cm)
$^{28}\mathrm{Si}(n,\gamma)^{29}\mathrm{Si}$	4.61	0.08	0.0037	2.88
$^{29}\mathrm{Si}(n,\gamma)^{30}\mathrm{Si}$	0.23	0.27	0.00062	0.49
$^{30}\mathrm{Si}(n,\gamma)^{31}\mathrm{Si}$	0.15	0.12	0.00018	0.14
			Total	3.51

$\xrightarrow{\text{2.6 hr}}$ P^{31}
β (1.5 MeV)

FAST NEUTRON KNOCK-ON RECOIL
(Average recoil energy of 50 keV assumed, $\nu = 25$ keV)

All	5	3 (avg)	0.15	3.8×10^3

the silicon isotopes, isotope concentrations, and cross sections
for thermal neutron capture[3] and for fast neutron scattering. The
probability of interaction (product of isotope concentration and
capture or scattering cross section) is listed in column 4. The
energies available from silicon recoils for producing atomic displacements are given in column 5. For thermal neutrons, these
energies were obtained from a product of the probability for
thermal neutron capture and the average recoil energy[3] of 780 eV.
Kirt and Greenwood[4] used 474 eV for the average silicon recoil
energy, but the lower energy only emphasizes further the dominance
of the fast neutrons for producing displacement damage. Energy
for displacements contributed by the β emissions and associated
^{31}P recoils is even less than that for the ^{31}Si recoil and is
neglected in the consideration of displacement damage. The details
of the damage produced by recoiling ^{31}Si and ^{31}P atoms may, however,
be important in determining the lattice location of ^{31}P introduced
by NTD. No experimental evidence is currently available on the
lattice location of ^{31}P prior to annealing, but it will most likely
be interstitial or substitutional with a trapped vacancy. If the
Fermi level can be suitably positioned, then EPR measurements
should be capable of determining if vacancy-substitutional-
phosphorus (E-center) defects[5] are present immediately after NTD
at reactor ambient temperature.

Column 5 of Table 1 shows that the energy available for displacements from an incident fast neutron is 10^3 times that from an
incident thermal neutron. Fast neutrons will, therefore, dominate
the displacement damage until thermal-to-fast ratios exceed 1000:1.
Thermal neutron capture cross-sections for germanium and gallium
arsenide[3] are much larger than those for silicon. Consequently,
displacement damage by thermal neutrons relative to fast neutrons
is expected to be more important in these materials than it is in
silicon.

2.2. Clustered Nature of Neutron Damage

An atom recoiling in a host material creates a high-defect
density (cluster of defects) along the recoil track. The upper
part of Fig. 1 illustrates clusters formed by recoil tracks calculated for a 50 keV silicon atom recoiling in silicon,[6] and for
a 10 keV germanium atom recoiling in germanium.[7] The track in
silicon is a side view and the authors[6] emphasize the high damage
density in subclusters expected near the end of range for each
recoiling silicon atom. These subcluster regions are similar to
the capture-recoil damage regions produced by thermal neutron
capture. Therefore, the kind of damage regions produced by thermal
neutrons is included in fast neutron damage. The track in germanium
is an end view where the open circles, which represent interstitials, are shown to be concentrated on the periphery of the track.

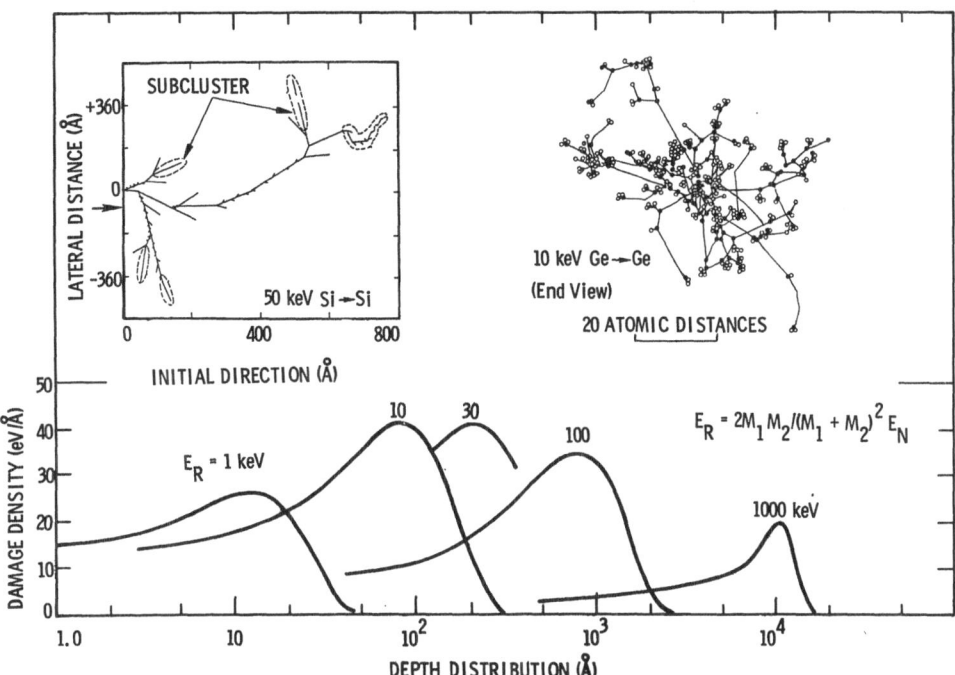

Fig. 1. Shown in the upper left is a track for a 50 keV Si atom recoiling in Si according to calculations by Van Lint, et al.[6] Upper right is a displacement cascade for a 10 keV Ge atom recoiling in Ge according to calculations by Yoshida,[7] where the open and solid circles represent interstitials and vacancies, respectively. The lower part of the figure shows the energy deposited per Å for different Si recoil energies, E_r, according to the formulations of Brice.[8]

The lower part of Fig. 1 shows results of calculating the depth distribution of displacement damage by averaging over a large number of tracks for silicon atoms recoiling with energy E_r in silicon.[8] The peak damage density first increases with E_r but then decreases because the damage clusters become more diffuse. Knock-on recoils in reactor-neutron irradiations have an average $E_r \approx 50$ keV. The areas under the curves give the energy spent in collision processes, and is ≈ 25 keV for $E_r = 50$ keV. The other half of E_r is spent in ionization.[8] Assuming 25 eV/displacement, there is sufficient energy to produce 1000 displacements.

We now consider the possibility of amorphous material formation in the recoil track. From ion implantation, it is known that
6×10^{23} eV/cm^3 into displacement processes at 80°K will produce an
amorphous silicon layer.[9-10] The damage density (eV/cm^3) is plotted
versus silicon recoil energy in Fig. 2. These results were obtained using the total energy deposited divided by the projected range
and lateral spread in area of the damage track. For a 50 keV

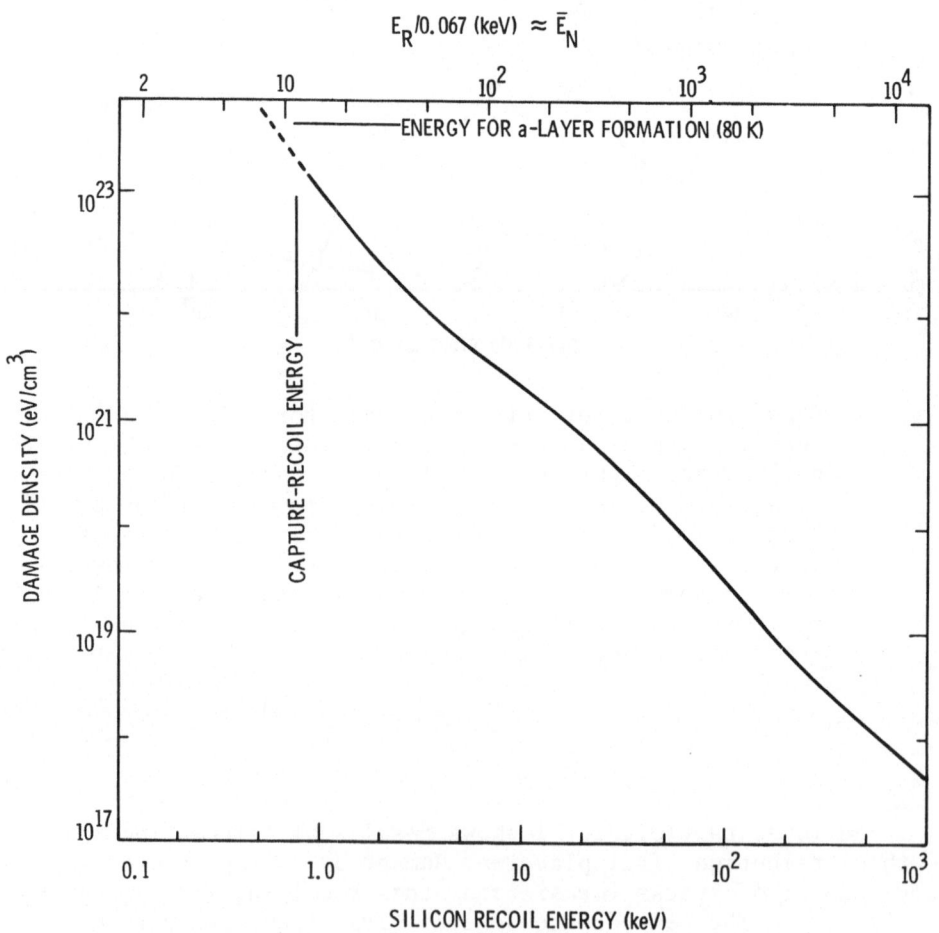

Fig. 2. Energy per unit volume deposited into collision processes
versus Si recoil energy, E_r. Energy required per unit
volume to form an amorphous layer is indicated. The
upper abscissa label is the relationship between E_r and
the neutron energy, E_N, assuming isotropic scattering.

recoil, the damage density is three orders of magnitude below that required for amorphous material formation. However, for the sub-clusters such as those sketched in Fig. 1 (Si recoil atoms with ≤ 10 keV), and for the capture recoil silicon atoms (780 eV), the damage density is approaching that for amorphous material formation. One must then consider the question of how much material is required to qualify as an amorphous zone. Defects with up to five vacancies[1] have been identified by EPR studies and described as defects in crystalline silicon. If 50 atomic displacements are assumed to be necessary for amorphous zone formation, then at 25 eV per displacement a recoil energy of 2 keV would be needed (assuming no loss to ionization), and, even assuming no annealing the damage density for a 2 keV recoil is still a factor of 10 below the threshold for amorphous material formation.

A further consideration is the possibility that overlap of damage tracks will cause amorphous zone formation. In Table 2 we consider the number of recoils required to fill one-half the material volume with damage tracks and therefore have a significant track overlap. For a 50 keV recoil track the estimate gives a fast neutron fluence of $5.1 \times 10^{16} \mathrm{n/cm^2}$ to create damage tracks in one-half of the silicon crystal, and assuming a 100:1 thermal to fast ratio, a phosphorous dopant concentration of $9.2 \times 10^{14} \mathrm{cm^{-3}}$ (5 ohm-cm) would be obtained. Therefore, for most applications of NTD silicon which are for resistivities > 5 ohm-cm there will be little overlap of damage tracks and certainly not enough to produce an amorphous zone on an entire track. Additional calculations were made assuming that the 50 keV recoil energy is all partitioned into 1 or 10 keV subclusters. The results listed in Table 2 indicate that the overlap of such subclusters is highly unlikely in usual NTD silicon processing, and therefore amorphous zone formation is not expected.

3. EXPERIMENTAL EVIDENCE FOR DAMAGE CLUSTERS

One of the most graphic early examples of defect clusters produced by neutron bombardment of semiconductors was obtained by Bertolotti, et al.[11] using etched surface replication transmission electron microscopy (TEM). Figure 3 is a sketch taken from such results obtained on 14 MeV neutron irradiated silicon.[12] Most people would agree that the central region is probably due to the core of the displacement-damage clusters. There is less agreement on the interpretation of the outer zone. Direct TEM measurements[13-14] indicate a strained region around the damage core so that strain-induced differential etching may have caused the outer zone observed by Bertolotti, et al.[11-12] A recent model by Nelson[15] postulates trapping of mobile defects by damage clusters so that the differential etching may have been caused by an excess of

Table 2. Estimates of fast neutron fluences and NTD dopant concentrations when displacement cascades occupy one-half of crystal volume.

Recoil Energy (keV)	Single Recoil		½ Crystal Occupation by Cascades		
	Recoil Range (Å)	Cascade Volume (cm^3)	Cascade Concentration (cm^{-3})	Fast Neutron Fluence (cm^{-2})	P-Dopant** Conc. (cm^{-3})
1	22	4.4×10^{-21}	2.3×10^{18}	1.5×10^{19}*	2.7×10^{17} (0.065 ohm–cm)
10	140	3.0×10^{-18}	1.6×10^{17}	1.1×10^{18}*	2.0×10^{16} (0.35 ohm–cm)
50	650	6.5×10^{-17}	7.7×10^{15}	5.1×10^{16}	9.2×10^{14} (5 ohm–cm)

* 50 keV recoil assumed but partitioned into 1 or 10 keV subclusters.

** Thermal/fast ratio of 100:1 assumed.

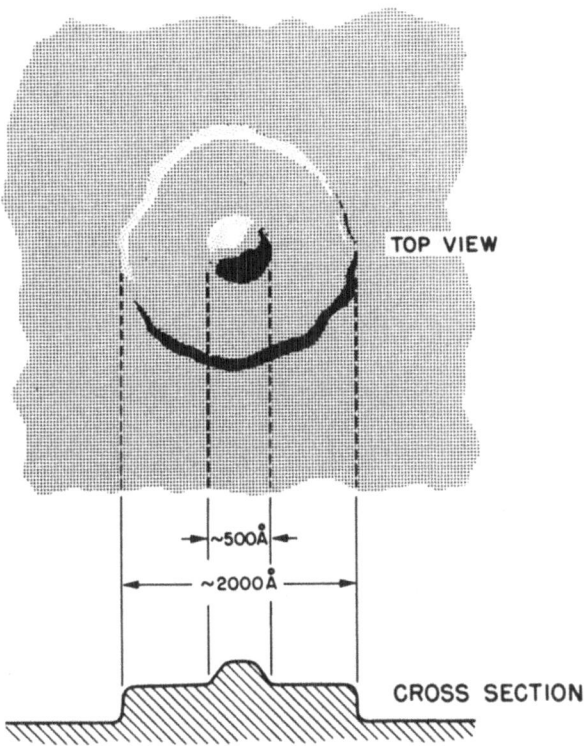

Fig. 3.   Sketch of a region observed on an etched surface of Si
          following 14 MeV neutron irradiation (after Bertolotti,
          et al.[12]). Central core (500 Å) is attributed to recoil
          damage.

trapped vacancies or interstitials. However, the interpretation
that has been most extensively used to explain experimental data
is due to Gossick,[16] and Crawford and Cleland.[17] In this model,
the outer zone represents a space charge region surrounding a
p-type damage cluster in n-type germanium. The damage cluster in
silicon is nearly intrinsic so that a space charge region would
be found in both n- and p-type silicon. The space charge model
has been used to interpret minority carrier lifetime data,[18]
changes in carrier concentration and carrier mobility,[19] the
light sensitivity of neutron-produced electrical changes,[19] photoconductivity,[20] and EPR observations on specific defects in neutron
irradiated silicon.[21]

The space charge model seems to be most applicable to the
initial defect cluster which undergoes significant annealing even
at room temperature. But residual clusters have been observed in
etched surface replication TEM,[22] and in direct TEM measurements[23]
after high temperature annealing. For example, Pankratz, et al.[23],
observed clusters with mean diameters of 40 Å and 22 Å in phosphorus doped and undoped neutron irradiated silicon, respectively,
after annealing at 700°C. These higher temperature clusters
apparently involve both impurities and atomic displacements.

4. ANNEALING AND DEFECT EVOLUTION

4.1. Temperatures < 500°C

Defects have characteristics such as electronic energy levels,
stress response, and temperatures or activation energies for
annealing which make it possible to correlate results from macroscopic electrical measurements with particular defects identified
in EPR or optical studies.[2] The more characteristics correlated,
the more positive the defect identification in macroscopic measurements. Although not the most positive for identification, the
defect characteristic most frequently used for correlation is the
thermal annealing temperature, and it is used in the present
discussion because it illustrates defect changes with increasing
temperature. Annealing results for carrier removal produced by
neutron irradiation at 80°K of 10 ohm-cm phosphorus doped float
zone silicon is shown in Fig. 4.[24] The difference in carrier removal measured with and without illumination has been attributed
to a reduction of the space charge zones by illumination.[24]
Similar measurements on electron irradiated silicon did not show
the illumination effect. The large recovery stage near 400°K
correlates with the vacancy-phosphorus annealing, the stage at
540°K correlates with divacancy annealing, and the stage near 620°K
correlates with oxygen-vacancy annealing.

Although crucible grown silicon is generally not used in transmutation doping due to high carbon and oxygen concentrations,
results from crucible grown silicon are presented because they aid
in illucidating the nature of neutron produced defects and defect-
impurity interactions. Results from measurements like those presented in Fig. 4 for float zone silicon are shown in Fig. 5 for
phosphorus doped crucible-grown silicon.[24] The E-center stage is
smaller in crucible-grown than in vacuum-float zone silicon because
the vacancy is more likely to be captured by an oxygen impurity
than by a phosphorus impurity. The reverse annealing peak near
620°K in crucible-grown material and the subsequent recovery are
presumably associated with oxygen and carbon in the material. A
residual amount of damage in both float zone and crucible-grown
silicon is still observed after annealing at 700°K.

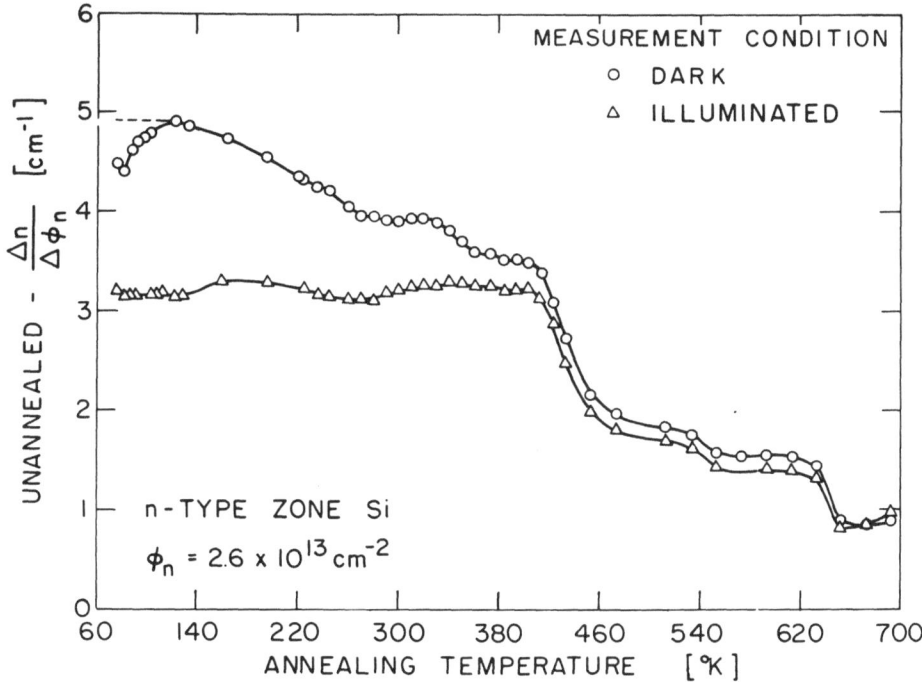

Fig. 4. Isochronal annealing of carrier removal produced by neutron
        irradiation of vacuum float zone phosphorus doped Si.
        Annealing stages for vacancy-phosphorus (400°K), divacancy
        (540°K) and oxygen vacancy (620°K) centers are identi-
        fiable. Illumination is believed to suppress space charge
        zones around defect clusters (Stein[24]).

        Figure 6 shows the annealing behavior observed by Whan[25] for
the oxygen-vacancy defect (A center) in crucible-grown silicon
following electron and neutron irradiation near 200°K.. In contrast
to the gradual annealing loss of the A center between 200 and 500°K
following electron irradiation, there is an increase in the A
center following neutron irradiation for the same annealing temperature. The fractional growth of the A center is compared in
Fig. 7 to the fractional loss of the light sensitivity of the
carrier removal shown in Fig. 5. The correlation is quite good,
consequently, the picture which emerges is that vacancies are
liberated from defect clusters (silicon recoil tracks) and the
vacancies are subsequently trapped by oxygen to form A centers.
The cluster becomes diffuse during annealing and the space charge
effects monitored by the illumination sensitivity decrease. An
increase in divacancy concentration upon annealing following low

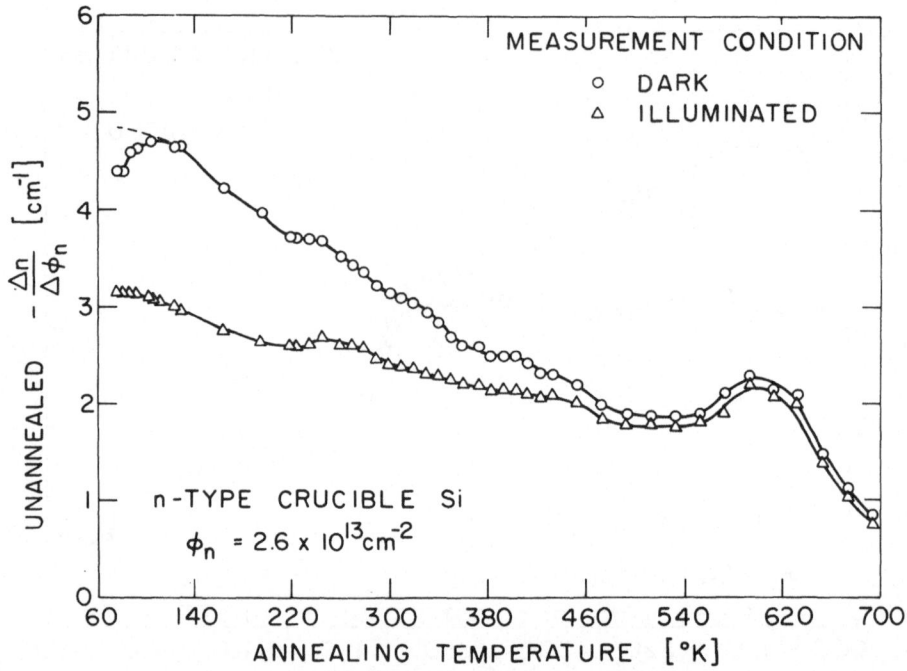

Fig. 5.   Isochronal annealing of carrier removal produced by neutron
          irradiation of phosphorus doped crucible-grown Si. Illu-
          mination effect is similar to that in float zone Si, but
          annealing stages for defects are growth method dependent
          (Stein[24]).

temperature neutron irradiation[26] or ion implantation[27] has also
been observed and attributed to vacancy liberation from clusters
and subsequent vacancy-vacancy interaction.

No direct observation of single vacancies has been made in
neutron irradiated silicon. Brower[28] performed EPR measurements
on silicon following 14 MeV neutron irradiation at 4°K in an
attempt to observe single vacancies but instead of single vacancies
observed distorted four vacancy (silicon-P3) centers (see Fig. 8).
It may be that neutron irradiation produces too much distortion in
the cluster region for detection of the single vacancy center.
This possibly could be explored experimentally by first producing
vacancies with electron irradiation and then performing neutron
irradiation with a fluence sufficiently large to produce recoil
tracks in a significant fraction of the crystal. The positive

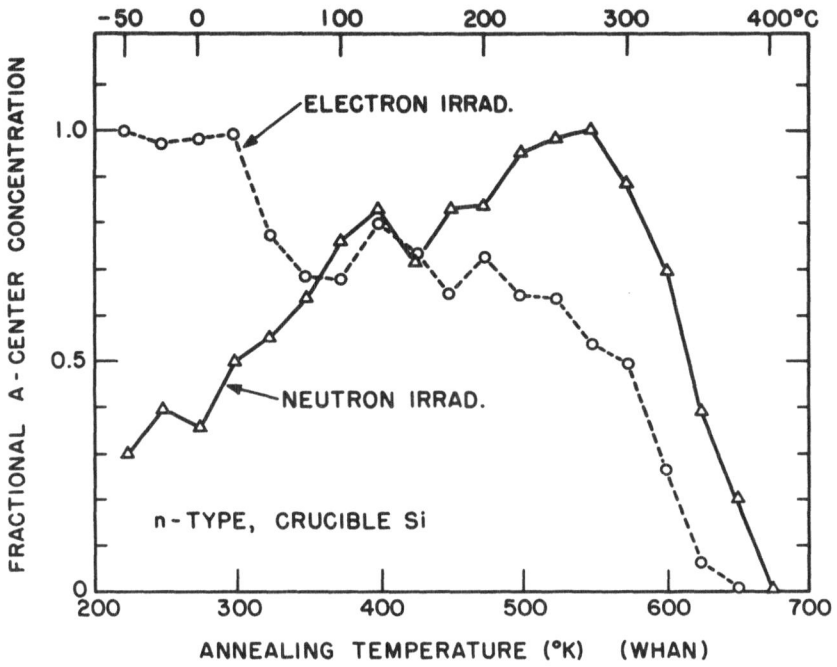

Fig. 6. Isochronal annealing behavior of oxygen-vacancy (A-
center) defects in electron and neutron irradiated
crucible-grown Si (Whan[25]).

divacancy (silicon-G6) was found to increase after neutron irradiation upon annealing between 100 and 400°K, however, an increase
in divacancy concentration cannot be distinguished from a Fermi
level effect in the EPR measurement.

The discussion thus far has centered around vacancy type
defects where more is known than about interstitial defects.
While less detailed than the evidence for vacancies, there is
strong experimental evidence for both impurity interstitials and
self-interstitials in irradiation produced defects in silicon at
temperatures < 500°C. Watkins[29] has shown that electron irradiation at 4°K displaces substitutional impurities to interstitial
sites and interstitial impurities have been identified in EPR
studies of annealing between 4 and 500°K.[1,29] A self-interstitial
defect in silicon observed by Brower[28] is shown in Fig. 8 to remain
in silicon to nearly 800°K. The split interstitial model determined
for the self-interstitial defect is sketched in Fig. 8.

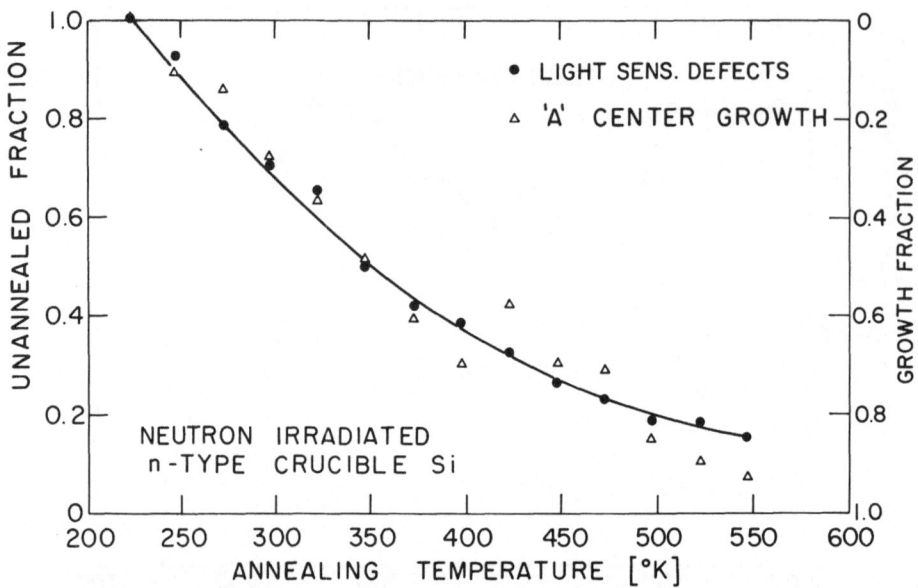

Fig. 7.   Correlation of annealing growth of A-centers following
          neutron irradiation (Fig. 6) and the annealing loss of
          illumination sensitivity from Fig. 5.

4.2. Temperatures > 500°C

Defects remaining at temperatures > 500°C are of major
importance in NTD, but few identifications have been made of the
residual defects at these high temperatures. Results from a number
of measurements on neutron-irradiated Si have been combined in Fig.
9a where the unannealed fraction of defects observed at temperatures
> 400°C are plotted as a function of annealing temperature.   Three
of seven defect levels observed by Guldberg[30] in deep level transient spectroscopy (DLTS) on NTD float zone silicon are represented
by solid segments connecting data points. The DLTS results show
very active defect annealing between 400 and 750°C. The light-
weight solid, dash-dot, dashed and dotted curves in Fig. 9a
represent results from EPR,[28,31] optical,[32] anomalous x-ray transmission,[33] and direct transmission electron microscopy (TEM),[23]
respectively. The split-self-interstitial[28] and the carbon-oxygen-
divacancy centers[31] are the highest temperature defects identified
in EPR studies. The 1124 cm^{-1} optical absorption band is one of a
group called "higher order bands" by Corelli, et al.[32] They
suggested clusters of interstitials and/or vacancies as the possible

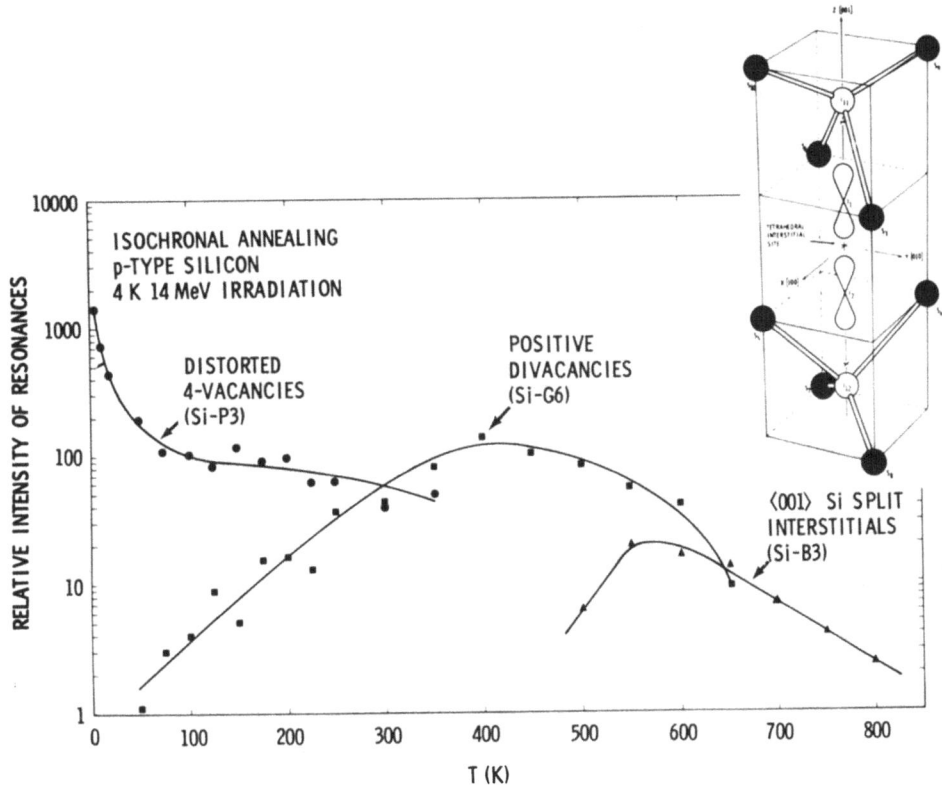

Fig. 8. Defects observed in EPR measurements on boron-doped Si
         after 14 MeV neutron irradiation at 4°K and subsequent
         isochronal annealing. Model for split interstitial
         defect is shown in the upper right of the figure
         (Brower[28]).

defects responsible for these higher order bands. The DLTS measurements on NTD silicon show defect annealing parallel to the annealing of centers observed in EPR and optical measurements. DLTS[30]
shows further defect annealing between 600 and 750°C, which is the
temperature required to optimize electrical activation for low
fluence NTD.[34] Annealing near 700°C was also found in TEM[23] and
x-ray transmission measurements[33] on neutron irradiated silicon.
Analyses of the images observed in TEM showed nearly equal concentrations of vacancy and interstitial clusters near 600°C, but
primarily extrinsic interstitial clusters at 700°C.[23] The x-ray
transmission measurement is especially sensitive to strain in the
crystals, and is presumably monitoring strain due to defect clusters.

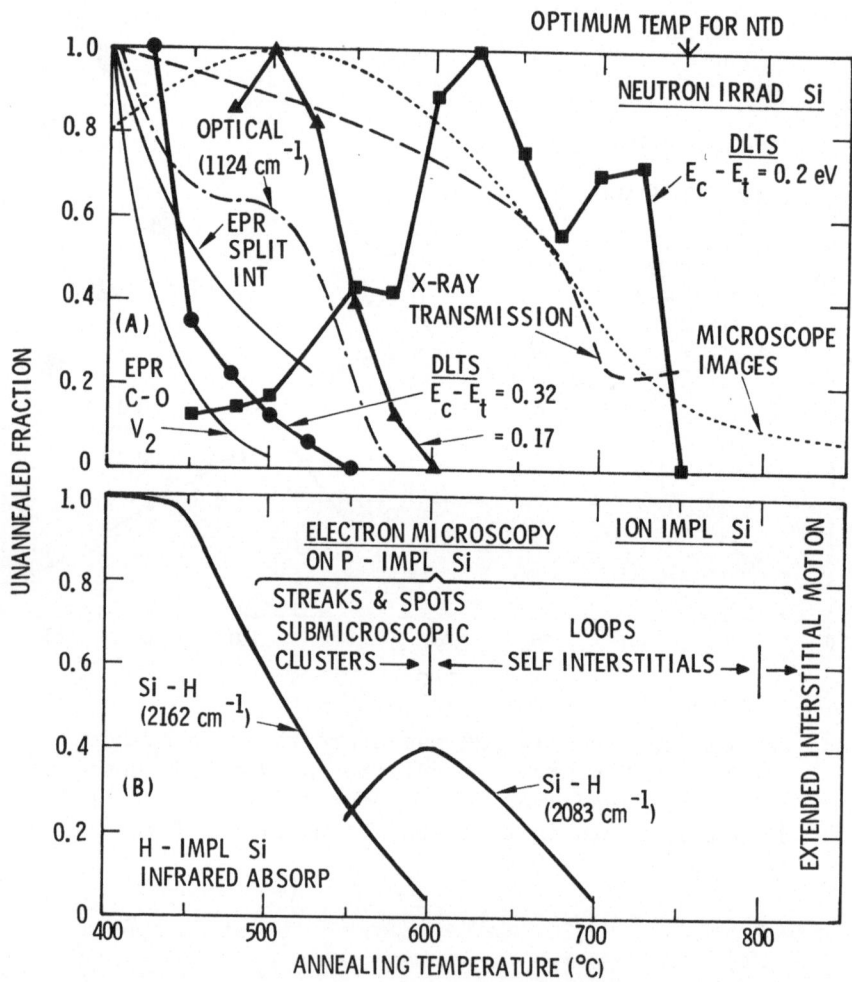

Fig. 9. Compilation of defect annealing data from a number of
different irradiation and ion implantation studies.
9a. Irradiations studies - EPR--carbon-oxygen-divacancy,[31]
EPR--split interstitial,[28] optical absorption--1124 cm^{-1}
band,[32] anomalous x-ray transmission--,[33] TEM--,[23] and
DLTS--.[30] 9b. Ion implantation studies - optical
absorption--Si-H bands for implanted hydrogen,[37] and TEM--
phosphorus implanted Si.[38]

Both the TEM and x-ray transmission measurements were made on LOPEX silicon which characteristically contains a significant amount of carbon.[35] Moyer and Buschert[35] have shown that the carbon content affects the lattice parameter change under electron irradiation with 750°C required for complete annealing. Infrared measurements have shown that once carbon is displaced from substitutional sites by irradiation it does not return to a normal substitutional site until annealing to $\sim$ 600°C.[36] Therefore, carbon clearly remains active in the high temperature defects. While the carbon concentrations in float zone silicon may be an order of magnitude lower than that in LOPEX silicon, it is sufficiently large to account for trap concentrations of $\sim 2 \times 10^{12} cm^{-3}$ observed in DLTS above 600°C, and Guldberg[30] found that carbon has an effect upon the annealing of the high temperature DLTS level.

Figure 9b shows the annealing behavior for two infrared absorption bands produced by hydrogen implantation of silicon.[37] These bands are vibrational modes for hydrogen bound to silicon with frequency perturbations caused by displacement defects. The annealing behavior for these Si-H centers shows a similarity to defect annealing in neutron irradiated silicon and suggests that ion implanted hydrogen can be used to monitor displacement-defect annealing. Guldberg[30] found that a 2 percent hydrogen environment during silicon growth affected the DLTS spectra in NTD silicon.

Annealing data obtained from TEM measurements on phosphorus implanted silicon by Seshan and Washburn[38] are also outlined in Fig. 9b. Beginning about 400°C, streaks and spots were observed which were attributed to submicroscopic clusters, and beginning about 600°C the streaks and spots transformed to resolvable clusters and loops. The loops observed at temperatures > 600°C were attributed to an excess of silicon interstitials assumed to have been displaced when implanted phosphorus became substitutional. The loops coarsen between 600° and 800°C and then anneal out. Motion of extended self-interstitials proposed by Seegar and Frank[39] was suggested to explain the annealing loss of the loops. In this case ion implantation is distinctly different from NTD because no excess atoms are introduced by NTD. Hydrogen implantation could also be used to achieve the condition of no excess atoms in displacement-damaged silicon because hydrogen escapes from silicon below 800°C.

Evidence was presented at this conference by Cleland, et al. and by Malmros that fast neutron damage aids lifetime recovery during annealing in contrast to electrical activation where higher damage levels require higher anneal temperature.[40-41] Minority carrier lifetimes are only relative to pre-irradiation values or to values in another crystal whereas target resistivities are determined by transmutations. Consequently, displacement damage gettering of crystal impurities by displacement damage may be playing a role in determining minority carrier lifetimes in NTD silicon.

5. SUMMARY

Calculations of energy deposition into displacement processes show that fast neutrons will dominate defect production in NTD processing of silicon unless thermal-to-fast neutron ratios exceed 1000:1. Defect clusters are produced by silicon-atom recoils from fast neutron collisions. Using an experimental value for the energy needed per unit volume to form amorphous material, it is argued that amorphous zone formation in silicon by NTD is highly unlikely.

The initial defect clusters produced by the silicon recoils from fast neutron encounters undergo extensive annealing at and below room temperature, yet defect clusters, vacancy loops and interstitial loops are also observed after annealing at temperatures > 500°C. Further work is needed to establish the relationships of the high temperature clusters and lattice imperfections to the initial clusters and to point defects produced by silicon recoils, and also the relationships to impurities in the host crystal.

ACKNOWLEDGEMENTS

The author wishes to acknowledge helpful discussions with K. L. Brower and J. M. Meese.

REFERENCES

*   This article sponsored by the U.S. Department of Energy under Contract AT(29-1)-789.
**   A U.S. Department of Energy Facility.
1)   J. W. Corbett and J. C. Bourgoin, Point Defects in Solids, Semiconductors and Molecular Crystals, Vol. 2, ed. by J. H. Crawford, Jr. and L. M. Slifkin (Plenum Press, New York, 1975), Chapter I.
2)   H. J. Stein, in Albany Conference on Radiation Effects in Semiconductors, ed. by J. W. Corbett and G. D. Watkins (Plenum Press, New York, 1971) p. 125.
3)   J. W. Cleland, in Proceedings-International School of Physics, Enrico Fermi - Course XVIII - Radiation Damage in Solids, ed. by D. S. Billington (Academic Press, New York, 1962) p. 384; see also J. H. Crawford, Jr. and J. W. Cleland, Int. J. Appl. Radiat. Isotopes 9, 189 (1960).

4) M. A. Kirk and L. R. Greenwood, this conference.

5) G. D. Watkins, (lot. cit. 1), Chapter 4.

6) V. A. J. van Lint, R. E. Leadon, and J. F. Colwell, IEEE Trans. Nucl. Sci. NS-19, 181 (1972).

7) M. Yoshida, J. Phys. Soc. of Japan 16, 44 (1961).

8) D. K. Brice, Radiat. Effects 11, 227 (1971).

9) F. L. Vook, in Reading Conference on Radiation Damage and Defects in Semiconductors, ed. by J. E. Whitehouse (Institute of Physics, London Conf. Series No. 16, 1973), p. 60.

10) J. R. Dennis and E. B. Hale, J. Appl. Phys. 49, 1119 (1978).

11) M. Bertolotti, T. Papa, D. Sette, V. Grasso, and G. Vitali, IL Nuovo Cimento 29, 4310 (1963).

12) M. Bertolotti, D. Sette, and G. Vitali, J. Appl. Phys. 38, 2645 (1967).

13) J. R. Parsons and C. W. Hoelke, IEEE Trans. on Nucl. Sci. NS-16, 37 (1969); see also M. L. Swanson, J. R. Parsons, and C. W. Hoelke, (loc. cit. 2), 359.

14) G. den Ouden, Phil. Mag. 19, 321 (1969).

15) R. S. Nelson, Radiat. Effects 32, 19 (1977).

16) B. R. Gossick, J. Appl. Phys. 30, 1214 (1959).

17) J. H. Crawford, Jr. and J. W. Cleland, J. Appl. Phys. 30, 1204 (1959).

18) B. L. Gregory, IEEE Trans. on Nucl. Sci. NS-16, 53 (1969).

19) H. J. Stein, IEEE Trans. on Nucl. Sci. NS-15, 69 (1968).

20) H. J. Stein, Appl. Phys. Letts. 15, 61 (1969).

21) D. F. Daly and H. E. Noffke, (loc. cit. 2), p. 359.

22) M. Bertolotti, in Santa Fe Conference on Radiation Effects in Semiconductors, ed. by F. L. Vook (Plenum Press, New York 1968), p. 311.

23) J. M. Pankratz, J. A. Sprague, and M. L. Rudee, J. Appl. Phys. 39, 101 (1968).

24) H. J. Stein, Phys. Rev. 163, 801 (1967).

25) R. E. Whan, J. Appl. Phys. 37, 3378 (1966).

26) C. E. Barnes, (loc. cit. 2), p. 203.

27) F. L. Vook and H. J. Stein, Radiat. Effects 6, 11 (1970).

28) K. L. Brower, in 5th Intl. Conf. on Ion Implantation in Semi-conductors, ed. by F. Chernow, J. A. Borders, and D. K. Brice (Plenum Press, New York, 1977), p. 427; Phys. Rev. B14, 872 (1976).

29) G. D. Watkins, in Radiation Damage in Semiconductors, ed. by P. Baruch (Dunod, Paris, 1965), p. 97.

30) J. Guldberg, Appl. Phys. Letts. 31, 578 (1977); Dissertation at Technical University of Denmark, "Micro-Defects in Float Zone Silicon," (1978).

31) Y. H. Lee, J. W. Corbett, and K. L. Brower, Phys. Stat. Sol. (a), 41, 637 (1977).

32) J. C. Corelli, D. Mills, R. Gruver, D. Cuddeback, Y. H. Lee,
     and J. W. Corbett, in Dubrovnik Conference on Radiation
     Effects in Semiconductors, ed. by N. B. Urli and J. W. Corbett
     (The Institute of Physics Conference Series No. 31, 1977), p.
     251.
33) T. O. Baldwin and J. E. Thomas, J. Appl. Phys. 39, 4391 (1968).
34) A. Senes, G. Sifre, and M. Breant, in Semiconductor Silicon
     1977 PV 77-2, ed. by H. R. Huff (The Electro. Chem. Soc.,
     Philadelphia, 1977), p. 106.
35) N. E. Moyer and R. C. Buschert, (loc. cit. 22), p. 444.
36) R. C. Newman and D. H. J. Totterdel, J. Phys. C8, 3944 (1975).
37) H. J. Stein, J. Electronic Mater. 4, 159 (1975).
38) K. Seshan and J. Washburn, Radiat. Effects (to be published).
39) A. Seeger and W. Frank, (loc. cit. 9), p. 262.
40) H. Herzer, (loc. cit. 34), p. 106.
41) H. J. Stein, Radiat. Effects 6, 175 (1970).

THE MINORITY CARRIER LIFETIME OF NEUTRON DOPED SILICON

Olof Malmros

TOPSIL A/S

Frederikssund, Denmark

ABSTRACT

A study was performed to establish the minority carrier lifetime of neutron doped silicon as a function of various parameters, e.g., the resistivity, the reactor, and the starting material. It was found that lifetime was strongly dependent on resistivity and purity and weakly on starting material and reactor.

1. INTRODUCTION

The minority carrier lifetime (τ) is an important parameter of silicon as it is a measure of the purity and the number of defects in the crystal.

In this paper a study of the lifetime of neutron doped silicon is presented. We will look at the values measured on the routine production at TOPSIL and correlate them with a number of parameters which might be suspected to influence them.

2. EXPERIMENTAL PROCEDURE

The material in this investigation is dislocation free float zone crystals with starting resistivities from 5 to 200 times the final resistivity (ρ). Both n- and p-type starting material were used with more than 90% being of the former type.

The crystals come from several sources. There are 3 process steps in the fabrication of the starting material, which interest

249

us. They are the growth of the polycrystalline rod and the two
zone passes. The investigation covers material from 5 different
companies, in the following denoted A to E. The crystals in this
investigation are therefore characterized by a 3-letter combination;
B-A-A , for instance, means that the poly rod was made by company
B, while the two zone passes were made by company A.

Prior to neutron doping, the crystals, measured for resistivity,
are cleaned according to the following procedure:
  a.   detergent cleaning in an ultrasonic bath,
  b.   rinsing in deionized water,
  c.   etching in an HNO_3/HF mixture,
  d.   rinsing in deionized water,
  e.   drying in hot air.
After cleaning, the crystals are packed in plastic bags which are
hermetically sealed and sent to one of several reactors. We have
used 3 different types of reactors as shown in Table 1. The
crystals irradiated at the Type III reactor, the swimming pool type,
were cleaned at the reactor before being sent back in order to remove any contamination from dissolved impurities in the water.
After returning from the reactor, all crystals were cleaned again
according to the procedure described above. After annealing, the
crystals were measured for resistivity and lifetime.

Table 1.   Characterization of the 3 different reactor types.

REACTOR	TYPE	PLACE OF IRRADIATION	CD-RATIO AT IRRADIATION POSITION	APPROXIMATE TEMPERATURE OF CRYSTAL SURROUNDINGS
I	GRAPHITE/ HEAVY WATER	DRY HOLE IN GRAPHITE AROUND CORE	1000-2000	185° C
II	GRAPHITE/ HEAVY WATER	DRY HOLE IN GRAPHITE AROUND CORE	20-100	40° C
III	SWIMMING POOL	LIGHT WATER CLOSE TO REACTOR CORE	5-10	30° C

The minority carrier lifetime is measured by the photo conductivity decay method using an AC-method. The light source consists
of a xenon flash lamp which illuminates approximately 3 cm^2 of the
crystal's cylinder surface. The lifetime value is displayed automatically on a counter through an electronic determination of the
time constant of the exponential decay of the photo conductivity.
It is possible to measure various parts on the decaying curve in
order to ensure it follows a true exponential.

The xenon flash can be supplied with an infrared filter consisting of a double-sided polished silicon wafer. By using this
filter it is possible to vary the effective penetration depth of
the light and thus ascertain that surface recombination has negligible influence.

The light intensity can be changed by altering the distance
between the flash lamp and the crystal.

Experience has shown that the decay curve normally is a true
exponential independent of surface recombination and light intensity.
Furthermore, the measurements have been correlated with DC-methods
several times and always with good correlation.

The issues we have investigated are the following:
a.   lifetime variation of individual crystals,
b.   lifetime distribution of a large number of crystals,
c.   lifetime versus crystal diameter,
d.   lifetime versus resistivity after irradiation,
e.   lifetime versus reactor type,
f.   lifetime versus crystal origin.
The answers to these questions have been found by looking at more
than 7000 individual measurements of τ. These have been taken on
statistically randomly selected crystals from our routine production
in the second half of 1977 and the first four months of 1978.

3.   EXPERIMENTAL RESULTS

The lifetime distribution within a crystal was established by
measuring τ at 8 points evenly distributed in a belt around the
middle of the cylindrical crystal. A measure of the variation was
calculated:

$$T = \tau_{max}/\tau_{min}.$$

The distribution of T on 112 crystals is given in Fig. 1. It is
clearly seen that the lifetime varies very much within a crystal.
The most common value of T is between 2 and 3, but the maximum is
as high as 30.

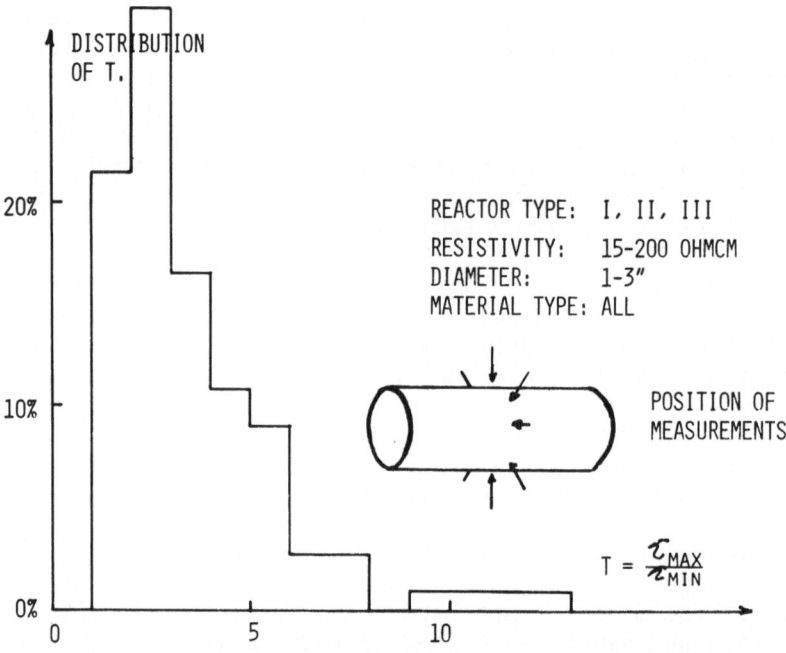

Fig. 1.   Lifetime distribution within a crystal.   The largest value
         found (not shown on the curve) is 30.

An example of the lifetime distribution as found from crystal
to crystal is shown in Fig. 2. As should be expected, we found
even higher divergences between the crystals than within them.

Figure 3 also shows an example of the way the lifetime distribution has evolved since the beginning of 1976. The very important
improvement is due to a number of minor process modifications, one
example being the installation of a new ultrasonic tank.

The lifetime was checked as a function of the crystal diameter
on the type A-A-A material. No correlation was found.

Lifetime versus final resistivity is shown in Fig. 4. On this
double logarithmic plot we find 2 intersecting straight lines, and
from these the position in the band gap of 2 different trapping
centers can be calculated:[1]

Center 1: 0.24 ± 0.02 eV below conduction band

Center 2: 0.37 ± 0.02 eV below conduction band.

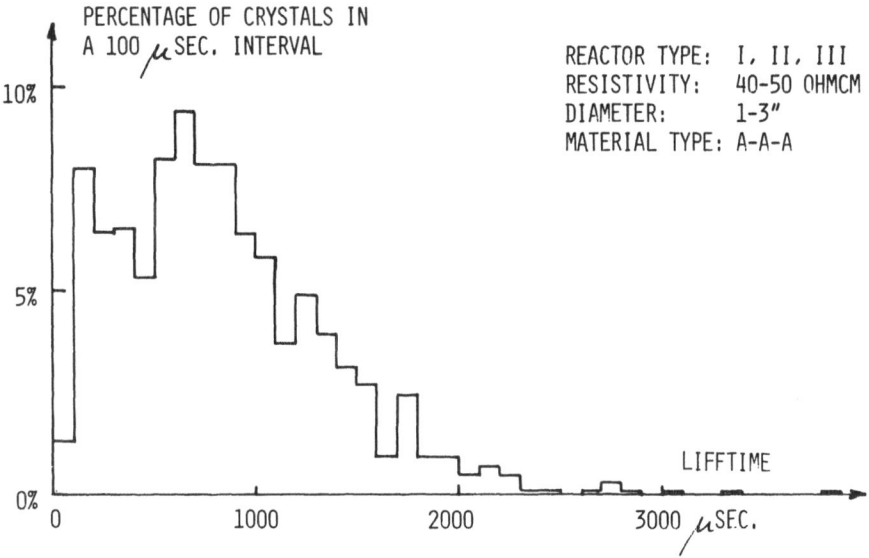

Fig. 2. Distribution of lifetimes.

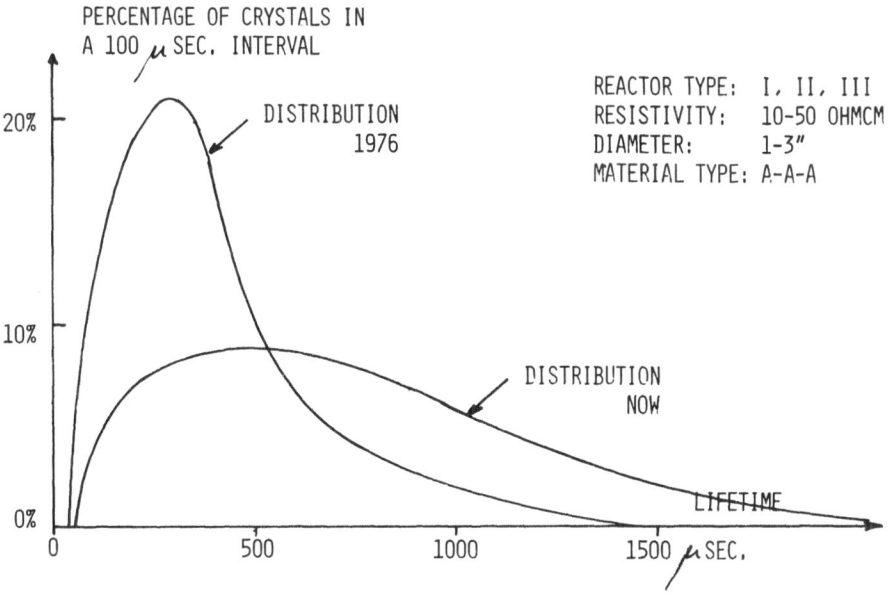

Fig. 3. Evolution of lifetimes from the beginning of 1976 and until now.

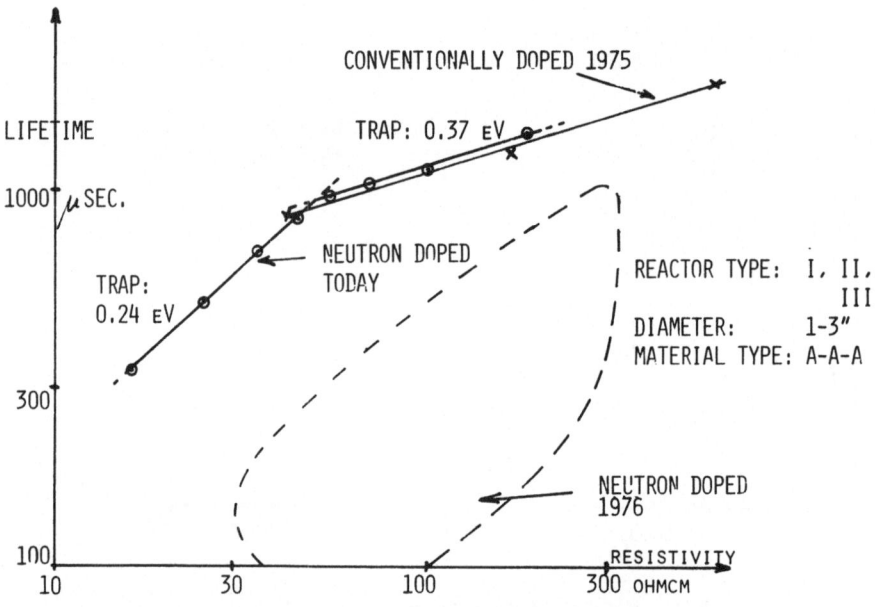

Fig. 4. Average lifetime versus resistivity.

Figure 4 also shows the lifetime results we obtained in 1976.[2]
Again, a very clear improvement (a factor 3 to 4) is seen. Shown
too, on Fig. 4, is the lifetime of conventionally doped n-type
crystals of type A-A-A grown in 1974-75. The similarity between
this curve and part of that for the neutron doped crystals is
obvious.

The lifetime versus reactor type is shown in Fig. 5. A distinct
but not very large difference between the 3 reactor types is seen.
The same form of lifetime versus resistivity dependence that led
to the two straight lines in Fig. 4 is found on all 3 reactor types.
From the data we calculate the proportion between the lifetimes:

    Reactor type   I: 0.73
    Reactor type   II: 0.87
    Reactor type III: 1.00

The lifetime versus material type is shown in Table 2. The
data have been found by taking all the τ-values and, with the aid
of Figs. 4 and 5, compensating for the influence of resistivity

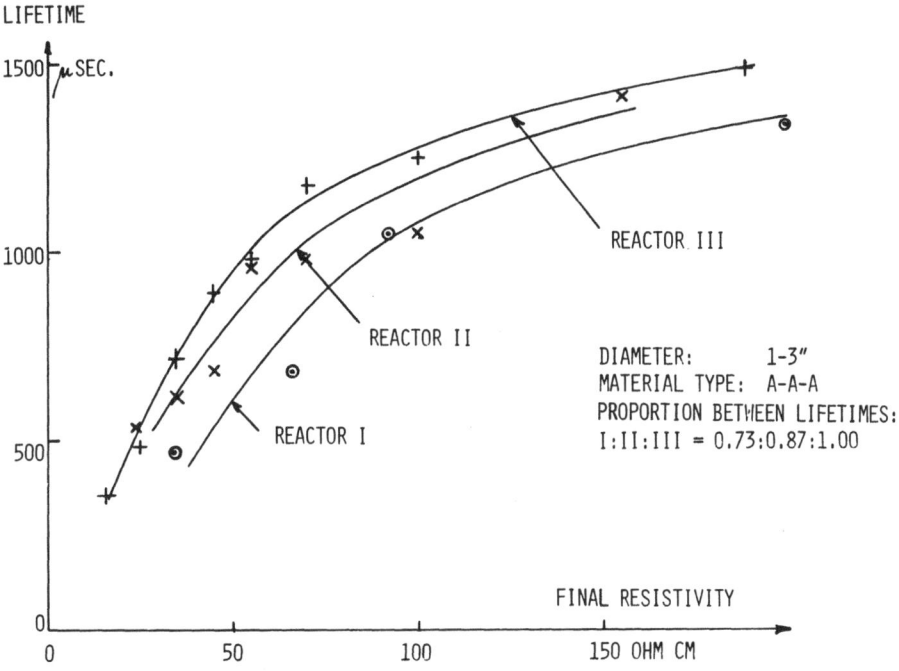

Fig. 5.   Average lifetime versus resistivity for the 3 reactor types.

Table 2.   Average lifetime for crystals 45 ohm-cm irradiated in
reactor Type III, as a function of origin of the undoped
material.

MANUFACTURER			AVERAGE LIFETIME
POLY	1. ZONEPASS	2. ZONEPASS	
A	A	A	885 μ SEC.
B	A	A	789 -
B	B	A	718 -
B	B	B	659 -
B	C	C	631 -
D	C	C	703 -
E	E	E	765 -

and reactor type. The lifetime obtained this way is the one which would be expected if an average 45 ohm-cm crystal had been irradiated in a Type III reactor.

A distinct, but not very large difference is seen between the individual crystal types. The most interesting of the data, showing the mixing of processes from producer A and B is plotted in Fig. 6. It is clearly seen how the lifetime increases as more and more of the processes are performed by company A.

4. DISCUSSION

It is most interesting in Fig. 4 to compare the lifetimes of conventionally doped and neutron irradiation doped crystals. It is seen that above 45 ohm-cm the neutron doped silicon actually has a slightly higher lifetime than the melt doped n-type silicon from 1974-75. It is especially interesting to notice that both materials show the same trap level at 0.37 eV below conduction band. It, therefore, seems safe to assume that this level has nothing to do with the irradiation process.

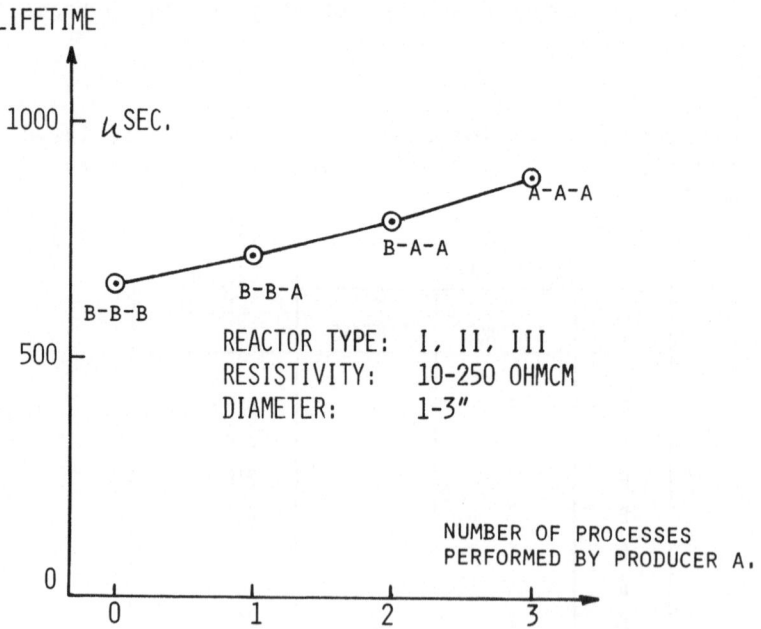

Fig. 6. Lifetime as a function of mixing processes from A and B.

It is not possible to compare lifetimes below 45 ohm-cm, as there are no data for the conventionally doped material. Therefore, we cannot say whether the 0.24 eV trap is characteristic of neutron doped silicon or if it also can be found in other n-type material.

Traditionally, "good" n-type float zone dislocation free silicon must have a lifetime in microseconds which is 5 times the resistivity in ohm-cm. From Fig. 4 it may be seen that the average lifetime for neutron doped as well as for conventionally doped silicon of the A-A-A type easily fulfills this criteria.

Even though we find a correlation between lifetime and resistivity, this does not necessarily mean that the concentration of the traps depends of resistivity. The traps may therefore be due to, for instance, a metal contamination, lattice defects, or a combination of both. Actually, the trap level calculation assumes no correlation between resistivity and concentration, and that is an assumption which has not been verified experimentally. Therefore, perhaps one should not worry too much about not having good candidates for the levels.

There are many facts pointing towards contamination as another very important parameter controlling the lifetime. First of all, the increase in average lifetime of neutron doped silicon from 1976 and to the present (Figs. 3 and 4) can only--with difficulty--be understood as anything but improved cleanliness. Next, the divergence in lifetimes between crystals of the same resistivity and origin (Fig. 2) also points strongly towards some sort of contamination. It is interesting to note that the A-A-A type material shows the same amount of lifetime scatter for conventionally doped silicon as for neutron doped. Finally, the lifetime variation within individual crystals (Fig. 1) also suggests a contamination.

The difference between the crystal types might be explained by the purity of the original material; otherwise the mixing of processes from company A and B (Fig. 6) cannot be easily understood. There exist, however, other possibilities, e.g., differences in crystallography quality, which for the time being, cannot be rejected.

With respect to the lifetime variation between the reactor types, it is puzzling that we observe the best lifetimes in crystals from the reactors with the highest number of fast neutrons. We might have expected the opposite situation: the greater the number of fast neutrons, the greater the amount of lattice defects, the more difficulty to annealing, and the lower the lifetime due to residual damage. This obviously is not the case.

We can construct several hypothetical explanations for the observed behavior:

a.   The cleanliness of the crystals following irradiation
     could depend on the reactor type. In this connection we
     remember that the swimming pool irradiated crystals are
     cleaned an extra time at the reactor.

b.   The detailed atomic configuration of the created lattice
     defects could depend on the irradiation temperature. Thus,
     a low temperature could create defect complexes or defect/
     contamination clusters easier to dissolve leading event-
     ually to higher lifetimes.

c.   The atomic configuration of the defects could depend on
     the composition of the elementary particles in the reactor.
     Thus, instead of getting more damage from the fast neutrons
     we would get more easily dissolved damage and thus higher
     lifetimes.

It is my guess that it is the extra cleaning rather than temperature
or fast neutrons which is the cause for the difference between the
3 reactor types.

5.   CONCLUSIONS

On the basis of minority carrier lifetimes measured on a large
number of neutron doped silicon crystals we have reached the following conclusions:

a.   The lifetime is to a very large extent influenced by
     spurious contaminations, but the average (in μsec) can
     now be held above 5 times the resistivity (in ohm-cm).

b.   The lifetime is correlated with resistivity as shown on
     Fig. 4. The higher the resistivity the higher the life-
     time.

c.   From Fig. 4 we calculate tentatively 2 trap levels at 0.24
     and 0.37 eV below conduction band. The latter level is
     also found in conventionally doped silicon and has there-
     fore hardly anything to do with the irradiation. No
     further information can be given on the 0.24 eV level.

d.   The lifetime depends to a small degree on the crystal
     origin. This is probably due to differences in purity
     of the various materials.

e.   The lifetime depends to a small degree on the atomic
     reactor which is used. Higher lifetimes are obtained in
     swimming pool reactors than in graphite/heavy water
     moderated reactors. The reason for this is not known.

ACKNOWLEDGEMENTS

I am indebted to Jens Guldberg for performing the calculations of the trap levels, to Ed Borbye for the lifetimes of conventionally doped silicon and to Noel de Leon for reading through the manuscript and suggesting valuable corrections.

REFERENCES

1)   Jens Guldberg, J. of Phys. D. To be published.
2)   Hans Janus and Olof Malmros, IEEE Trans., ED-23, 797 (1976).

ELECTRICAL PROPERTY STUDIES OF NEUTRON TRANSMUTATION DOPED

SILICON*

J. W. Cleland, P. H. Fleming, R. D. Westbrook,

R. F. Wood, and R. T. Young

Oak Ridge National Laboratory, Solid State Division

Oak Ridge, TN 37830

ABSTRACT

Irradiation of semiconducting materials in a nuclear reactor[1] was initiated about 30 years ago at the Oak Ridge National Laboratory (ORNL). The primary interest at that time was the study of radiation damage effects, but transmutation doping was used to determine isotopic cross sections, introduce a known dopant concentration, and study impurity banding effects. Research on neutron transmutation doped (NTD) silicon was resumed at ORNL approximately four years ago. In this paper results of several studies of electrical properties conducted since then will be reviewed and discussed.

1. INTRODUCTION

NTD silicon offers four clearly identifiable advantages over silicon doped by more traditional techniques. These advantages are (1) areal and spatial uniformity of dopant distribution; (2) precise control of doping level; (3) elimination of dopant segregation at grain boundaries in poly-crystalline silicon; and (4) superior control of heavy-atom contaminants. Uniformity of doping has proved to be of importance in a number of device applications, particularly high power rectifiers and thyristors.[2] Precise control of doping levels is important in any application such as avalanche or infrared detectors which require high resistivity material. Transmutation doping of epitaxial layers deposited on low resis-

261

tivity n-type[3] or p-type[4] silicon substrates has also been demonstrated, and is of interest because conventional chemical vapor deposition and diffusion techniques do not provide well-controlled and uniform dopant concentrations in large area epitaxial layers, especially at low dopant concentrations.

Capture of thermal neutrons converts ^{30}Si to ^{31}P thus resulting in transmutation doping of the material. The anticipated rate of introduction of ^{31}P can be calculated for NTD silicon from the ^{30}Si isotopic abundance, the ^{30}Si thermal neutron absorption cross section, and the thermal neutron flux, $n_{th}/cm^2 sec$. It has long been recognized that, because of multiple atomic displacements produced by recoiling silicon atoms, a lattice defect concentration many orders of magnitude greater than the anticipated ^{31}P concentration is created. This is the case even for samples irradiated with a low total neutron fluence in a reactor locale with a very large thermal-to-fast (t/f) neutron ratio.[5] Because of this recoil damage, even heavily doped n- and p-type silicon is converted to nearly intrinsic conditions as a consequence of such irradiations.

The annealing conditions required to obtain the anticipated resistivity, carrier concentration, and carrier mobility have been reported[6] for float zone refined (FZ) single crystal silicon of low oxygen content ($\leq 3 \times 10^{16} cm^{-3}$). Most of the simple recoil-induced defects were removed by annealing up to 400°C. Annealing for 30 minutes at 750°C was sufficient to remove virtually all lattice damage in FZ NTD silicon irradiated up to a total fluence of approximately $10^{21} n\ cm^{-2}$ in different reactor locales with t/f ratios of from 1700 (low total fluence) down to 3-10 (high total fluence).

The corresponding annealing requirements for obtaining the anticipated electrical properties in pulled or Czochralski (Cz) grown single crystal NTD silicon that contains a moderate ($5 \times 10^{17} cm^{-3}$) to large ($4 \times 10^{18} cm^{-3}$) oxygen concentration have not been clearly established. Most of the oxygen in Cz silicon after initial crystal growth may be present as electrically inactive interstitial atoms (O_i), but some of it may be present in electrically inactive oxygen clusters, and some as electrically active clusters that act as n-type donors. Commercially available Cz silicon is usually limited to about 50 ohm-cm resistivity (10^{14} donors or 3×10^{14} acceptors cm^{-3}) unless a special pedestal-type growth technique is employed. Most producers anneal after crystal growth to reduce any donor concentration due to clustered oxygen. The post annealing procedure may vary somewhat among producers, but is normally about 1 hr at 700 to 750°C for ingots up to two inches in diameter and this is followed by quick quenching (2-3 min) to below 200°C to minimize any additional oxygen donor formation.

Heat treatments of Cz silicon after crystal growth and post annealing, which may be required during device processing or to remove lattice damage in NTD Cz silicon, can still form oxygen donors at 400 to 500°C, electrically inactive oxygen clusters at 600 to 900°C, and electrically inactive oxygen precipitates (SiO$_2$) near 1000°C. Reheating near the melting point can dissociate most of these clusters and precipitates but some re-form during sub-sequent cooling. Recent experiments[7] have indicated that the apparent interstitial oxygen concentration in Cz silicon, as determined by infrared absorption (IR) techniques, cannot be reduced below about 10 ppm (5 x 10^{17}cm^{-3}) even by extended anneal-ing in a gettering atmosphere. In fact, because of some remaining clusters and precipitates, the actual oxygen content may be greater by a factor of from 2 to 5 than that indicated by IR techniques.[7]

The annealing requirements to remove lattice defects in Cz NTD silicon were first investigated by Mordkovich et al.[8] They used resistivity data to estimate the ^{31}P concentration after irradiation at 50°C with 10^{17} - 10^{19} thermal neutrons cm^{-2}, and after 30 min isochronal annealing in 20°C steps from 100 to 1000°C. Their results for FZ NTD silicon agree almost identically with their calculated values after annealing at 800°C; but the apparent ^{31}P concentration in their Cz NTD silicon after annealing at 1000°C was larger than that predicted by about a factor of 2. This discrepancy increased with both initial oxygen content and total neutron fluence. They did not state if their Cz silicon was post annealed. They concluded that highly stable neutron-induced lattice defects are tightly bound to oxygen atoms, but they did not suggest a model for a defect-oxygen complex that could serve as a donor in Cz NTD silicon.

Liaw and Varker[9] used resistivity, Hall coefficient, and IR measurements to investigate both FZ and Cz (not post annealed) NTD silicon that has been subjected first to a single irradiation cal-culated to introduce about 3.5 x 10^{14} ^{31}P cm^{-3}, and then to 30 min. isochronal or extended isothermal annealing from about 450 to 950°C. They obtained fairly reasonable agreement with their cal-culated value for the ^{31}P concentration in FZ NTD silicon after annealing for 1-2 hrs at 680°C. They stated that excess donors were generated in their Cz NTD silicon to such an extent that annealing for 1 hr at 950°C was required to remove them, and thus obtain reasonable agreement with their calculated value for the ^{31}P concentration. They also do not suggest a model for a defect-oxygen complex that serves as a donor in Cz NTD silicon, and requires annealing at 950°C.

Previous investigators[10-12] have shown that lithium can be introduced into silicon during crystal growth or by subsequent diffusion. Lithium, which is highly mobile in silicon at ambient temperatures, can serve as an electrically active donor unless precipitated and it can interact with other impurities, defects already present and defects introduced by irradiation to form additional donors.[13-15] Because of this and for other reasons, the properties of lithium in NTD silicon have been of particular interest to us at ORNL, and some of our recent results in this area will be discussed.

2. IRRADIATION FACILITIES AND EXPERIMENTAL PROCEDURES

Figure 1 is a drawing of a new thermal neutron facility (D_2O tank) recently installed in the pool of the 2 MW Bulk Shielding Reactor (BSR) at ORNL. This facility will accommodate ingots or wafers up to 6-inches in diameter in several locales, and samples can be raised, lowered, and rotated as required to maximize the radial and axial dopant uniformity. Some of the sample tubes terminate 8 feet above the tank to permit sample exchange under water without any requirement for reduced reactor power or shutdown. Any of the tubes except those against the BSR core can be equipped with an "S" shaped extension for dry irradiations, and any empty tubes, including those against the BSR core, can be plugged with graphite, magnesium, or D_2O-filled cylinders to increase the thermal neutron fluence and t/f neutron ratio in the other facilities. Preliminary results indicate a thermal neutron flux of $\sim 10^{13}$ and $10^{12} cm^{-2} sec^{-1}$ for the 4-inch diameter core holes and 6-inch tank holes closest to the BSR core, respectively. The corresponding fast ($\geq$ 1 MeV) neutron flux has been estimated from nickel wire dosimeter techniques as 1.3 x 10^{12} and $10^{10} cm^{-2} sec^{-1}$ for the above holes, respectively.

Samples from single crystal silicon ingots were characterized by conventional van der Pauw and IR techniques to determine the ingot type, resistivity, initial apparent carrier concentration, and initial apparent oxygen content as indicated in Table 1. Isothermal annealing experiments were carried out in a quartz tube furnace in a helium, argon, or nitrogen atmosphere to·study the introduction and dissociation of donors due to clustered oxygen in Cz silicon samples prior to irradiation. Ingot sections or slices of FZ and Cz silicon were irradiated in the D_2O tank where the temperature did not exceed 50°C to introduce $\sim 10^{12}$ to 10^{16} ^{31}P cm^{-3} in estimated t/f ratios of 100 and 10. Other ingot sections were irradiated in various locales in the Oak Ridge Reactor (ORR), the High Flux Isotopes Reactor (HFIR) at ORNL, and the National Bureau of Standards Reactor (NBSR) to introduce up to 10^{18} ^{31}P cm^{-3} in a t/f ratio of 3-10. Isochronal (30 min) annealing at 750 to 1000°C was used to remove the radiation-induced defects in both FZ and Cz NTD silicon samples, and additional annealing experiments were carried out to study the introduction and dissociation of donors due to oxygen or of defect-donor complexes in selected Cz NTD silicon samples.

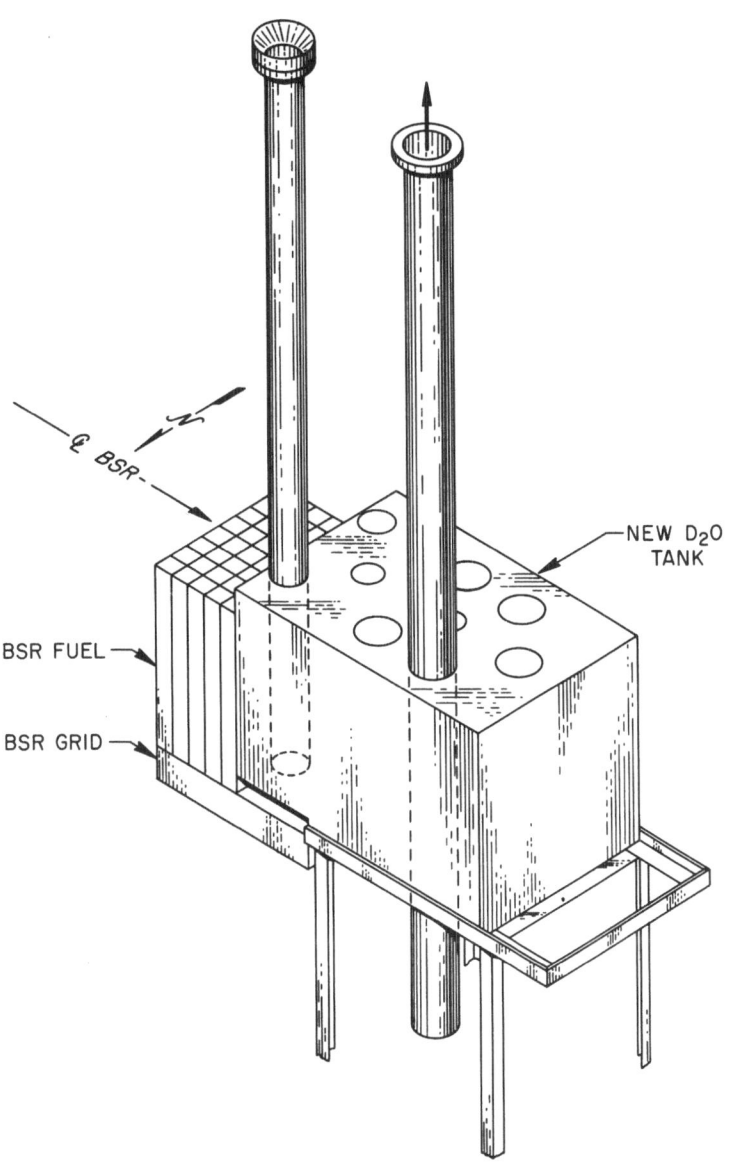

Fig. 1. Thermal neutron facility (D_2O tank) of the Bulk Shielding
Reactor.

Table 1.   Ingot characteristics of silicon samples studied.

Ingot Number	Ingot Type	Dopant	Growth Technique	Post Anneal	Resistivity (ohm-cm)	Carrier Cont. (cm^{-3})	Oxygen Content (cm^{-3})
1	-	-	Cz	No	50	2.5×10^{14}	2.4×10^{18}
2	n	P	Cz	Yes	30	3.0×10^{14}	1.3×10^{18}
3	n	P	Cz	Yes	194	1.8×10^{13}	1.7×10^{18}
4	n	P	Cz	Yes	41	8.3×10^{13}	8.5×10^{17}
5	n	P	Cz	Yes	149	2.5×10^{13}	7.5×10^{17}
6	p	B	Cz	Yes	50	3.5×10^{14}	1.3×10^{18}
7	p	^{10}B	Cz	No	0.42	5.5×10^{16}	8.0×10^{17}
8	n	P	FZ	No	1000^{+}	$1\text{-}5 \times 10^{12}$	$\leq 3.0 \times 10^{16}$

Isotopic ^{10}B (19.8% normal abundance) has a very large
($\sim 3{,}840 \times 10^{-24} cm^2$) thermal neutron absorption cross section and
fissions to introduce ^4He and ^7Li. Since the properties of lithium
in NTD silicon have been of interest, a special Cz ^{10}B doped
single crystal silicon ingot (No. 7 - Table 1) was obtained,[16]
and samples were subjected to a variety of irradiation and annealing experiments to study the behavior of lithium in this material.
Yet another experiment was carried out in which samples of high
purity FZ silicon were irradiated, annealed to remove lattice
damage, and diffused with lithium from an oil bath suspension to
investigate any potential lithium-phosphorus interactions.

3.   EXPERIMENTAL RESULTS

3.1.   Annealing of Float Zone NTD Silicon

It was stated above that annealing for 30 minutes at 750°C is
sufficient to obtain the anticipated electrical activity in FZ
NTD silicon irradiated to introduce up to 6×10^{16} ^{31}P cm^{-3} irrespective of the t/f neutron ratio. These measurements have very
recently been extended to FZ NTD silicon irradiated in a core
position of the HFIR with a t/f ratio of no more than 3-10 to
introduce up to 3.5×10^{17} ^{31}P cm^{-3}, and identical annealing

results were obtained. Previous investigators and we agree that
one can predict a phosphorus concentration in FZ NTD silicon from
the isotopic abundance, cross section, and thermal neutron fluence
data and obtain that value after annealing for 1-2 hours at 700
to 800°C.

Recovery of the minority carrier lifetime (MCL) in FZ NTD
silicon requires a higher annealing temperature than that required to obtain the anticipated carrier concentration and mobility,
and the recovery is also dependent on the time at temperature,
furnace atmosphere, and rate of cooling.[17] Studies of the MCL in
FZ NTD silicon at ORNL have led to annealing conditions, not yet
optimized, that restore lifetimes in the milliseconds in lightly
irradiated material. The annealing cycle is one hour at 1000°C
in a gettering atmosphere, cooling at 20°C hr^{-1} to 650°C to permit
vacancy stabilization, and then rapid cooling to room temperature
to minimize oxygen clustering. The process has been used to increase the MCL in commercial NTD silicon and to restore the MCL in
samples whose lifetimes were thermally degraded intentionally.

Figure 2 is a graph of the MCL as a function of epithermal
neutron fluence for various samples of FZ NTD silicon which were
irradiated in a number of different reactor locales to an epithermal
neutron fluence as indicated, and after the partial but standard
annealing schedule just outlined. The annealing schedule was later

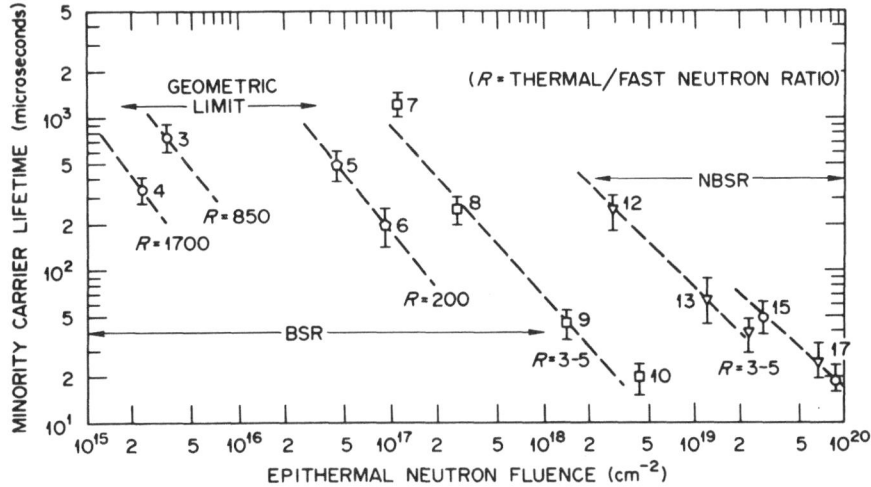

Fig. 2. Minority carrier lifetime vs. epithermal neutron fluence
         for selected FZ NTD silicon samples after irradiation and
         partial annealing.

extended first to 7 hrs at 1000°C with some improvement in the
MCL, and then to 1 hr at 1100°C with little additional improvement.
Note that the recovery of the MCL is not dependent on the total
fluence or t/f ratio in a manner which might have been anticipated.
All of the samples had an identical annealing sequence, but the
neutron fluence for sample No. 12 was ten times that for sample
No. 8 and thirty times that for No. 6. The fact that the amount
of recovery of the MCL in these samples was nearly identical would
seem to indicate that the presence of fast neutron induced lattice
damage may actually enhance the rate of lifetime recovery in FZ
NTD silicon. There is also an evident difference between the
neutron energy spectrum of the BSR and the NBSR reactors that was
not indicated by any dosimeter data that was available to us.

3.2. Annealing of Unirradiated Czochralski Silicon

Figure 3 is a graph of the interstitial oxygen concentration
determined by IR spectroscopy, vs. the time of annealing at
different temperatures for samples from ingot 1 of Table 1 (not
post annealed).[16] This data, taken on a coarse time grid, indicates
that $\sim$ 100 hrs at 450°C were required to introduce $\sim$ 2.3 x 10^{16}
donors cm^{-3} due to clustered oxygen, with no significant decrease
in the O_i content. Annealing for $\sim$ 100 hr at 750°C, in contrast,
had little if any apparent effect on the donor concentration,
although a major fraction of the O_i was clustered. This data would
seem to suggest that annealing for 1-2 hrs at 700-800°C, which is
required to remove lattice damage in FZ NTD silicon, would not be
expected to significantly alter the O_i concentration or the donor
concentration due to clustered oxygen in Cz NTD silicon. However,
we shall see in the next paragraph that the situation is not that
simple.

Figure 4 is graph of the donor concentration vs. the time of
annealing at 450°C for samples of Cz silicon from ingot 1 (not
post annealed) and ingots 2, 5, and 6 (post annealed of Table 1,
and this data indicates that formation of donors by oxygen should
not be ignored in any isochronal or isothermal annealing studies
of Cz NTD silicon that includes temperatures around 400 to 500°C,
irrespective of any post annealing. Annealing for 1 hr at 450°C
was sufficient to introduce $\sim$ 1.6 x 10^{15} donors cm^{-3} in the sample
from ingot 1 (not post annealed) and $\sim$ 1-4 x 10^{14} donors cm^{-3} in
the samples from ingots 2, 5, and 6 (post annealed). The sample
from ingot 6, initially p-type, was converted to n-type after
annealing for 1 hour. Another sample from ingot 1 (not post
annealed) had an apparent p-type acceptor concentration of 3.4 x
10^{13} cm^{-3} after a post anneal of 1 hour at 750°C, and an apparent
n-type donor concentration of 2.3 x 10^{14} cm^{-3} after 1 hour at 450°C.
All of the samples used to obtain the data of Fig. 3 were reannealed

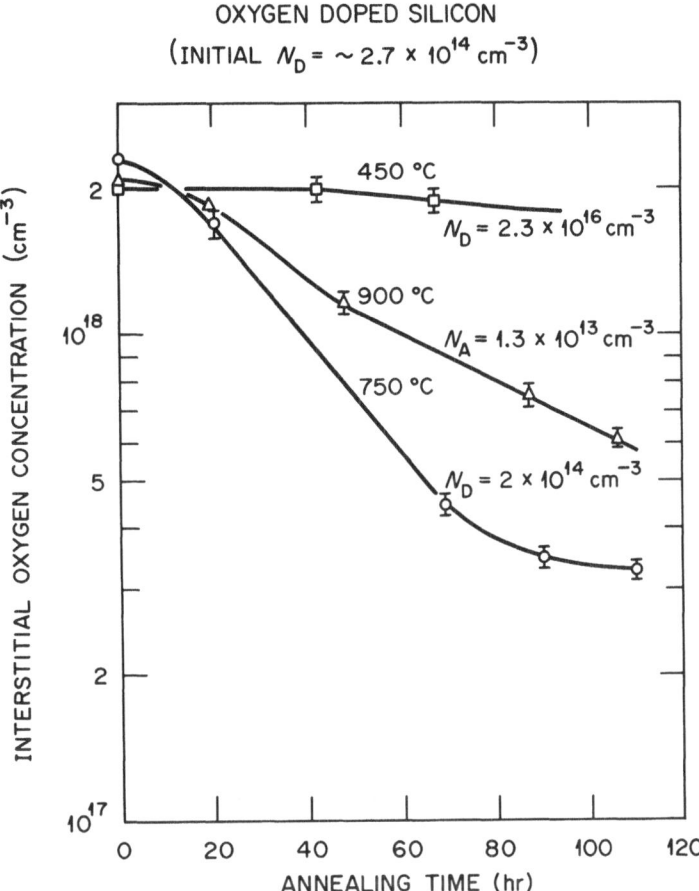

OXYGEN DOPED SILICON
(INITIAL $N_D = {\sim}2.7 \times 10^{14}\,cm^{-3}$)

Fig. 3. Interstitial oxygen concentration vs. time of annealing
at indicated temperatures for samples from ingot 1 of
Table 1 and the donor or acceptor concentration after
annealing.

for 1 hour at 900°C, and the apparent donor concentration was reduced to approximately the initial concentration listed in Table 1.
The subsequent rate of formation of donors as a consequence of
additional annealing at 450°C was redetermined, and the final donor
concentration after extended annealing was within about 10% of that
indicated for all of the samples in Fig. 3.

The most significant point from these results can be summed
up as follows. Previous investigators[8-9] examined Cz NTD silicon

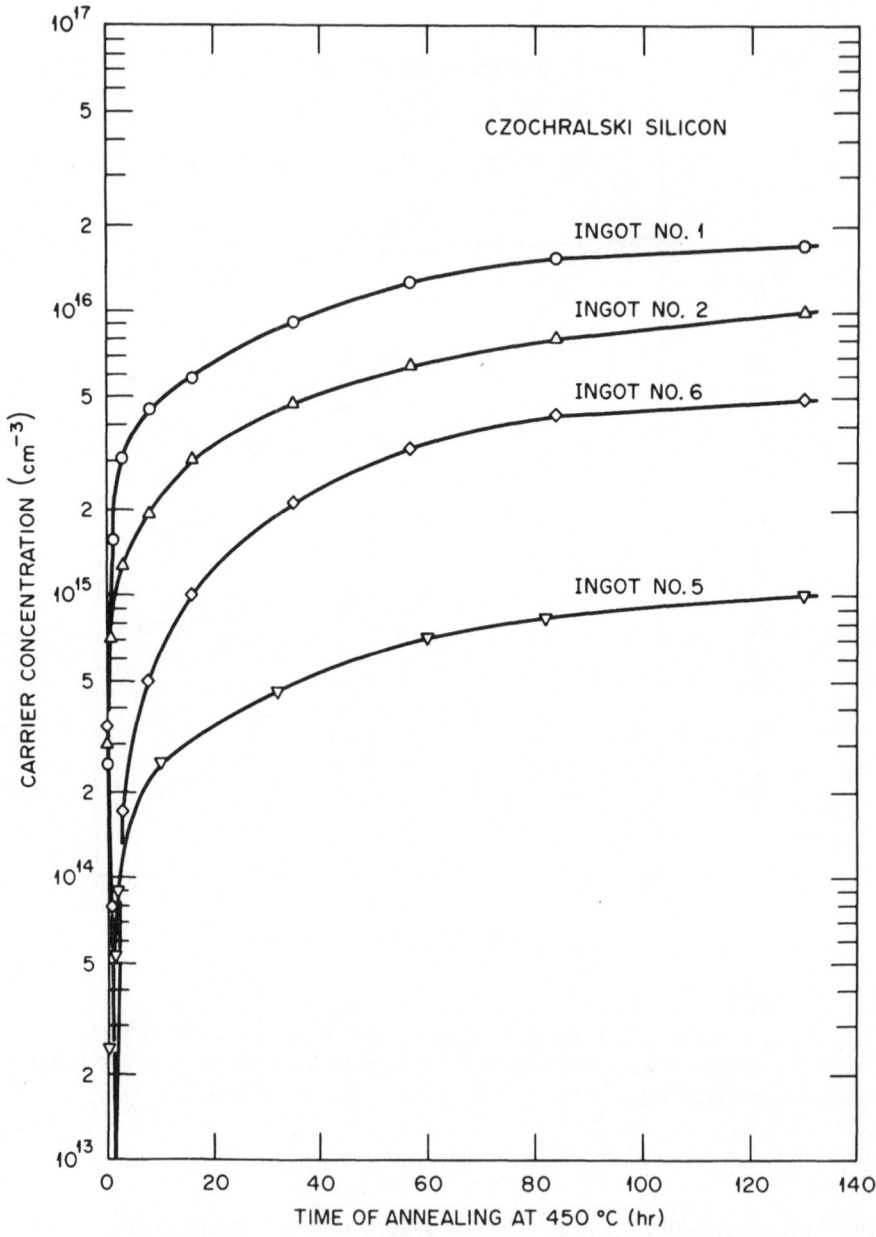

Fig. 4. Carrier concentration vs. time of annealing for samples
from ingots 1, 2, 5, and 6 of Table 1.

that was irradiated to introduce only 3×10^{14} to 2×10^{15} phosphorus cm^{-3} and they used both isochronal and isothermal annealing between 100 and 1000°C. They did not clearly differentiate between donors as introduced by phosphorus and those introduced by oxygen, and they concluded that some type of defect-oxygen donor complex was produced by irradiation and/or annealing that is stable up to 1000°C[8] or requires extended annealing at 950°C.[9] The present data indicates that annealing for 1 hour near 400-500°C is sufficient to introduce $\sim 10^{15}$ donors cm^{-3} in unirradiated commercial Cz silicon, irrespective of any post annealing up to at least 900°C. Clearly, the separate effect of these oxygen-induced donors must be considered in any subsequent annealing experiments that involve Cz NTD silicon.

3.3. Comparison of Annealing Results for FZ and Cz NTD Samples

Figure 5 shows a graph of the calculated phosphorus concentration vs. the thermal neutron fluence (solid line) and the experimental points are the measured phosphorus concentration for various samples after irradiation and annealing for 30 minutes at 750°C. This data indicates that the calculated and measured ^{31}P concentration were equal within about 5% for all concentrations in excess of about 10^{16} ^{31}P cm^{-3} for both FZ and Cz NTD silicon. The fact that the apparent donor concentration was larger in the Cz NTD silicon at lower total ^{31}P concentrations may be due to the presence of oxygen donors or defect-oxygen complexes, and will be discussed separately.

Subsequently annealing for 30 minutes at 1000°C has little if any effect on the apparent donor concentration in any of the FZ or Cz NTD silicon samples shown in Fig. 5 that contained $\geq 10^{16}$ phosphorus cm^{-3}. The Cz NTD silicon samples that had an apparent donor concentration of $\sim 5 \times 10^{15}$, 10^{16}, and $3 \times 10^{16} cm^{-3}$ after irradiation and annealing at 750 and 950°C were reannealed for 100 hours at 450°C, and the apparent donor concentration increased by 1.42, 1.69, and $1.26 \times 10^{16} cm^{-3}$ respectively in the three samples. The fact that one can introduce an almost equivalent concentration of donors by oxygen clustering at 450°C in Cz NTD silicon after extended irradiation, and after annealing at 750 and 950°C to remove lattice defects, shows that it is not necessary to invoke the formation of a special defect-oxygen complex that requires higher temperature annealing. Of course, such a complex may be present but its effects masked at these high donor concentrations.

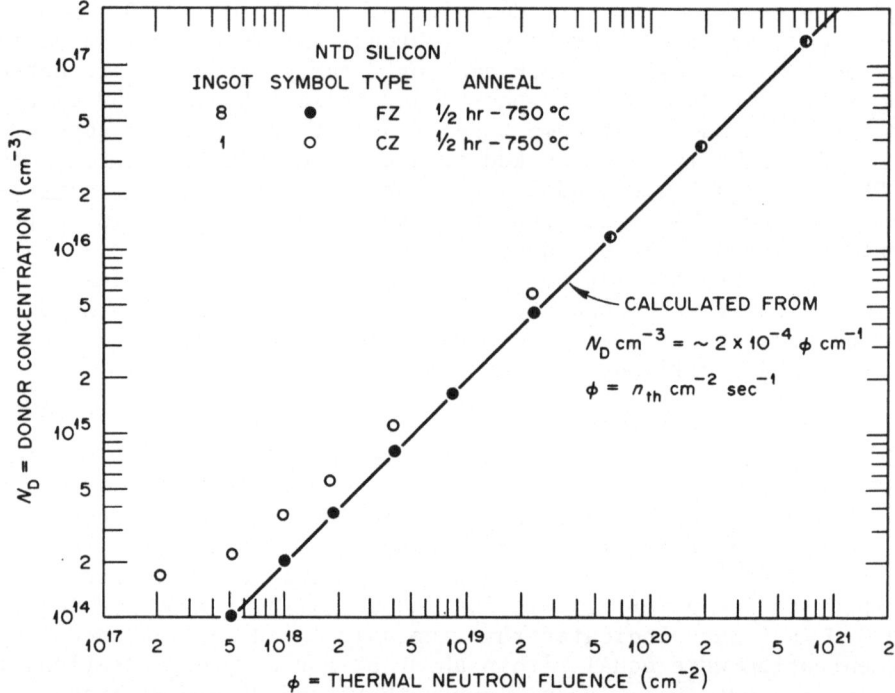

Fig. 5. Calculated and measured phosphorus concentration vs.
thermal neutron fluence for selected samples of FZ and
Cz NTD silicon after irradiation and annealing.

3.4. Comparison of Annealing Results for FZ and Cz NTD Samples
After a Low (10^{16} - 10^{18} cm^{-2}) Neutron Fluence

Previous experiments[6] have indicated that most of the simple
(n,γ) recoil type defects can be removed in FZ NTD silicon by
annealing up to ∿ 400°C, but the data of Figs. 4 and 5 indicate
that oxygen donor formation may mask any attempt to determine the
annealing of radiation-induced defects in Cz NTD silicon around
400°C. Since it has already been demonstrated[6] that annealing for
1/2 hr at 750°C is sufficient to remove the radiation-induced
lattice defects in FZ NTD silicon, it was decided to irradiate
companion FZ and Cz samples, anneal them for 1/2 hr at 750°C,
and then continue the annealing experiments at 750 and 950°C.
Since the anticipated phosphorus concentration was obtained for
the FZ NTD samples after the first anneal, and since extended
annealing at 750 and 950°C had no apparent additional effect, that
data was used to estimate the actual ^{31}P concentration in the
companion Cz NTD samples.

Figure 6 is a graph of the measured donor concentration vs. the time of annealing first at 750 and then at 950°C for samples from ingot 1 of Table 1 after irradiation, and after an initial anneal of 1/2 hr at 750°C. The anticipated ^{31}P concentrations obtained from the FZ NTD samples are as shown. This data indicates that the measured donor concentration for the Cz NTD samples was larger than anticipated at the outset, particularly for the low dose irradiations, and may have been even larger than indicated if the probable effect of a reduction in initial donor concentration due to post annealing is included. Note that the apparent donor concentration was increased by $\sim$ 1-3 x 10^{15}cm^{-3} in each of the six Cz NTD samples as a consequence of annealing for 14 hrs at 750°C, even though the initial lattice defect concentration was in the ratio of $\sim$ 1 to 6,000 for these samples. It is evident that some type of donor formation due to oxygen, or to defect-oxygen complexes, can occur in Cz NTD silicon as a consequence of extended annealing at 750°C after a low total dose irradiation. Some of this enhanced donor concentration remains after additional annealing at 950°C, but the present data does not indicate any obvious correlation between the number of defects present and the potential number of defect-oxygen complexes, unless that number depends almost entirely upon the initial oxygen concentration. Almost identical data to that in Fig. 6 was obtained on samples from ingot 4 of Table 1, which contained less oxygen and had been post annealed. Here the apparent donor concentration was increased by $\sim$ 3-6 x 10^{14} cm^{-3} as a consequence of extended annealing at 750°C, irrespective of the number of defects present.

Perhaps the major point to be made from the results of these low total fluence irradiation experiments is that $\sim$ 1-3 x 10^{15} and $\sim$ 3-6 x 10^{14} donors cm^{-3} were introduced into Cz NTD silicon samples of high (ingot 1) or moderate (ingot 4) oxygen content by extended annealing at 750°C after irradiation. Some fraction of that additional donor concentration remained after additional annealing at 950°C, but that fraction was $\leq$ 10% of the number of donors that can be introduced by clustered oxygen in these same samples, irrespective of any defect concentration due to irradiation. There may be some type of special defect-oxygen donor complex in Cz NTD silicon that can only be removed by extended high temperature annealing, but the actual donor content is apparently not neutron fluence dependent, and the total donor contribution is minimal ($\leq$ 10%) compared to that which may be introduced by clustered oxygen alone.

3.5. Irradiation Experiments on Heavily Doped Silicon

Potential defect-impurity interactions between an NTD epitaxial silicon layer and its substrate may be of importance in a

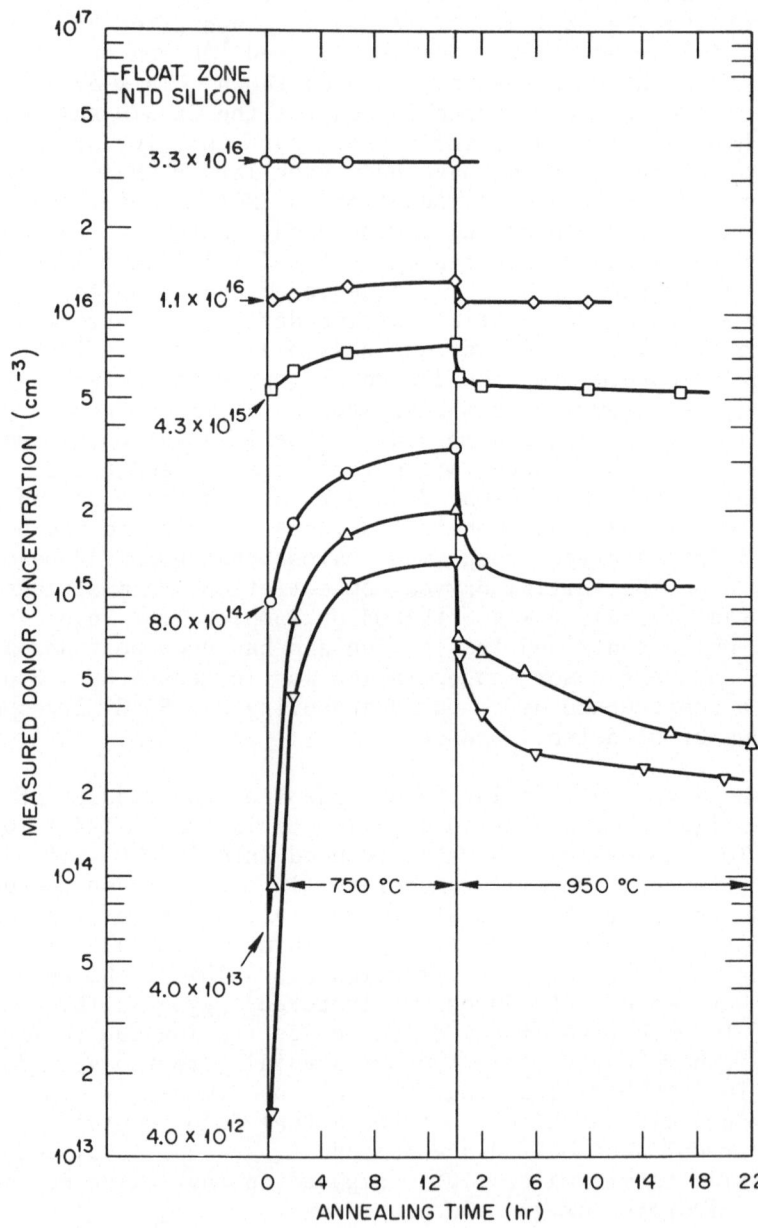

Fig. 6. Carrier concentration vs. time of annealing at 750 and 950°C for Cz NTD silicon samples. The anticipated ^{31}P concentration obtained from companion FZ NTD silicon samples are also indicated.

number of applications. The published data on thermal neutron cross sections, half-lifes of radioactive decay, and transmutation end products can be used to select a preferred silicon substrate dopant. The standard n-type silicon dopants are phosphorus, arsenic, and antimony, but antimony has a large cross section and long half-life, and phosphorus has extensive autodoping tendencies during any subsequent device fabrication. The standard p-type dopants are boron, indium, or gallium, but indium is difficult to introduce during initial crystal growth, and isotopic ^{10}B fissions to introduce ^{4}He and ^{7}Li. The above would suggest n-type arsenic or p-type gallium as the preferred substrate dopant impurities.[3-4]

It was also stated that lithium can act as an electrically active donor in silicon, and can interact with other impurities or lattice defects. One can use cross section data to calculate that the rate of introduction of ^{31}P (donor) and the rate of removal of ^{10}B (acceptor) by fission should be about comparable for NTD silicon initially doped with $\sim$ 5 x 10^{16} ^{10}B or $\sim$ 2.5 x 10^{17} normal B cm^{-3}. A special Cz silicon single crystal ingot (No. 7 - Table 1) that was doped with $\sim$ 5.5 x 10^{16} ^{10}B cm^{-3} from a stable isotopic source in which ^{10}B was $\sim$ 94.5% abundant was provided for our experiments.[16] The initial oxygen concentration, determined by IR spectroscopy, was somewhat less than that for some of the other Cz ingots in Table 1, but isothermal annealing for 100 hr at 450°C decreased the apparent p-type acceptor concentration in a sample from ingot 7 of Table 1 by about 7.5 x 10^{15}cm^{-3} by addition of donors due to clustering. This value is comparable to that obtained for the other Cz ingots on the basis of their relative oxygen content.

Samples of the ^{10}B doped Cz silicon were irradiated with FZ control samples, and the results on the impurity concentrations are given in Table 2. One must calculate the amount of ^{10}B removed by fission, since the total concentration is reduced by extended irradiation, but some ^{11}B remains. The ^{31}P concentration was determined after annealing the FZ NTD control samples for 1/2 hr at 750°C. The net concentration (line 5) is the anticipated value if the only change is that of addition of ^{31}P donors and removal of ^{10}B acceptors. The calculated concentration of ^{7}Li (line 6) should be identical with the calculated removal of ^{10}B (line 3), and this value was used to estimate the net concentration (line 8) if ^{7}Li is active as a donor. The measured concentration was obtained (line 9) for the Cz NTD ^{10}B silicon after annealing for 1/2 hr at 1000°C to remove lattice damage introduced by (n,γ) recoils, fast neutrons, and fission, since annealing for 1/2 hr at 750°C was not sufficient to stabilize the apparent impurity concentration in every sample.

It is evident that ^{7}Li is present as a donor in the low total dose samples in Table 2 that remained p-type, and in the 21 day

Table 2.   Transmutation reactions in isotopic ^{10}boron-doped silicon.

REACTOR	BSR	BSR	NBSR	NBSR	NBSR
Irradiation (days)	20	40	7	21	84
Calc. ^{10}B Removed	0.34	0.52	1.05	2.57	4.74
Total B Remaining	5.16	4.98	4.45	2.93	0.76
Measured ^{31}P Added	-0.34	-0.52	-1.21	-3.66	14.00
Net Concentration	4.82	4.46	3.24	-0.73	-13.24
Calculated ^{7}Li Added	-0.34	-0.52	-1.05	-2.57	- 4.74
Net Concentration	4.48	3.94	2.19	-3.30	-17.98
Measured Concentration	4.18	3.14	1.57	-2.35	-14.40

All data are in units of 10^{16}cm^{-3}
Minus sign indicates n-type donor concentration

NBSR sample that was converted to n-type. The 84-day NBSR sample required six months for radioactive decay, and a major portion of the ^{7}Li may have precipitated in that time period. An alternative explanation is that the ^{7}Li interacts with boron to form a stable donor, and the remaining boron concentration was insufficient in this sample to retain such a donor activity. Previous experiments have indicated that annealing for 1/2 hr at 750°C was sufficient for both FZ and conventionally doped Cz NTD silicon to remove the lattice damage and obtain the anticipated carrier concentration out to a total ^{31}P concentration of 10^{18}cm^{-3}. The ^{10}B-doped Cz NTD silicon was different in that annealing up to 1000°C was required to stabilize the carrier concentration in the low total dose, 7-day and 21-day NBSR samples, whereas annealing at 750°C was sufficient for several samples from the 84-day NBSR ingot section. Another major difference is that reannealing at 450°C was used to introduce additional donors due to oxygen clustering after irradiation in conventional Cz NTD silicon, but there was little if any indication of oxygen clustering in any of the ^{10}B-doped Cz NTD silicon samples as a consequence of extended annealing at 450°C after irradiation. The higher temperature annealing requirement may be due to more extensive lattice damage as a consequence of fission, and the absence of donor formation at 450°C may be due to some type of interaction between the ^{7}Li and oxygen.

Yet another experiment was carried out in which samples of high purity FZ silicon were irradiated, annealed for 1/2 hr at 750°C to remove lattice damage, and diffused with lithium from an oil

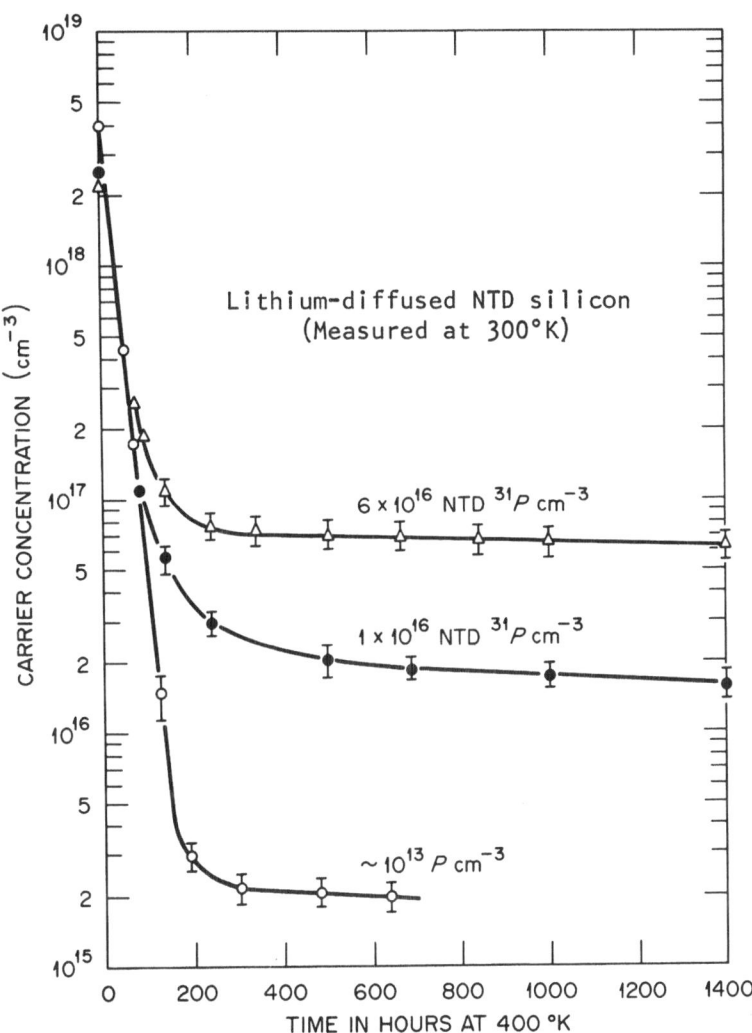

Fig. 7. Carrier concentration vs. time of annealing after lithium
         diffusion in FZ NTD silicon samples and a FZ silicon
         control sample.

bath suspension at 600°C. Figure 7 is a graph of the carrier concentration as a function of time of accelerated annealing at 400°K.
This data indicates there is no evident interaction between the
lithium and ^{31}P in NTD silicon, and that the lithium precipitates
to a final solubility that may depend on the actual oxygen concentration in FZ NTD silicon.

4. DISCUSSION

The primary result of these various experiments is that a
better understanding of the annealing requirements to remove
lattice damage and observe the anticipated ^{31}P concentration in FZ
and Cz NTD silicon has been obtained. It has been demonstrated
that annealing for 1/2 hour at 750°C is sufficient to obtain the
anticipated resistivity, carrier concentration, and carrier mobility
in both FZ and Cz NTD silicon irradiated in a low t/f ratio to
introduce ∿ 10^{16} to 10^{18} ^{31}P cm^{-3}, but the annealing requirements
to recover the anticipated MCL are still under investigation. The
electrical role of clustered oxygen and of any defect-oxygen
complex in Cz NTD silicon was investigated, but any separate
effects between these two types of potential donor sources were
not clearly identified. It has also been shown that one can produce an NTD epitaxial layer on a heavily doped n- or p-type
substrate, but careful choice of substrate dopant is recommended.
The use of boron will result in the introduction of ^{7}Li that may
migrate and alter the desired dopant concentration in the epitaxial
layer. There is no evident interaction between lithium introduced
by diffusion and ^{31}P introduced by irradiation, but there may be
some type of pairing reaction between ^{7}Li introduced by ^{10}B fission
and any remaining boron, since samples that contain an equivalent
concentration require a higher annealing temperature than those in
which most of the ^{10}B was removed by fission. The ability to add
a known concentration of ^{7}Li in silicon by the technique of ^{10}B
fission should lead to a better understanding of the interaction
between lithium and oxygen in silicon, if FZ ^{10}B-doped samples of
low or controlled oxygen content can be made available.

ACKNOWLEDGMENTS

The authors wish to thank L. E. Katz of the Bell Telephone
Laboratories, Allentown, PA, for providing both the heavily oxygen-
doped and ^{10}B-doped ingots.

REFERENCES

*   Research sponsored by the Division of Materials Sciences,
      U.S. Department of Energy under contract W-7405-eng-26 with
      the Union Carbide Corporation.

1)   J. W. Cleland, K. Lark-Horovitz, and J. C. Pigg, Phys. Rev.
      78, 814 (1950).

2)   Third International Symposium on Silicon Materials Science
      and Technology: Semiconductor Silicon/1977, PV 77-2, Electro-
      chemical Society, 1977. IEEE Trans. Electron Devices, ED-23,
      8 (1976), (special issue on high power semiconductor devices.)

3)   S. Prussin and J. W. Cleland, J. Electrochem. Soc. 125, No. 2,
      350 (1978).

4)   C. W. Pearch, Western Electric Co., Allentown, PA. Private
      communication.

5)   R. F. Wood, R. D. Westbrook, R. T. Young and J. W. Cleland,
      Proceedings of the Solar Cell High Efficiency and Radiation
      Damage Conference, NASA Conference Publication 2020, p. 109.

6)   R. T. Young, J. W. Cleland, R. F. Wood and M. M. Abraham,
      J. Appl. Phys. 49, 4752 (1978).

7)   L. E. Katz and D. W. Hill, J. Electrochem. Soc. 125, No. 7,
      1151 (1978). G. A. Rozgonyi and C. W. Pearce, Appl. Phys.
      Letters 32 (11), June 1, 1978.

8)   V. N. Mordkovich, S. P. Solov'ev, E. M. Temper and V. A.
      Kharchenko, Sov. Phys. Semicond. 8, No. 1, July (1974).

9)   H. M. Liaw and C. J. Varker, Semiconductor Silicon/1977,
      PV 77-2, Electrochemical Society, p. 116 (1977).

10)   E. M. Pell, Solid State Physics in Electronics and Communi-
      cation (Academic Press, Inc., New York, 1960), Vol. 1, p. 261.

11)   C. S. Fuller, J. Phys. Chem. Solids 19, 18 (1961).

12)   J. W. Cleland, F. J. James and R. D. Westbrook, IEEE Trans.
      Nucl. Sci. ND-19, No. 6, 224 (1972).

13)   V. S. Vavilov, I. V. Smirnova, and·V. A. Chapnin, Sov. Phys.
      Solid State 4, 830 (1962).

14)   G. J. Bruckner, Phys. Rev. 183, 712 (1969).

15)   L. C. Kimerling, P. J. Drevinsky, and C. S. Chen, Defects in
      Semiconductors (Reading, England, 1972), Conf. Series No. 16,
      P. 182.

16)   We are indebted to L. E. Katz, Bell Telephone Laboratories,
      Allentown, PA, for the careful preparation of ingots 1 and 7.

17)   R. D. Westbrook and J. W. Cleland, Extended Abstracts, J.
      Electrochem. Soc. 125, No. 3, 144C (1978).

DEFECT ANNEALING STUDIES IN NEUTRON TRANSMUTATION DOPED SILICON*

B. C. Larson, R. T. Young and J. Narayan

Oak Ridge National Laboratory, Solid State Division

Oak Ridge, TN 37830

ABSTRACT

Electrical, optical, x-ray and electron microscopy techniques have been used to investigate lattice damage and annealing in NTD silicon. Dislocation loops of 10-100 Å diameter were identified in both the as irradiated state and after thermal anneals of up to 900°C by the electron microscopy and x-ray techniques. A correlation of these results with the infrared absorption and electron mobility recovery stage between 600 and 750°C suggests that defects < 10 Å are also contributing to the recovery processes at this temperature.

1. INTRODUCTION

Neutron transmutation has been shown to be a useful method of producing a controlled, uniform phosphorus doping in silicon. The lattice damage associated with the recoiling atom in the (n,γ) transmutation process and the damage resulting from recoiling silicon atoms following direct hits by fast neutrons must, however, be removed before the electrical properties of the phosphorus doping can be realized. This lattice damage is in general quite complicated,[2] consisting of vacancy complexes, divacancies, vacancy clusters and interstitials presumably in complexes as well as larger clusters and dislocation loops. The concentrations of these defects can be expected to vary depending on the ratio of the fast neutron flux to the thermal neutron flux present during the irradiation and on the energy spectrum of the fast neutrons and the temperature of the silicon in the particular reactor locale.

As a result of the complex defect structure and the difficulty in specifying or controlling all of the parameters in the irradiation our understanding of the lattice damage and its impact on the electrical properties of the doped silicon is incomplete. In this paper, we report x-ray and electron microscopy measurements along with electrical and infrared absorption measurements on rather highly neutron transmutation doped silicon in order to relate the presence of structural defect clusters to the electrical and optical properties of the material.

2. EXPERIMENTAL

One inch diameter ingots of float zone refined ($\sim$ 700 Ω-cm) n-type silicon were neutron irradiated at temperatures < 100°C at the National Bureau of Standards Reactor (NBSR), the Oak Ridge Research Reactor (ORR) and the Bulk Shielding Reactor (BSR) at Oak Ridge for neutron transmutation (phosphorus) doping. Table 1 summarizes the neutron fluences and the reactor locales utilized. After irradiation, wafers ($\sim$ 0.5 - 1.0 mm) were cut from these ingots, lapped and then chemically polished with CP6 solution to obtain mirror like surfaces.

These specimens were then studied by infrared absorption in the 1-6 µm wavelength range (Perkin Elmer 21) at room temperature both in the as-irradiated state and after 20 or 30 minute isochronal anneals. The annealing was carried out between 150°C and 900°C in a quartz tube furnace with a helium flow atmosphere.

X-ray diffuse scattering and transmission electron microscopy measurements were carried out on the ORR irradiated specimens in the as-irradiated state and after selected annealings in order to study structural defects such as small defect clusters and dislocation loops and to relate them to the optical data. The x-ray data were collected, using CuKα radiation, in the form of integral diffuse scattering measurements[3-4] near the 333 Bragg reflection. Data were collected as a function of the rocking angle $\Delta\Theta$ relative to the 333 Bragg reflection angle. Crystals with [111] surface normals were used for this purpose and the neutron irradiation induced diffuse scattering was separated from the usual background scattering (Compton and thermal diffuse scattering) through measurements on unirradiated specimens. The irradiation induced scattering intensity is reported in the form of the net symmetrical scattering (averaged over plus and minus $\Delta\Theta$ for each crystal setting $\Delta\Theta$) as described in detail elsewhere.[4]

Direct observations of the irradiation induced defects were made in the electron microscope after chemically thinning 3 mm discs cut from the irradiated samples. Micrographs were made using

Table 1. Irradiation conditions and measured divacancy concentra-
          tions for the silicon samples studied.

Sample #	Reactor Locale	Phosphorus Conc. (cm^{-3})	Fast Neutron Fluence (n/cm^2)	Divacancy Conc. (cm^{-3})
1	NBSR G-2	1.4×10^{15}	1.5×10^{18}	1.4×10^{18}
2	NBSR G-2	3.0×10^{16}	3.2×10^{19}	1.1×10^{19}
3	NBSR G-2	6.0×10^{16}	6.4×10^{19}	1.6×10^{19}
4	ORR	5.0×10^{16}	1.0×10^{20}	3.5×10^{19}
5	BSR SW	2.0×10^{15}	1.0×10^{16}	--------

bright and dark field (including weak beam) imaging under both
kinematical and dynamical diffraction conditions. The loop sizes
were determined on bright field dynamical micrographs where the
width of the black-white transition region in the defect image
was taken as the measure of the loop diameter.[5-6]

3. RESULTS

The irradiation conditions for the silicon used in this study
are shown in Table 1 where the reactor, reactor locale, the
phosphorus doping level and fast neutron fluences (E > 1 MeV) are
indicated. In addition, the divacancy concentrations as measured
by the 1.8 μm infrared absorption band are given.

Figure 1 shows the infrared near edge absorption coefficient
measured at 1.5 μm for silicon irradiated in the NBSR and ORR
facilities. The absorption coefficient is shown in the as-irradiated state and after annealing from 150 to 750°C. For each dose,
at least 90% of the recovery of the absorption coefficient occurs
between 150 and 450°C and in the two higher dose cases where
measurements were carried out to 750°C a second well defined
recovery state was found between 600 and 700°C. From the divacancy
concentrations listed in Table 1 for samples 1, 3 and 4, it can be
seen that the number of divacancies increases less than linearly
with the neutron fluences.

Turning now to the x-ray results the irradiation induced
diffuse scattering in the as-irradiated and annealed specimens from
the ORR (sample 4) crystal are shown in Fig. 2 where the form of

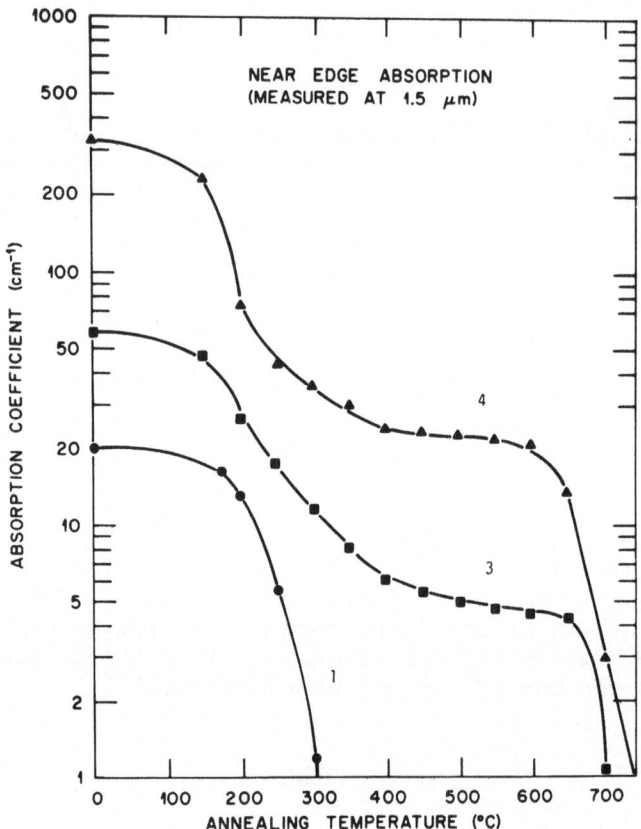

Fig. 1. Near edge infrared absorption measurements as a function
         of isochronal annealing for ORR and NBSR irradiated
         silicon. (Samples Nos. refer to Table 1.)

the intensity as a function of $\Delta\Theta$ indicates[4] the presence of
rather small defect clusters. It is further seen though that the
annealing of these clusters (as monitored by the relative decrease
in the magnitude of the diffuse intensity) has proceeded rather
slowly compared to the recovery of the 1.5 μm near edge absorption
coefficient in Fig. 1. There is only about a 30% decrease in the
diffuse scattering after annealing to 550°C whereas a greater than
90% recovery in the absorption coefficient was observed. Larger
changes in the diffuse scattering were observed upon annealing to
650 and 700°C where the decrease in the scattering intensities tend
to correspond more nearly to the recovery of the near edge absorption measurements. The average cluster sizes inferred from the

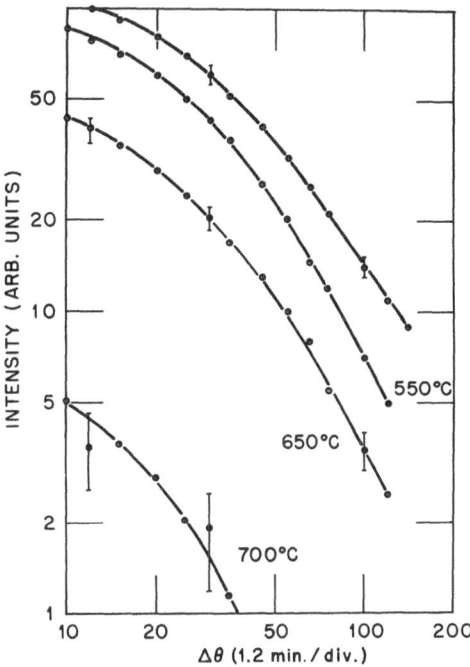

Fig. 2.   Integral x-ray diffuse scattering measured on the ORR
          (sample 4) irradiated specimen. The upper curve corre-
          sponds to the as-irradiated state.

x-ray measurements by determining the $\ln(\Delta\Theta)$ intercept of an intensity vs. $\ln(\Delta\Theta)$ plot[4] of the data in Fig. 2 are 21, 25, 30, and
65 Å diameter for the as-irradiated, 550, 650, and 700°C annealed
specimens, respectively. Although this size determination is
weighted by the fourth moment of the sizes if a distribution of
sizes is present, these will be referred to as average sizes in
the absence of information on possible distributions.

The electron micrographs shown in Fig. 3 indicate directly
that defect clusters in the form of small dislocation loops exist
in the ORR specimen (sample 4). These micrographs show the loops
to have sizes 100 Å diameter and initial checks indicated that both
vacancy and interstitial type loops were present. The average
diameter of the loops in the as-irradiated specimen was measured
to be 40 Å and the number density was $3 \times 10^{16} cm^{-3}$. After annealing to 450°C, the average loop size and number density were found
to be 46 Å and $1.5 \times 10^{16} cm^{-3}$, respectively. The loop size changed
to 60 Å after annealing to 650°C with a number density of

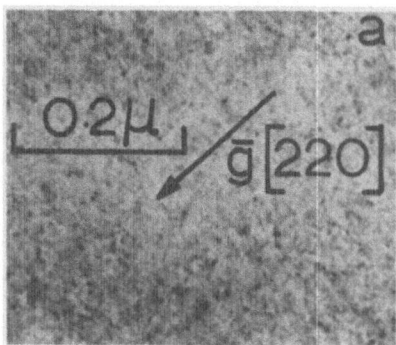
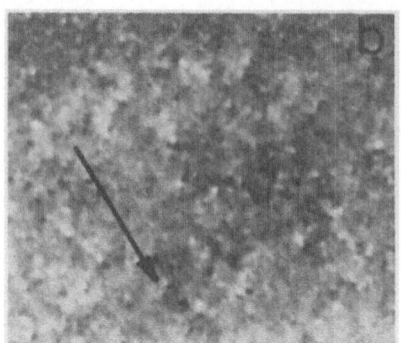

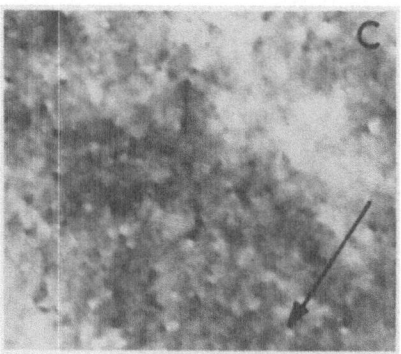

Fig. 3. Bright field electron micrographs under 2-beam conditions
on the ORR (sample 4) specimen showing dislocation loops.
(a) as irradiated, (b) 450°C anneal, and (c) 650°C anneal.
The arrow indicates the direction of the 220 diffraction
vector used here.

$1 \times 10^{16} cm^{-3}$ and after 750°C the number density of loops decreased
to about $3 \times 10^{15} cm^{-3}$ with an average diameter of about 64 Å.
The decrease in the density of dislocation loops after annealing
to 650 and 750°C is in qualitative agreement with the infrared
absorption and the recovery of x-ray diffuse scattering at these
temperatures. The increases in the average sizes of the loops
remaining at these temperatures are also in qualitative agreement
with the x-ray results although the average sizes estimated from
the x-ray measurements are about a factor of two lower initially.

These results can be compared with the electrical mobility
recovery for samples 3 and 4 as presented in Fig. 4 where the
electron mobility of these samples is seen to recover rather sharply
in the temperature range of 600 to 750°C. Sample 5 in Fig. 4 had
nearly the full mobility before annealing above 500°C and shows
only a few percent increase upon annealing to 750°C. From Table 1
we see, however, that this sample was irradiated in the BSR in a
position which had the benefit of a D_2O moderator leading to
dramatically ($\sim 10^3$) lower fast neutron fluxes than the other two
samples.

4. DISCUSSION

Comparing the recovery of the near edge absorption coefficient
in Fig. 1 with the annealing results for the x-ray diffuse scatter-

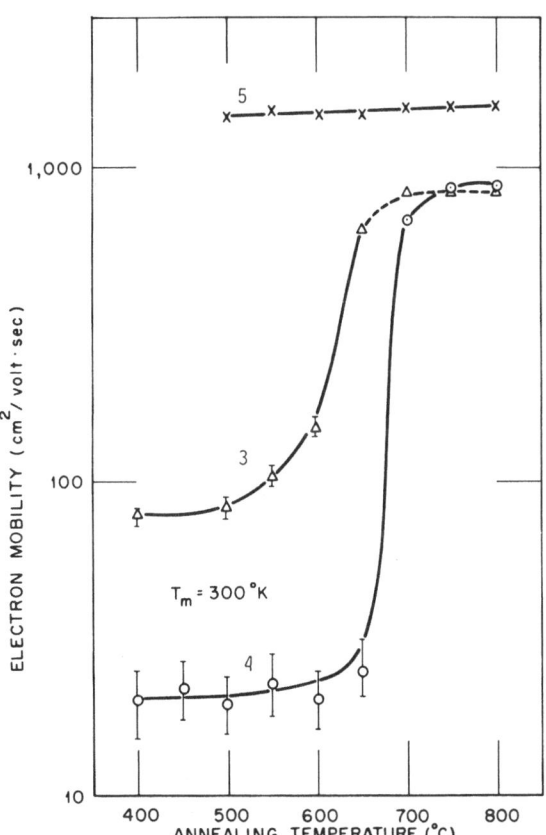

Fig. 4. Recovery of the electron mobility as a function of
isochronal annealing for ORR, NBSR and BSR irradiated
silicon.

ing data in Fig. 2 and the electron microscopy data listed in the previous section we find that there is a correlation between all three measurements only with regard to the annealing stage between 600 and 750°C. Since the results of the electron microscope and the diffuse x-ray scattering measurements reported here are really sensitive only to defects in the form of clusters and small aggregates, the absence of a significant annealing stage between 150 and 450°C seems to indicate that the annealing at these temperatures is largely carried on by very small defects such as divacancies and small vacancy complexes or aggregates. In light of the fact that there is not a one-to-one correspondence between the as-irradiated divacancy concentrations (listed in Table 1) and the as-irradiated 1.5 μm near edge absorption coefficient in Fig. 1, it does not appear that the first stage annealing can be ascribed only to divacancies for these doses. Some combination of divacancies and small vacancy aggregates would seem more probable.

Looking closely at the x-ray and electron microscopy results for the temperature range above 450°C, we find that there is a much faster rate of decrease in the diffuse scattering intensity and the dislocation loop number density (and in the total point defect densities calculated from the loop sizes and concentrations) after annealing above 650°C than for anneals below this temperature. Although the rate of decrease in the dislocation loop number density does not appear to be as fast as the recovery of the diffuse scattering and optical absorption, the accelerated rate of change in the loop densities indicates some correlation exists between the measurements. Without definite information on the sensitivity of the near edge absorption to the larger clusters imaged in the electron microscope relative to the small defects such as vacancy aggregates and complexes, it is difficult to establish a direct correspondence between the optical absorption and electron microscopy results. On the other hand, the x-ray scattering sensitivity to clusters of varying sizes and types can be calculated[3] and the smaller average sizes extracted from the x-ray data could, for instance, arise as a result of a component of defects too small to be imaged in the microscope but large enough to play a role in the x-ray scattering. Some evidence for this exists in EPR[7-10] studies which have identified small vacancy aggregates or complexes which were found to anneal at about 550°C. In addition, the larger increases in the cluster sizes as a function of annealing obtained from the x-ray measurements compared to that for the electron microscopy results would be consistent with the breakup of clusters too small to be imaged in the microscope.

The processes discussed above are, of course, of significance because the annealing takes place in the same temperature range that the electron mobility recovers in NTD silicon. Even though the mobility seems to recover completely by about 750°C, it has been observed[11] that the minority carrier lifetimes do not recover until $\sim$ 1000°C in these crystals. Since electron microscopy observations in this study have indicated that dislocation loops 50-100 Å in diameter with number densities $\sim 10^{14} cm^{-3}$ remain even up to 900°C for sample 4, it is possible that these loops play a role in the trapping of the minority carriers.

Finally, it is important to emphasize that in neither the x-ray diffuse scattering nor the electron microscopy measurements was there any evidence found for dislocation loops with diameters greater than 100 Å. This result is similar to that reported by Pankratz[12] et al. and indicates that for the irradiation conditions used here very large loops such as these suggested by Truell,[13] Bertolotti[14] and Ghosh and Sargent[15] do not appear.

ACKNOWLEDGMENTS

The authors would like to thank P. H. Fleming and T. A. Stephanson for their assistance during this work and useful discussions with J. W. Cleland, R. D. Westbrook, and R. F. Wood are gratefully acknowledged.

REFERENCES

*   Research sponsored by the Division of Materials Sciences, U.S. Department of Energy under contract W-7405-eng-26 with the Union Carbide Corporation.

1)   R. T. Young, J. W. Cleland, R. F. Wood and M. M. Abraham, J. Appl. Phys. 49, 4752 (1978).

2)   O. L. Curtis, Jr., Point Defects in Solids (Plenum Press, New York and London, 1975) Vol 2, p. 257.

3)   B. C. Larson, J. Appl. Phys. 45, 514 (1974).

4)   B. C. Larson and F. W. Young, Jr., Z. Naturforsch. 28a, 626 (1973).

5)   K.-H. Katerbau, Phys. Stat. Sol. (a) 38, 463 (1976).

6)   K.-H. Katerbau, Proceedings of Ninth international Congress on Electron Microscope (The Imperial Press Ltd., 1978) Vol. 1, p. 184.

7)   Y. H. Lee, Y. M. Kim and J. W. Corbett, Rad. Eff. 15, 77 (1972).

8)   Y. H. Lee and J. W. Corbett, Phys. Rev. 8, 2810 (1973).

9)   Y. H. Lee, P. R. Brosious, and J. W. Corbett, Rad. Eff. 22, 169 (1974).

10) Y. H. Lee, N. N. Gerasimenko and J. W. Corbett, Phys. Rev. 14, 4506 (1976).

11) J. W. Cleland, P. H. Fleming, R. D. Westbrook, R. F. Wood, and R. T. Young, this conference.

12) J. M. Pankratz, J. A. Sprague and M. L. Rudee, J. Appl. Phys. 39, 101 (1968).

13) R. Truell, Phys. Rev. 116, 890 (1959).

14) M. Bertolotti, Radiation Effects in Semiconductors (Plenum Press, New York, 1968), ed. F. L. Vook, P. 311.

15) S. Ghosh and G. A. Sargent, Radiation Damage and Defects in Semiconductors (Reading, England, 1972) p. 165.

ISOCHRONAL ANNEALING OF RESISTIVITY IN FLOAT ZONE AND CZOCHRALSKI NTD SILICON*

Paul J. Glairon and J. M. Meese

University of Missouri, Research Reactor Facility

Columbia, MO 65211

ABSTRACT

Resistivity and thermal probe measurements have been used to characterize the isochronal annealing, type conversion, and phosphorus electrical activation of NTD-Si over the temperature range of 20 to 850°C in undoped float zone and pulled silicon. Similar experiments are reported for Ga doped float zone for [P]/[B] ratios of 1 to 3. The annealing associated with fast neutron damage has been isolated using boron shielding during irradiation. The effects of β^- recoil damage have been isolated by rapid pre-annealing to 850°C before appreciable $^{31}Si \rightarrow {}^{31}P + \beta^-$ decay had occurred. Type conversion annealing peaks (p → n) occur at 650 - 750°C in NTD float zone, at 400°C in NTD Czochralski and at 350°C (n → p) in β^- recoil damaged silicon. A dominate acceptor defect dominates the conductivity in the annealing temperature range of 575-650°C in both float zone and Czochralski. Several reverse annealing peaks from 20-550°C are also observed in most samples and heavily irradiated float zone exhibits many annealing features similar to Czochralski.

1. INTRODUCTION

Neutron transmutation doping (NTD) in silicon is a new technology which is being used for many commercial applications by the semiconductor industry. This doping process is well known[1] and has recently been reviewed,[2] however, relatively little is known about the details of the radiation damage which accompany the process. During the doping irradiation, silicon crystals will become damaged by the high energy particles which are present in a fission reactor. There are roughly two types of damage we would expect in these types

291

of irradiations:

1.   Defects due to high energy collisions (fast neutrons, high
     energy gammas, fast charged particles).
2.   Defects due to recoils during nuclear reactions (n,γ), (n,p),
     (n,α) reactions and recoils due to beta decay).

This paper presents a survey of the annealing of electrical
resistivity and majority carrier type in silicon. The technique of
isochronal annealing was used to characterize this recovery as a
function of temperature. A similar technique has been used previously by Kharchenko et al.[3-5] and Herzer[2] but over a more limited
fluence range. Experiments were performed in an attempt to re-
produce and to extend these results and to determine which damage
mechanisms were responsible for the various annealing stages
observed. The isochronal annealing of NTD high purity detector
grade silicon is presented here for the first time.

2.   EXPERIMENTAL PROCEDURES

Irradiation was performed at the University of Missouri Research
Reactor Facility (MURR) in Columbia, Missouri. This facility is a
10 megawatt U-235 source in a light water pool with beryllium and
graphite reflectors. Samples were irradiated in a variety of
positions in both the graphite reflector and in the pool to give
cadmium ratios of between 10:1 and 30:1. Neutron doses were determined using self-powered detectors and cobalt doped aluminum flux
wires.

Isochronal annealing was performed in a Spectrosil quartz tube
using an argon atmosphere flowing at a rate of about 15 cc/min.
Annealing steps were 15 minutes long. The temperature was increased
25°C for each step. This annealing was done using a Hoskins electric
furnace and an Omega 49-818 temperature controller with a chromel-
alumel thermocouple. A temperature stability of ±5°C over the
annealing range of 50°C - 850°C was attained using this apparatus.

Electrical resistivity was determined using a 4-point probe
(Signatone SP4 625-180 TC SR) with a Keithley 225 current-source
and a Hewlett-Packard 3465A Digital Multimeter. For resistivities
in excess of 10^4 ohm-cm, a Keithley 602 Electrometer was used as a
unity gain amplifier to increase input impedance and reduce the time
constant. Conduction type determinations were made using the thermal
probe method and utilizing an Isotip soldering iron and a Keithley
602 Electrometer.

Thermal neutrons were shielded from the sample in some experiments to investigate the effects of fast neutron damage by using
a boron rich sample container. Beta recoil damage was isolated

by irradiation in a pneumatic tube facility which provided fast
sample retrieval, followed immediately by a 15 minute anneal at
850°C. The sample was then allowed to cool to room temperature
and the remaining β^- activity used to self-redamage the sample.

3. RESULTS AND DISCUSSION

An examination of the dependence of isochronal annealing on
thermal neutron dose is shown in Fig. 1. All three samples were
cut from the same high purity float zone ingot (5250 ohm-cm, p-
type), and received doses of 1.67×10^{16} n/cm^2 for (0); 3.56×10^{17} n/
cm^2 for (Δ); and 1.57×10^{19} n/cm^2 for ($\bullet$). In order to perform
these irradiations over three orders of magnitude, it was necessary
to use positions of different flux and different thermal-to-fast
ratios. The cadmium ratios of those positions used for sample
irradiation were 30:1, 12:1, and 10:1 respectively.

All three samples of Fig. 1 were p-type at the start of the
anneal and converted to n-type at some higher temperature. In fact,
a thermal probe check of 50 ingots irradiated to a variety of doses
always indicated p-type after irradiation in contrast to Ref. 3.
The type conversion annealing temperature depended upon the neutron
dose and, therefore, was dependent upon the amount of transmuted
phosphorus added. Sample (0) which was irradiated to balance the
boron concentration with phosphorus, never fully type converted,
but attained mixed conduction type above 800°C. Samples (Δ) and
($\bullet$) type converted at 725°C and at 425°C respectively. This
suggests that the transmuted phosphorus does not become electrically
active until higher annealing temperatures and that this temperature
is lower for larger phosphorus concentrations.

At about 600°C an extremum in the resistivity can be observed
in Fig. 1. We identify this extremum with the production of
acceptors resulting from damage plus annealing and will call these
defect structures "600°C-acceptors." Since the initial resistivity
was 5250 ohm-cm ($p = 2.4 \times 10^{12}$ cm^{-3}), the resistivity minimum in
samples (0) and (Δ) corresponds to an acceptor concentration of
$\sim 6.3 \times 10^{13}$ cm^{-3} or an increase of about 6×10^{13} cm^{-3}. This concentration is at least an order of magnitude higher than any
electrically active impurity concentrations in the starting material.
Sample ($\bullet$) is not as easily analyzed since it is n-type at 600°C,
but if we assume that the acceptor has balanced the donor concentration at 600°C to raise the resistivity to the intrinsic value,
we need only approximate the resistivity we would expect if the
acceptor were absent. A straight line drawn from the resistivity
value at 475°C to the value at 675°C yields a resistivity value at
600°C of about 90 ohm-cm or a donor concentration of about 5×10^{13}
cm^{-3}. Since the "600° acceptor" has balanced these donors, the
concentration of the "600° acceptor" is also about 5×10^{13} cm^{-3}.

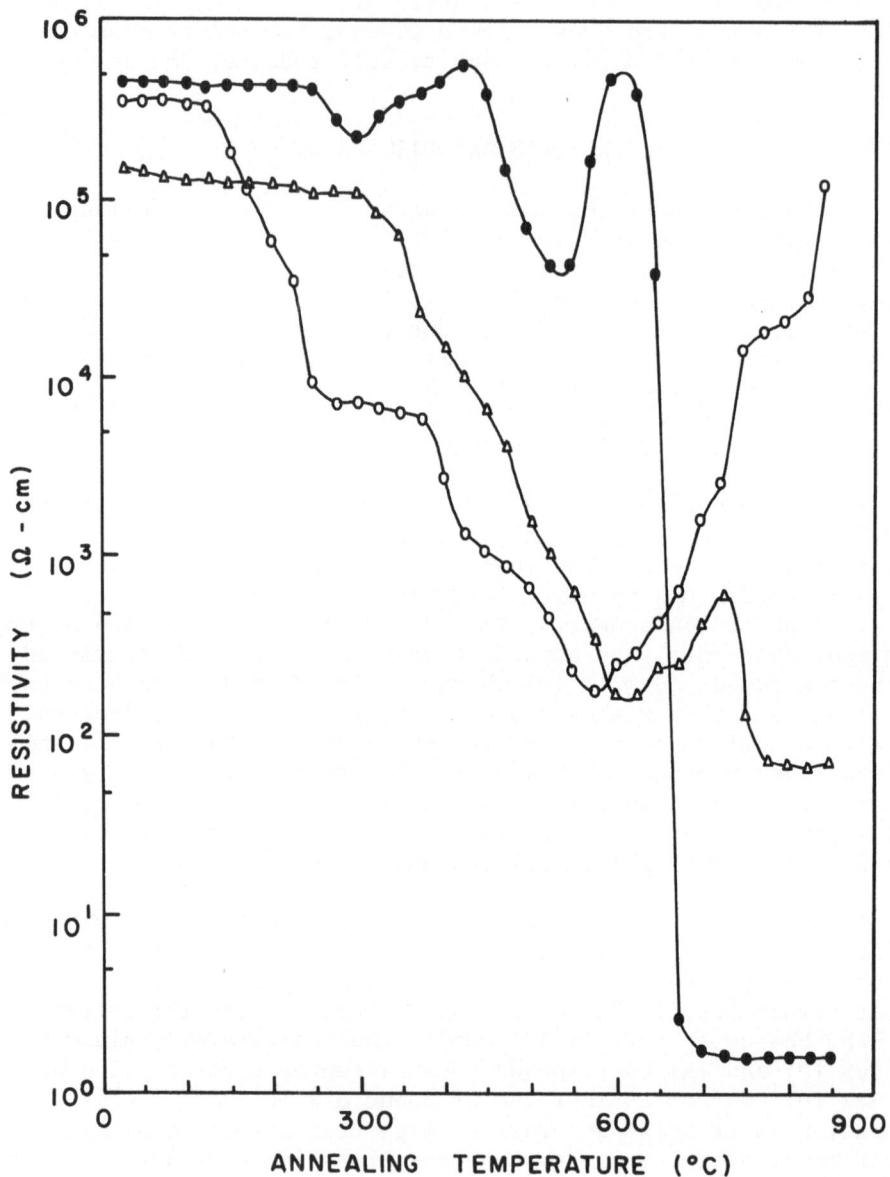

Fig. 1.   Dose dependence of isochronal annealing of detector grade
          NTD float zone silicon. Initial resistivities before
          irradiation were approximately 5250 Ω-cm p-type. Thermal
          neutron doses and type conversion (p→n) temperatures were
          as follows: (O)=1.67×10^{16}n/cm^2, mixed type at 800°C; (Δ)=
          3.56×10^{17}n/cm^2 725°C; (●)=1.57×10^{19}n/cm^2, 425°C.

This rather crude analysis has shown that the "600°C acceptor" is
not strongly dependent on a neutron dose. In samples from other
sources, we have observed that the concentration of "600°C acceptors"
is an order of magnitude higher suggesting that this defect complex
is sample dependent rather than dose dependent.[10] Shallow acceptors
resulting from annealing to these temperatures, in concentrations
of between 10^{12} and $10^{15} cm^{-3}$, have been observed by other authors.[11-13] It should also be noted that the isochronal annealing
for these float zone samples shows considerably more annealing
structure than has been previously reported for float zone.[2-5]

The question of cadmium ratio was examined by irradiating three
samples in the same position, with and without thermal neutron
shielding. Figure 2 shows the results for samples cut from the
same ingot of float zone silicon (5500 ohm-cm, p-type). This ingot
differs from that used for the experiments shown on Fig. 1. Samples
(0) and (Δ), which were shielded from thermal neutrons, remained
p-type during the entire anneal, while the sample which was not
boron shielded type converted from p-type to n-type between 700°C
and 725°C apparently due to transmuted phosphorus. Two qualitative
observations can be made concerning the anneals of Fig. 2. First,
the major annealing features, other than a slight shift in the
curves, do not seem to depend upon the presence of thermal neutrons
during irradiation. Secondly, the temperature at which the resistivity minimum occurs is lower for smaller fast neutron doses.
Again the "600°C acceptor" concentration is $\sim 6 \times 10^{13} cm^{-3}$ and is
apparently due to fast neutron damage. It should be noted that a
study of damage due to recoil effects would be difficult in the
presence of the fast neutron damage since this damage tends to
dominate the annealing characteristics.

Figure 3 shows the comparison of isochronal annealing of a
Czochralski silicon sample (50 ohm-cm, n-type), thermal neutron
dose = $3.80 \times 10^{18} n/cm^2$ to the heavily irradiated float zone sample
shown previously in Fig. 1. The two anneals are surprisingly similar
both in the type conversion point from p-type to n-type at 400°C
and in the rise in acceptor concentration to about $5 \times 10^{13} cm^{-3}$ at
600°C. We can directly compare these results to those of Kharchenko
et al.[3-5] While the results for Czochralski silicon agree quite
well, those for float zone are very different. Furthermore, their
float zone sample was initially n-type after irradiation while ours
are p-type.[3] Since our results show only a weak cadmium ratio
dependence above 600°C, we believe these differences are possibly
due to differences in irradiation temperature.

In measuring the resistivity after each annealing step, a rise
in the resistivity as a function of time was noted. A measurement
of this rise was taken for the heavily irradiated float zone sample
of Fig. 1. Data for the first 60 minutes after cooling the sample
to room temperature is shown in Fig. 4 and is arranged in order of

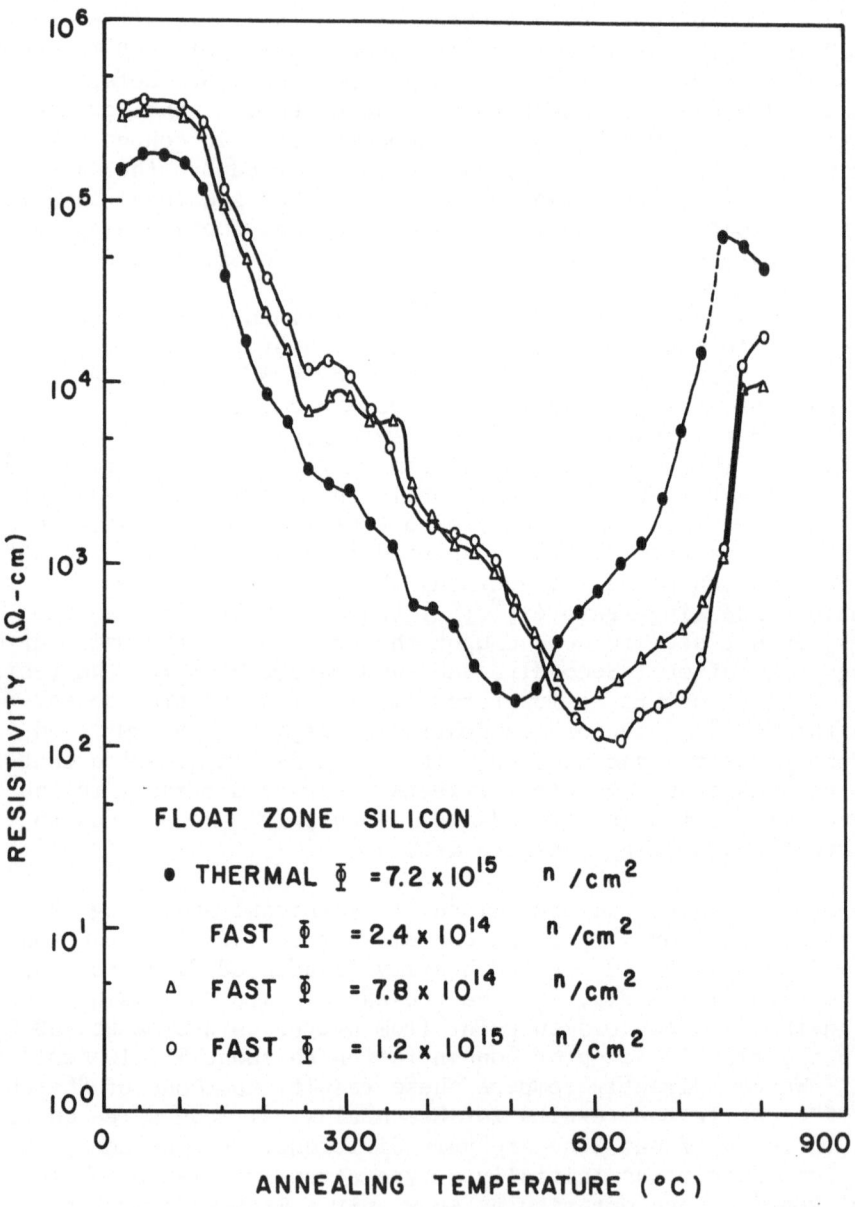

Fig. 2. Comparison of the effects of various fast-to-thermal neutron ratios on the isochronal annealing of detector grade
silicon. Initial resistivity was approximately 5500 Ω-cm,
p-type. Boron shielding was used for samples (O) and ($\triangle$)
while no boron shielding was used to irradiate sample ($\bullet$).
Sample ($\bullet$) type converted at $\sim$ 715°C.

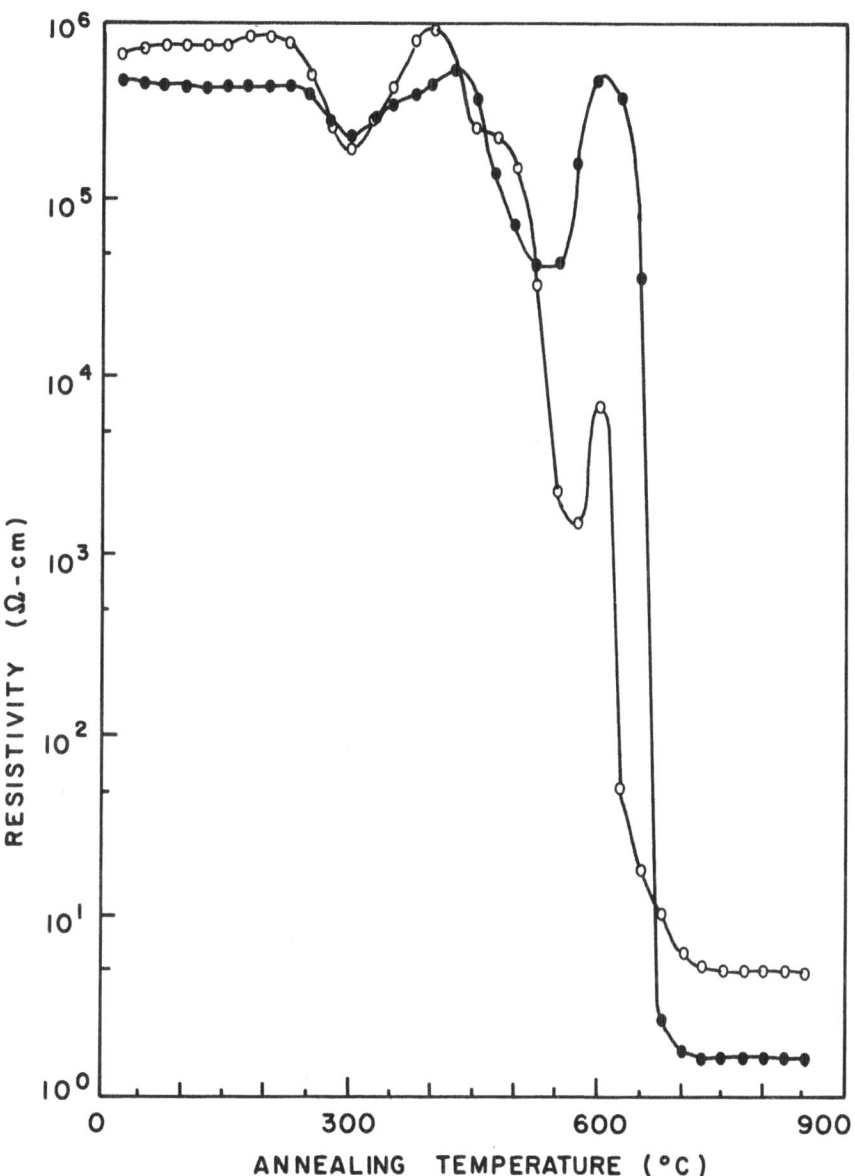

Fig. 3. Comparison of isochronal annealing of heavily neutron
doped float zone (●) and Czochralski (0). Type conversion of both samples occurred in the 400°C annealing peak.

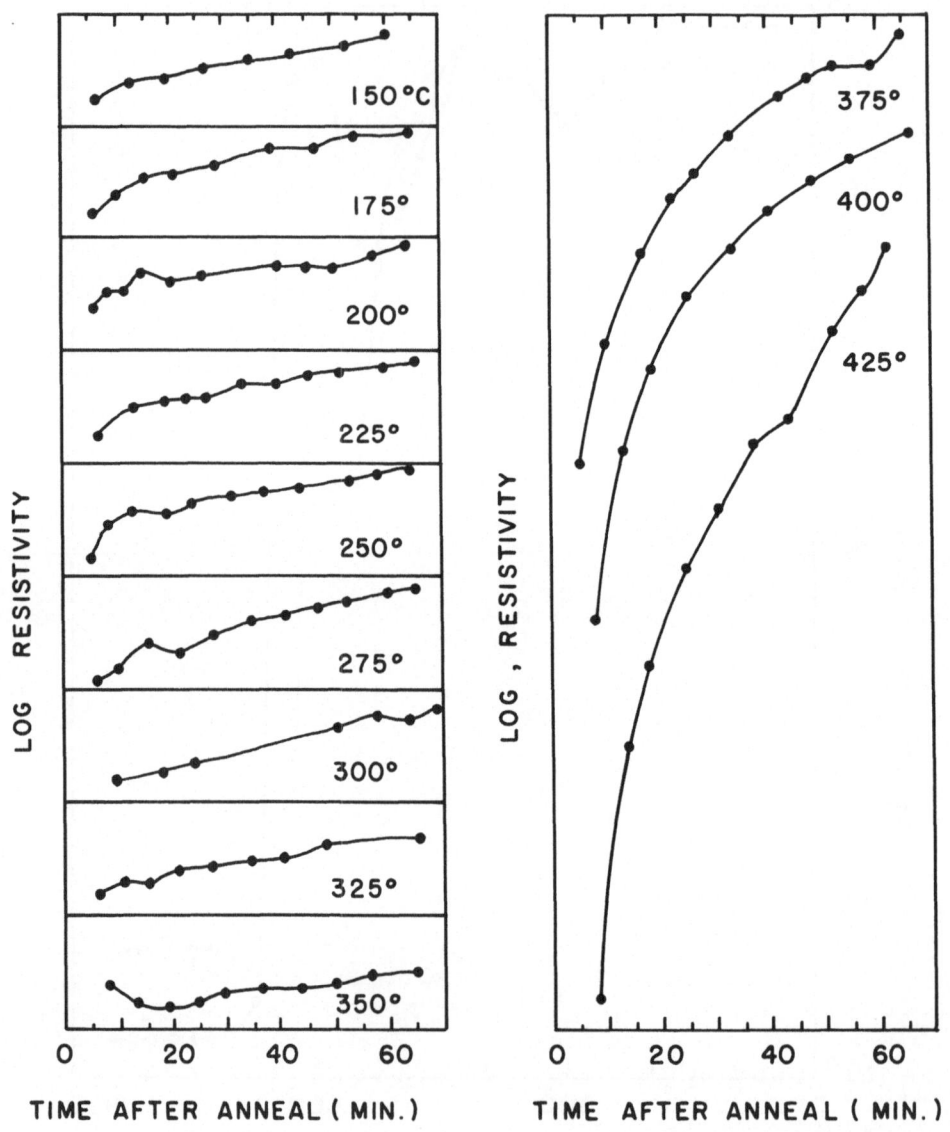

Fig. 4a. Resistivity instability as a function of time after anneal-
          ing in heavily irradiated float zone Si ($\Phi=1.57 \times 10^{19} n/cm^2$;
          $150°C \leqslant T \leqslant 425°C$).

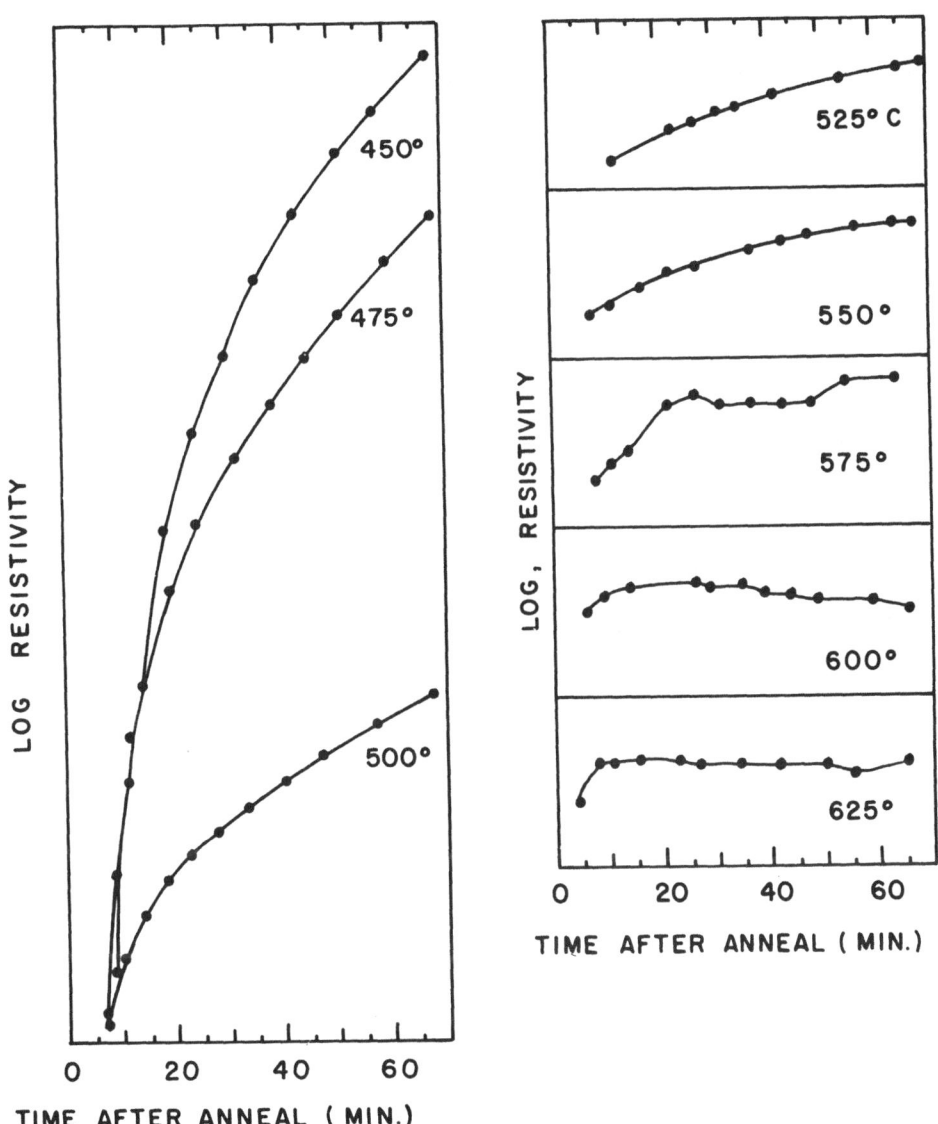

Fig. 4b. Resistivity instability as a function of time after anneal-
          ing in heavily irradiated float zone Si ($\Phi=1.57 \times 10^{19} n/cm^2$;
          $450°C \leqslant T \leqslant 650°C$).

annealing temperature (this figure does not indicate the absolute change in resistivity between annealing steps, however, all relative changes are to the same scale). A large effect can be noted between 375°C and 500°C which corresponds to the type conversion peak.

While this effect may have something to do with the initiation of phosphorus electrical activity, such a conclusion cannot be drawn without direct examination of energy levels in the band gap. The mechanism for this resistivity instability is not understood at this time. We have observed this effect in samples using soldered contacts as well as probes and in etched samples as well as lapped. It should also be noted that this effect is very large from 375 - 500°C but is negligible in the annealing range of 550-650°C even though the resistivity is similar in magnitude in these two anneal- ing peaks. This effect might be due to carrier trapping at defects with extremely small capture cross section, to very slow room temperature migration of defects freed from complexes during the anneal, or even due to electrical activation of donor oxygen com- plexes which are normally only observed in Czochralski. Further experimentation will be required to identify the proper mechanism.

Isochronal annealing of more highly doped samples ($N_A \geq 10^{16}$ cm^{-3}) was performed on gallium doped float zone silicon irradiated to [P]/[B] ratios of about 1:1, 2:1, and 3:1 in order to compensate the residual boron concentration ($N_B \sim 5 \times 10^{12} cm^{-3}$) without un- duly compensating the gallium concentration. The annealing results are shown in Fig. 5. Thermal neutron doses were $3.22 \times 10^{16} n/cm^2$ for (0); $6.44 \times 10^{16} n/cm^2$ for (Δ); and $9.66 \times 10^{16} n/cm^2$ for ($\bullet$). These anneals showed essentially no structure and had no indication of a "600°C acceptor." While acceptors have been observed at 600°C using Hall effect versus temperature techniques on similar gallium doped samples,[11] room temperature resistivity is not sensitive to these defects in purposely doped material due to concentration masking.

Figures 6 and 7 are isochronal anneals of two float zone silicon samples which had been beta recoil damaged by the technique described in Section 2. The sample of Fig. 6 (1200 ohm-cm, p-type) was irra- diated to a thermal neutron dose of $3.84 \times 10^{16} n/cm^2$ making the boron concentration approximately equal to the concentration of transmuted phosphorus. In Fig. 7, the sample (5250 ohm-cm, p-type) had a dose of $1.92 \times 10^{17} n/cm^2$, making the phosphorus to boron ratio about 10:1. In contrast to previous samples which were p-type as a result of fast neutron damage, both of the samples were n-type after the recoil damage was completed. Since both samples are n- type at the start of the anneal, we can make an estimate of the donor defect concentration due to beta recoil events by assuming that the mobility has not drastically changed. The starting mate- rial in Fig. 6 has $N_a \simeq 1 \times 10^{13} cm^{-3}$, and after beta recoil damage $n \simeq 1.8 \times 10^{12} cm^{-3}$, so $N_d \simeq N_a + n = 1.18 \times 10^{13} cm^{-3}$. In Fig. 7

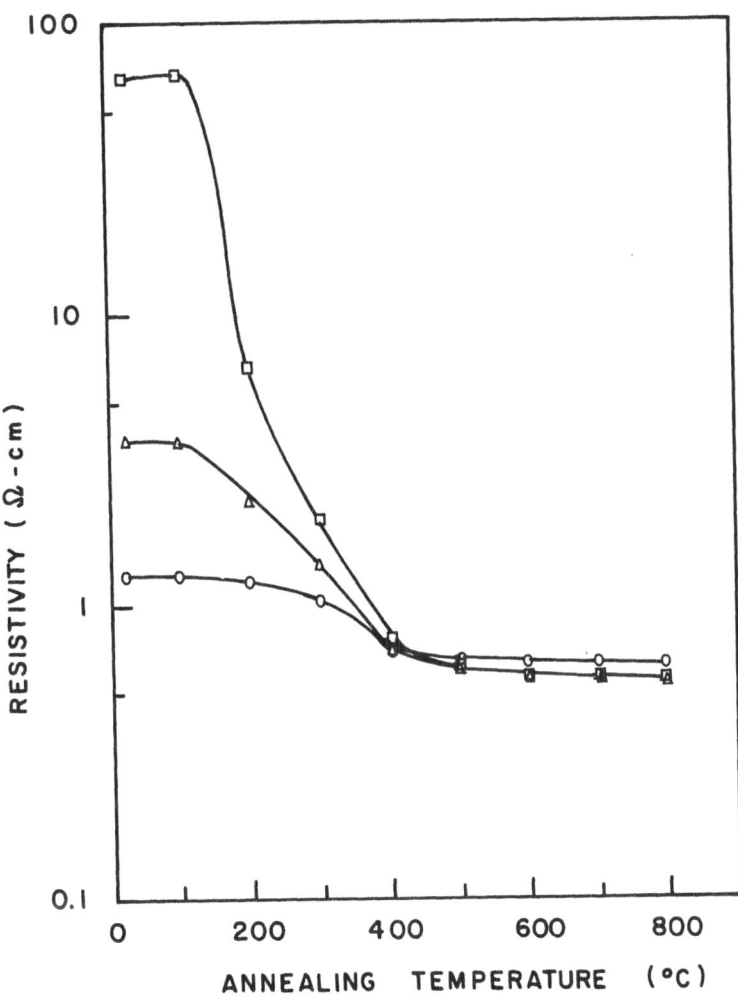

Fig. 5.   Isochronal annealing of Ga doped ($\sim 10^{16}\text{cm}^{-3}$) float zone
neutron doped to P/B ratios of 1:1=(0), 2:1=(Δ) and 3:1
for ($\bullet$). The boron concentration in these samples was
approximately $5 \times 10^{12}\text{cm}^{-3}$.

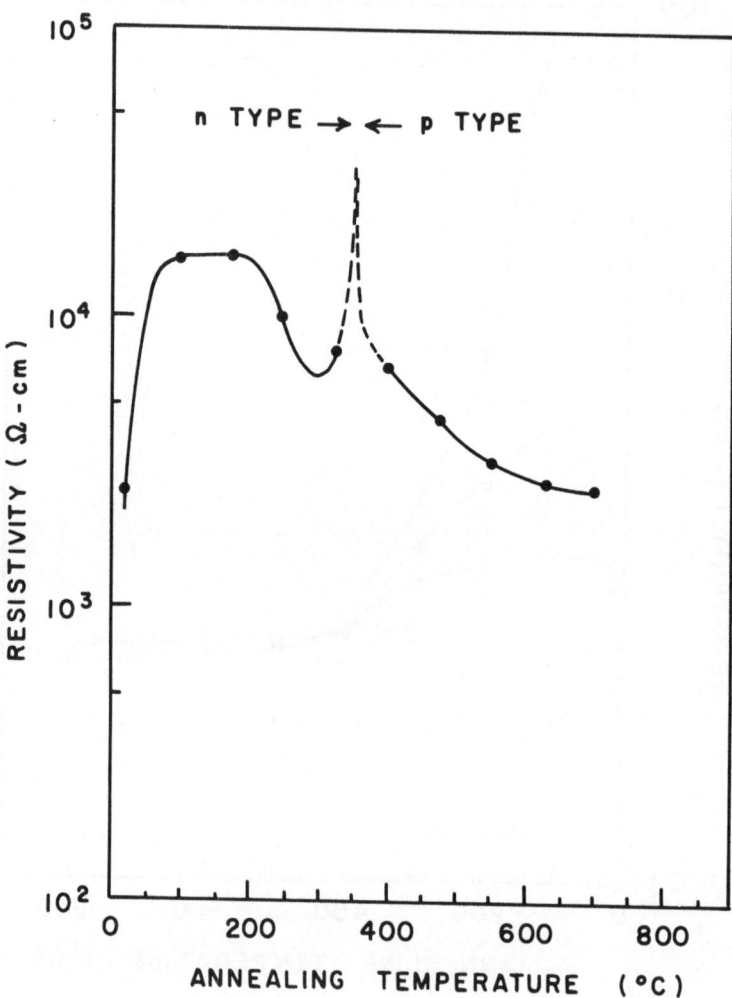

Fig. 6. Isochronal annealing of beta recoil damaged float zone
with an initial acceptor concentration of $\sim 1 \times 10^{13} \mathrm{cm}^{-3}$ and
a thermal neutron dose of $\sim 3.84 \times 10^{16} \mathrm{n/cm}^2$.

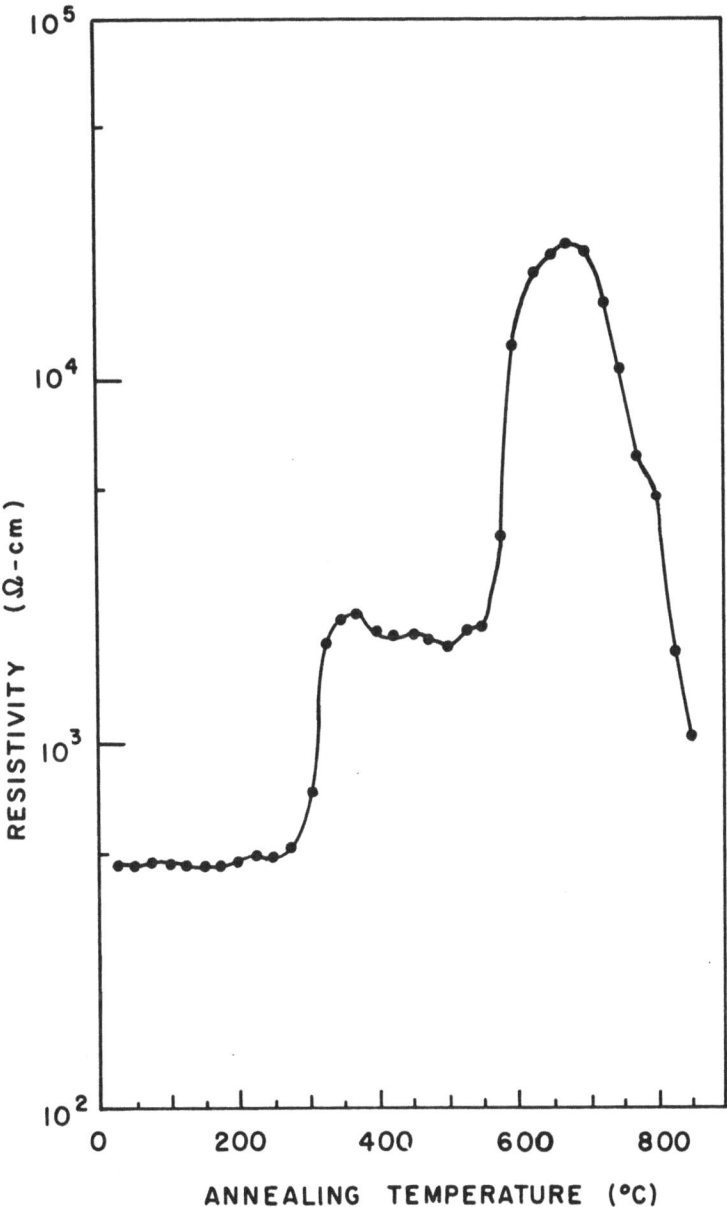

Fig. 7. Isochronal annealing of beta recoil damaged float zone
with an initial acceptor concentration of $\sim 2.4 \times 10^{12} cm^{-3}$
and a thermal neutron dose of $\sim 1.92 \times 10^{1.7} n/cm^{2}$. This
sample remained n-type throughout the anneal.

$N_a \simeq 2.4 \times 10^{12} cm^{-3}$, $n \simeq 8.95 \times 10^{12} cm^{-3}$, so $N_d \simeq 1.14 \times 10^{13} cm^{-3}$.
Therefore, a donor concentration of about $10^{13} cm^{-3}$ seems to be pro-
duced by the beta recoil damage in both samples. Other features of
these anneals are not as directly comparable and apparently depend
upon either the boron to phosphorus ratio or dose. We can see,
however, the appearance of an acceptor or the disappearance of the
initial donor in annealing between 300°C and 400°C in both samples.

4. SUMMARY AND CONCLUSIONS

In summary, we have surveyed the isochronal annealing of both
float zone and Czochralski NTD silicon under a variety of irradiation
conditions. A discrepancy with some of the previous literature[3] has
been found for NTD float zone both as to the carrier type after
irradiation but before annealing and in the amount of annealing
structure and final annealing temperatures.

We have found that the dominate fast-neutron-damage plus-
annealing effect, in terms of the concentration of electrically
active defects, is the production of the "600°C acceptor." Although these acceptors have been reported by others in float zone
material, we have been able to show that this defect also occurs
in Czochralski, that its concentration is reasonably independent of
dose, and hence phosphorus concentration, and that the concentration
of those defects appears to be sample dependent. Since this defect
occurred in high purity silicon in concentrations of $5 \times 10^{13} cm^{-3}$ or
higher, it is clear that these defects do not involve impurities
other than perhaps oxygen or carbon.

We have also observed an instability in the resistivity as a
function of time after anneal which is most pronounced in the
temperature range of 375-500°C. We have suggested several possible
mechanisms for this resistivity instability, however, further experimentation will be required to determine which, if any, of these
models is true.

Finally, we have been able to, for the first time, isolate the
damage and annealing effects of beta recoil damage from the other
damage mechanisms in the NTD process. This damage mechanism clearly
produces donors before annealing. By comparing two samples with
different initial doping conditions, we have been able to show that
the "beta recoil donor defect" concentration is of the order of
$10^{13} cm^{-3}$, again about an order of magnitude larger than the largest
of the electrically active impurities in detector grade material.

REFERENCES

*   Sponsored in part by Air Force Materials Laboratory, Air Force Systems Command, United States Air Force, Wright-Patterson AFB, Ohio 45433 under Contract No. F33615-76-C-5230.

1)   M. Tanenbaum and A. D. Mills, J. Electrochem. Soc. 108, 171 (1961).

2)   H. Herzer, Semiconductor Silicon 1977, ed. by H. R. Huff and E. Sirtl, The Electrochemical Society, Princeton, N.J. (1977), p. 106.

3)   V. A. Kharchenko and S. P. Solov'ev, Izv. Akad. Nauk USSR, Neorgan. Mat. 7, 2137 (1971).

4)   V. A. Kharchenko and S. P. Solov'ev, Sov. Phys. Semicond. 5, 1437 (1972).

5)   V. N. Modkovich, S. P. Solov'ev, E. M. Temper, V. A. Kharchenko, Sov. Phys. Semicond. 8, 139 (1974).

6)   H. A. Herrmann and H. Herzer, J. Electrochem. Soc. 122, 1568 (1975).

7)   E. W. Haas and M. S. Schnoller, IEEE Trans. on Electron Devices ED-23, 803 (1976).

8)   H. J. János and O. Malmros, IEEE Trans. on Electron Devices ED-23, 797 (1976).

9)   M. J. Hill, P. M. Van Iseghem and W. Zimmerman, IEEE Trans. on Electron Devices ED-23, 809 (1976).

10) J.M. Meese, Silicon Detector Compensation by Nuclear Transmutation, Technicial Report AFML-TR-77-178, Air Force Materials Laboratory, Wright-Patterson AFB, Ohio 45433, (February 1978).

11) M. H. Young, O. J. Marsh and R. Baron, Bull. Am. Phys. Soc. 22, 1241 (1977).

12) R. T. Young and J. W. Cleland, Bull. Am. Phys. Soc. 20, 328 (1977).

13) J. Goldberg, Appl. Phys. Lett. 31, 578 (1977).

ELECTRON SPIN RESONANCE IN NTD SILICON

L. Katz and E. B. Hale

University of Missouri-Rolla

Rolla, MO 65401

ABSTRACT

Neutron transmutation doped silicon has been studied and characterized by electron spin resonance. Two samples were irradiated
for the same final target resistivity (25 ohm-cm) in the University
of Missouri Research Reactor. One was in a position with high flux,
and the second was exposed to a lower flux. The room temperature
ESR spectrum as a function of anneal temperature for both samples
has been obtained from the irradiation temperature up to 600°C.
A number of ESR centers have been found. The lattice defect configuration corresponding to some of these centers has previously been
identified and is discussed. The data indicates that ESR measurements are a non-destructive and convenient way to obtain the
sample's temperature during irradiation.

1. INTRODUCTION

When silicon is irradiated with neutrons, a variety of point
defect structures are produced in the crystal. The microscopic
structure of these defects can be determined by electron spin resonance[1] (ESR). (ESR has been an essential technique for understanding many important radiation damage centers and processes in
silicon.[2-5]) Although ESR can only measure defects containing one
or more electron(s) with an unpaired spin, such defects are rather
common in silicon. In fact, the detailed analysis of ESR data has
permitted the assignment of lattice configurational models to a
variety of rather complex defects, several of which will be discussed in Section 4.1. In this paper, the ESR signals observed from
NTD silicon are reported both after irradiation and after various

annealing stages. Signals from a number of different defect
centers have been found, and previously reported models for these
centers are discussed. In addition, of special interest to several
attendees at this conference was the discussion (Section 4.2) on
how ESR measurements can be used to determine the irradiation temperature variations at various positions and under various conditions in a reactor.

2.   EXPERIMENTAL DETAILS

Two irradiated samples were kindly provided by Dr. Jon Meese
of the University of Missouri Research Reactor Facility. Both
samples were float zone silicon, grown by Topsil. Each was placed
in an aluminum can during irradiation and exposed to both thermal
and fast neutrons. Both were irradiated to a total fluence of
$1.05 \times 10^{18} n/cm^2$, corresponding to a target (high temperature)
resistivity of 25 ohm-cm. The difference between the samples was
in their irradiation positions.

Sample #1 was exposed to a higher flux in the reactor.

Sample #2 was exposed to a lower flux just out beyond the
reflector in the pool (S-basket position).

The flux on Sample #1 was about $3 \times 10^{13} n/cm^2/sec$ with a Cd ratio
of thermal to fast neutrons of about 10:1. The flux on Sample #2
was about $10^{12} n/cm^2/sec$ with a Cd ratio a factor of two or three
higher.

Several days after irradiation ESR samples were cut from the
boules. They had approximate dimensions of 3 mm x 3 mm x 10 mm.
ESR measurements were made at room temperature using a standard X-
band Varian Associates EPR spectrometer. The samples were rotated
about the long dimension which was cut as the $[1\bar{1}0]$ axis so that
the [001], [111], and [110] axes could all be aligned along the
magnetic field which was perpendicular to the $[1\bar{1}0]$ rotation axis.
The signal intensity at each anneal temperature was normalized to
a ruby standard in the ESR cavity.

Anneals were carried out in a quartz tube with flowing argon.
All anneals were for ten minutes at the specified temperature, which
does not include the five minute warm-up time or the cool-down time.

3.   RESULTS

As the magnetic field was swept, numerous ESR lines of various
intensities were seen. In addition, the position of these lines in
the spectra changed substantially with small changes in the sample
orientation. Often, many lines belong to one particular defect

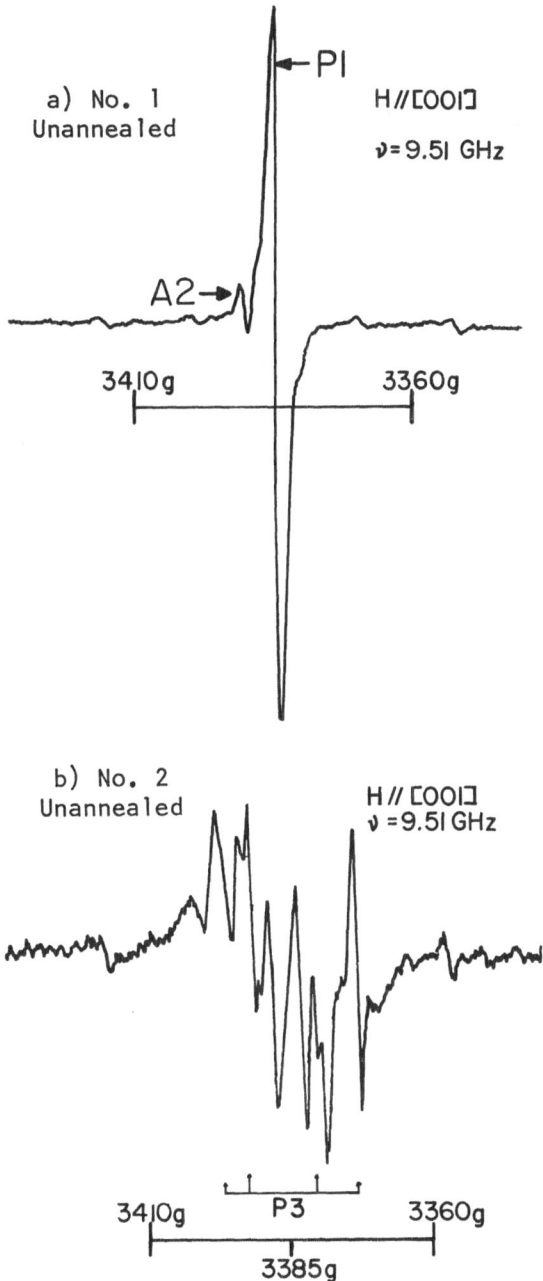

Fig. 1.   ESR spectra for unannealed samples with H along [001] axis.
(a) The spectrum from sample #1 is dominated by a line due
to center P1 with a small line due to center A2.   (b) The
spectrum from sample #2 is quite complex, but the strongest
lines are from center P3.

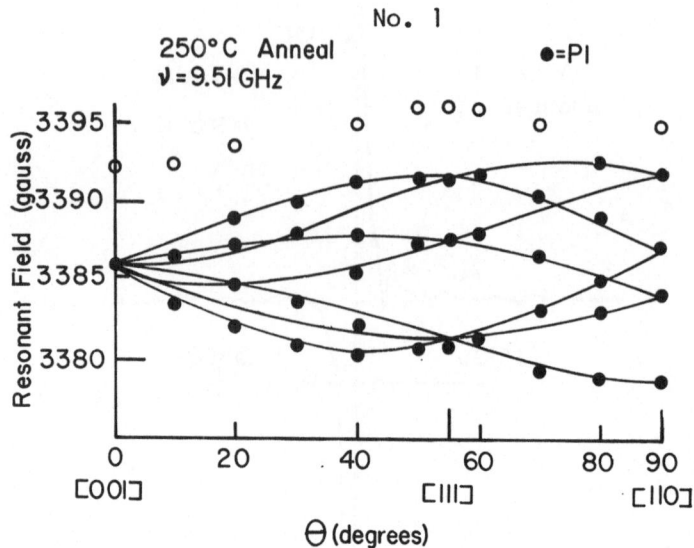

Fig. 2. Angular dependence of ESR spectra from sample #1. The
solid curved lines are computed from the g-tensor known
for P1. (See Ref. 7.)

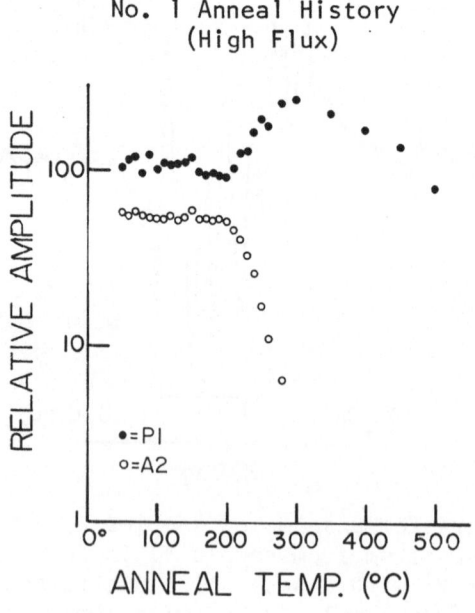

Fig. 3. Anneal history of sample #1. Large changes in the ESR
amplitudes from both the P1 and A2 centers occur after
the 200°C anneal.

since there can be a number of symmetrically equivalent orientations
of the defect in the crystal which are no longer equivalent once
the magnetic field has established its direction. Fortunately,
along the three principal directions, i.e., [001], [111], and [110],
many of the lines become degenerate because of the higher crystal
symmetry. Typically, the least number of lines is seen along the
[001] direction.

Sample #1 showed a rather simple unannealed spectra along [001]
which is shown in Fig. 1a. The complete angular spectra was taken
and it shows the strong line at [001] breaks up into several lines
at other angles as in Fig. 2. From this angular pattern of this
strong line spectra one can determine a unique g-tensor, which is
actually a characteristic fingerprint for this point defect in the
crystal. This strong line g-tensor has been previously observed in
neutron irradiated silicon and has been designated P1, (originally,
N-center[6]). The solid lines in Fig. 2 are the theoretically predicted line positions assuming the previously measured[6] g-tensor
for P1, and confirm that our signals are from the P1 defect center.
(The microscopic model for this defect will be discussed in the
next section.)

Sample #2 showed a considerably more complex unannealed
spectra, even along the [001] axis as shown in Fig. 1b. This
spectra basically remains unchanged for the various anneals up
through the 170°C anneal with the strongest lines in Fig. 1b no
longer being observed. The higher temperature anneals yielded
spectra the same as for sample #2.

An angular dependent plot, such as in Fig. 2, was also made
for the unannealed sample #2. From this plot it was determined
that most of the lines are from a defect designated P3, (originally[7] center (II, III)). In addition, low concentrations of the
P1 center and another defect designated[8] A2 were found. (Center
A2 was also found in sample #1 as the "shoulder" line in Fig. 1a.)

Once the various centers have been identified, the change in
relative concentrations with anneal temperature can be easily
determined. Figures 3 and 4 are the anneal histories we have
determined by plotting the relative amplitudes of the ESR lines.
These amplitudes have been normalized to take into account symmetry
degeneracies. The amplitude of P3, a spin one center, is the sum
of both branches. These amplitudes are roughly proportional to
defect concentration, but are not precise concentrations because of
different linewidths and hyperfine interactions.

Figure 4 shows that the irradiation temperature of sample #2
was between 170°C and 190°C, while Fig. 3 shows that the irradiation
temperature of sample #1 was between 200°C and 210°C. Hence,

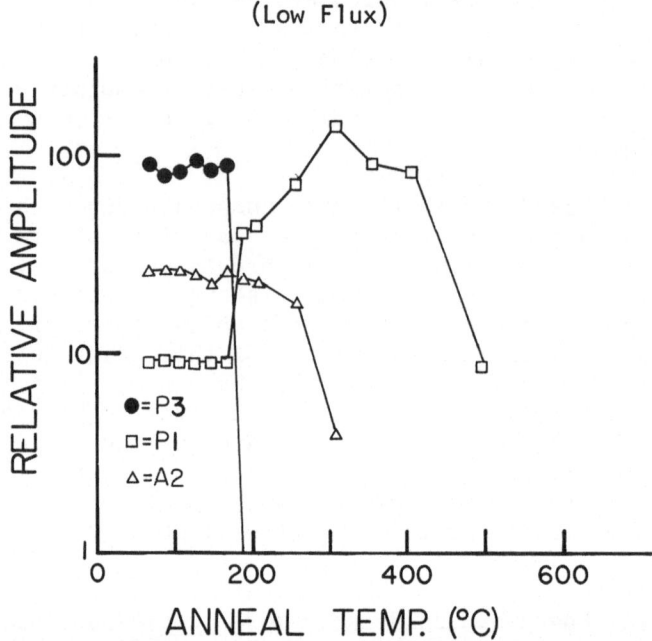

Fig. 4.   Anneal history of sample #2. Drastic changes in the ESR
          spectra occur at the 190°C anneal. The P3 spectra is no
          longer detected, and amplitude changes in the P1 and A2
          center also start to occur.

sample #1 which was exposed to the higher flux nearer the core was
at a higher irradiation temperature, as might be expected. In the
next section we estimate the temperature difference between the two
samples to be approximately 30°C.

4.   DISCUSSION

4.1. Microscopic Model

     Our results are consistent with others[6-9] which show that the
P3 center is the dominant paramagnetic defect at lower irradiation
temperature ($\leq$170°C), while the P1 center dominates at higher
irradiation temperatures. Configurational models for the primary
defects have been proposed[10-11] since they are generically related
to a group of defects. All the defects in this family have for
their basic structure a chain of missing silicon atoms.[7,3,10] The
vacancies and missing {111} bonds in this chain form a "sawtooth"
arrangement as shown in Fig. 5a. (In some members of the family,

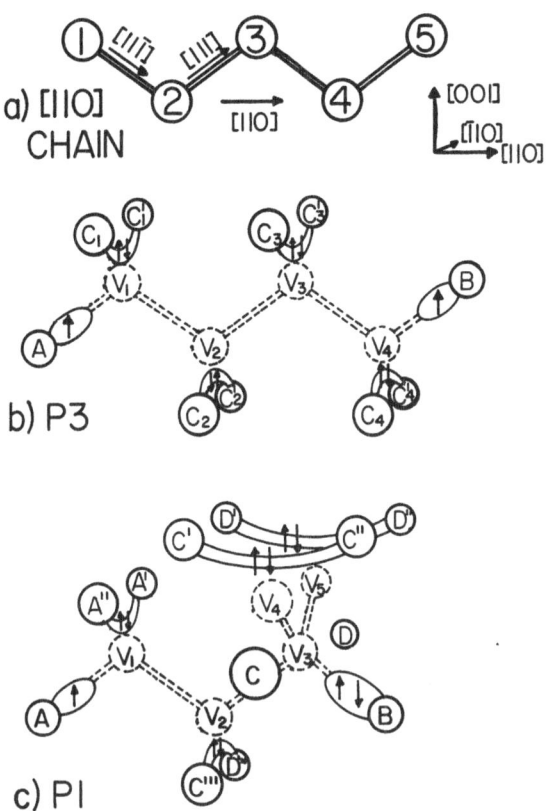

Fig. 5.   Lattice defect models. (a) Defects produced in neutron
          irradiated silicon often occur in a sawtooth chain con-
          figuration of vacancies. (b) The P3 center is a tetra-
          vacancy chain with two electrons having parallel spins
          located principally in orbitals at the end of the chain.
          It is the dominant paramagnetic defect for irradiation
          and anneal temperatures < 170°C. (c) The P1 center is a
          pentavacancy cluster and the dominant paramagnetic defect
          for irradiation and anneal temperatures > 170°C. It can
          be considered a trivacancy chain (V_1, V_2, V_3) with two
          side linked vacancies (V_4, V_5). It is a spin one-half
          defect with the unpaired electron primarily associated
          with the A atom at the end of the chain. (The extended
          orbital between the C and D atoms is not shown in the
          figure.)

an oxygen atom is located close to a vacancy site.) In addition, the electron configuration near the chain changes substantially since many electrons can no longer pair up in the tetrahedral bonds which were attached to the now missing atoms. About half the electrons attached to the nearest neighbor silicon atoms surrounding the chain remain associated with the defect for approximate charge neutrality. Some of these electrons pair off to form a "bent bond" spanning a region where two tetrahedral bonds had previously existed. Other electrons pair off at greater distances into various "extended bonds." The unpaired electron(s) studied by ESR remain predominately in orbital(s) associated with the end silicon atom(s) and extending into the vacancy chain. If only one unpaired electron exists, then a spin one-half center is observed with a g-tensor which reflects the lattice symmetry about the defect. If two unpaired electrons exist, then a spin one center is observed with not only a g-tensor, but also a spin coupling D-tensor. This tensor is important because it not only reflects the defect symmetry, but also yields the separation distance between the atoms at the ends of the chain.[10]

The smallest defect* that is a member of this family is the S1 center,[15] a spin one center which is an excited state of a single vacancy with a nearby oxygen. The two parallel spin electrons are in bent orbitals which replace the two tetrahedral bonds previously along the chain. (In Fig. 5a the vacancy would be at site 2 with the two unpaired electrons attached to the silicon atoms at sites 1 and 3.) The spin one-half divacancy in two charge states (the G6 and G7 centers[16]) have also been reported. The spin one P2 center[7] is believed to be a divacancy with oxygen.[10]** The spin one P4 center[7] and P5 center[7] are believed to be trivacancy chains with oxygen.[10]** Of importance here is the tetravacancy chain[10] which is associated with the dominant defect in the low temperature irradiated NTD silicon, the spin one P3 center.[7] The model for this center is shown in Fig. 5b with its four vacancies and the two unpaired electrons "dangling" into the chain from the two end silicon atoms.

Finally, the most complex defect so far identified is the dominant high temperature NTD defect, the spin one-half P1 center.[10] The model[11] is a pentavacancy with three vacancies forming the chain and the two other vacancies "attached" as symmetric side links. This is shown in Fig. 5c. At low measuring temperatures (< 100°K), the unpaired electron is "dangling" from the chain.[11] At higher measuring temperatures, this electron is thermally excited into a more extended orbital which encompasses the silicon atom at the side-linked end of the chain.[11]

If the above models are correct, then the major annealing stage near 170°C would be attributed to liberation of vacancies from the tetravacancy chain, i.e., P3 center. The vacancies then

become stabilized in a more clustered configuration, i.e., the P1
center, rather than in a more extended chain configuration. The
clustered vacancies, such as the P1 center or voids in amorphous
silicon eventually anneal near 500-600°C.

4.2. Irradiation Temperature

It is clear from Fig. 3 that sample #1 does not have a
spectral change until the 210°C anneal, and hence its irradiation
temperature was about 200°C. Similarly, Fig. 4 shows sample #1
had an irradiation temperature less than 190°C and perhaps as low
as 170°C. This last temperature can be determined by comparing
the signals we observe and their relative amplitude, with the
measurements of others.[7-9] By this comparison, the irradiation
temperature is calculated as 165 ± 5°C since P3 and P1 have
comparable amplitudes only over a small temperature interval.
(At lower temperature, the P3 concentration is substantially
greater, and P1 is not observed at all.) We thus believe our
irradiation temperature indicators. Even in temperature regions
where the dominant center concentration is rather stable with
temperature, there are other less concentrated centers which may
be annealing, such as A2 or A3 centers.[9] In such a case, the ESR
data could be taken under measurement conditions which would
optimize the detection of these lesser centers.[8]

We would like to acknowledge the assistance of Dr. Jon Meese
in obtaining and preparing the samples, and also for his interest
in this project.

REFERENCES

*   In some sense, the single vacancy, which has been observed in
      several charge states (the G1 center[12] and G2 center[13]), and
      also when occupied by an oxygen (the B1 center[14]), is the
      simplist member of the family.
**   The P2, P4, and P5 centers are found at the higher anneal
      temperatures (> 250°C) in crucible grown NTD silicon.
1)   R. S. Alger, Electron Paramagnetic Resonance (John Wiley,
      New York, 1968).
2)   G. Lancaster, Electron Spin Resonance in Semiconductors
      (Plenum Press, New York, 1967).
3)   J. W. Corbett in Solid State Physics, ed. by F. Seitz and D.
      Turnbull (Academic Press, New York, 1966), Supp. 7, Chapt.
      III, Sect. 11.
4)   G. D. Watkins, IEEE Trans. on Nuc. Sci. NS-16, 13 (1969).
5)   G. D. Watkins in Lattice Defects in Semiconductors, ed. by
      F. A. Huntley (Inst. of Physics, London, 1975), p. 1.
6)   M. Nisenoff and H. Y. Fan, Phys. Rev. 128, 1605 (1962).

7) W. Jung and G. S. Newell, Phys. Rev. 132, 648 (1963).

8) Y. H. Lee, Y. M. Kim, and J. W. Corbett, Radiat. Effects, 15 77 (1972).

9) E. B. Hale in Semiconductor Research Semiannual Report, ed. by H. Y. Fan. (Purdue University, Lafayette, 1965), p. 19. In this technical report ESR measurements in neutron damaged, float zone silicon annealed at temperatures in the range 100°C to 500°C were presented. A new center, designated P7, was also discovered. This center is the same as center A3 reported in Ref. 8.

10) K. L. Brower, Radiat. Effects 8, 213 (1971).

11) Y. H. Lee and J. W. Corbett, Phys. Rev. B8, 2810 (1973).

12) G. D. Watkins, J. Phys. Soc. Japan 18, Supp. II, 22 (1963).

13) G. D. Watkins in Radiation Damage in Semiconductors, ed. by P. Baruch (Dunod, Paris, 1965), P. 97.

14) G. D. Watkins and J. W. Corbett, Phys. Rev. 121, 1001 (1961) and references therein.

15) K. L. Brower, Phys. Rev. B4, 1968 (1971).

16) G. D. Watkins and J. W. Corbett, Discussions Faraday Soc. 31, 86 (1961). Also J. W. Corbett and G. D. Watkins, Phys. Lett. 7, 314 (1961).

DEFECT LEVELS CONTROLLING THE BEHAVIOR OF NTD SILICON DURING ANNEALING

B. Jayant Baliga and Andrew O. Evwaraye

General Electric Company, Corporate R and D Center

Schenectady, NY   12301

ABSTRACT

The variation of the resistivity and minority carrier lifetime has been examined during the annealing of neutron transmutation doped silicon. During each annealing step deep level transient spectroscopy has also been used to detect the presence of deep lying energy levels within the silicon energy gap. By the use of both p- and n-type crystals with doping levels higher than the phosphorus concentration created by the neutron transmutation process, deep levels lying in both the upper and lower half of the energy gap have been measured.

Although the resistivity anneals to its final value at temperatures around 600°C, it has been found that during isochronal annealing, the minority carrier lifetime continues to increase up to annealing temperatures around 700°C. Under these annealing conditions, four majority carrier trap levels have been found in the n-type wafers and five majority carrier trap levels have been found in the p-type wafers. Measurements of the minority carrier lifetime as a function of ambient temperature indicate that the lifetime controlling recombination center lies at 0.30 eV below the conduction band in the p-type silicon and at 0.30 eV above the valency band in n-type silicon.

1.   INTRODUCTION

Although the neutron transmutation doping (NTD) process was experimentally demonstrated by Tannenbaum and Mills in 1961[1], it has only recently been developed into a viable process for the fabri-

cation of homogeneously phosphorus doped silicon.[2-4] Both high
voltage power devices and detectors fabricated using the homogeneously doped NTD silicon have been shown to exhibit improved
characteristics when compared with conventionally doped silicon.[5-6]
However, during the neutron transmutation process a high density of
defect levels is created in the silicon due to (a) atom recoil after
thermal neutron capture, (b) β-radiation due to the decay of the
^{31}Si isotope after the transmutation reaction, and (c) fast neutron
collisions. It is well known that immediately after radiation, the
silicon has a resistivity in excess of 10^5 ohm-cm and that the
minority carrier lifetime is extremely low. To remove this radiation
damage, high temperature annealing procedures have been studied. Due
to the lack of a sensitive deep level measurement technique, earlier
studies have been confined to the measurement of either the resistivity or the minority carrier lifetime.[7-11] Recently, deep level
transient spectroscopy (DLTS) has been developed as an improved
sensitive technique for the measurement of deep levels.[12] This
technique has also been recently applied to the measurement of deep
levels during the annealing of NTD silicon.[13] However, these measurements were confined to majority carrier trap levels in the upper
half of the silicon band gap. In this paper, measurements of the
resistivity, minority carrier lifetime, as well as the deep levels,
has been conducted for both p- and n-type silicon as a function of
the annealing conditions after neutron transmutation doping. By
the use of both p- and n-type silicon crystals with doping levels
higher than the phosphorus concentration created by the NTD process,
deep levels lying in both the upper and the lower half of the silicon
energy gap have been measured by deep level transient spectroscopy.
By forming p-n junction diodes before neutron irradiation, it has
been possible to detect minority carrier trap levels for the first
time. The use of these junction diodes also allows the measurement
of lifetime with much greater sensitivity and precision than is
feasible with the photoconductivity measurement technique used in
the earlier studies.

2. EXPERIMENTAL PROCEDURE

To evaluate the behavior of neutron transmutation doped silicon
during annealing, three types of starting material were used: (a)
5 ohm-cm, n-type, phosphorus doped, Czochralski grown crystal, (b)
7 ohm-cm, p-type, boron doped, Czochralski grown crystal, and (c)
2000 ohm-cm, n-type, phosphorus doped, float zone grown crystal.
Wafers were cut and polished from each of these crystals prior to
the neutron irradiation. Junctions were formed on some wafers from
the 5 ohm-cm, n-type and the 7 ohm-cm, p-type crystals by the diffusion of boron and phosphorus. All the wafers were then simultaneously irradiated at the M.I.T. Research Reactor to a dosage of
3.5×10^{17}n/cm^2. The irradiation was performed at a location in
the reactor with a thermal-to-fast neutron flux (Cd Ratio) of 250.

After a post-irradiation cooling period of several weeks, the wafers were thoroughly cleaned with solvents and acids to remove any surface contamination.

Both isothermal and isochronal annealing experiments were conducted over a temperature range of 400° to 1000°C in an argon ambient. During each annealing step, the following samples were included: (1) 5 ohm-cm, n-type wafer after NTD, (2) 7 ohm-cm, p-type wafer after NTD, (3) 2000 ohm-cm, n-type wafer after NTD, (4) 5 ohm-cm, n-type wafer with junction after NTD, (5) 7 ohm-cm, p-type wafer with junction after NTD, (6) 5 ohm-cm, n-type wafer without NTD, (7) 7 ohm-cm, p-type wafer without NTD. Prior to conducting any of the annealing experiments, samples of the 5 ohm-cm, n-type and the 7 ohm-cm, p-type crystals with junctions were annealed at 1100°C for several hours in the furnace to check the quality of the annealing procedure. Diodes from these wafers were used for both lifetime and DLTS measurements. The lifetime was found to be about 40 to 60 μsec after this annealing cycle, and the DLTS ·spectra of these devices showed no electron or hole traps indicating that the concentrations of the deep levels were below the detectability of the system $(<10^{10} \text{cm}^{-3})$. This indicated sufficient cleanliness in our annealing procedure to allow accurate measurement of the deep levels arising from the NTD process without interference from the introduction of deep levels during either the annealing or the diode fabrication procedure.

The resistivity measurements were performed by using the four-point probe technique and verified on some samples by Van der Pauw measurements.[14-15] The lifetime measurements were conducted using the reverse recovery technique with the diodes fabricated from the 5 ohm-cm, n-type and the 7 ohm-cm, p-type wafers.[16] These diodes were also used for the measurement of the majority carrier trap and minority carrier trap spectra by deep level transient spectroscopy. The results of these measurements are presented in the next section.

3. EXPERIMENTAL RESULTS

The resistivity data obtained from isochronal annealing (60 min. steps) are shown in Fig. 1 for a temperature range of 400° to 900°C. From Fig. 1, it can be seen that the neutron irradiation produced a phosphorus concentration of about $7 \times 10^{13} \text{cm}^{-3}$ corresponding to a fully annealed resistivity of about 70 ohm-cm for the 2000 ohm-cm, n-type sample. It can also be seen that the higher resistivity (2000 ohm-cm) sample annealed slower than the lower resistivity samples (5 and 7 ohm-cm). This is to be expected because under the same annealing conditions all the samples are expected to contain approximately the same concentration of the deep levels, and the compensation effect by these deep levels is stronger when the shallow dopant level concentration is smaller.

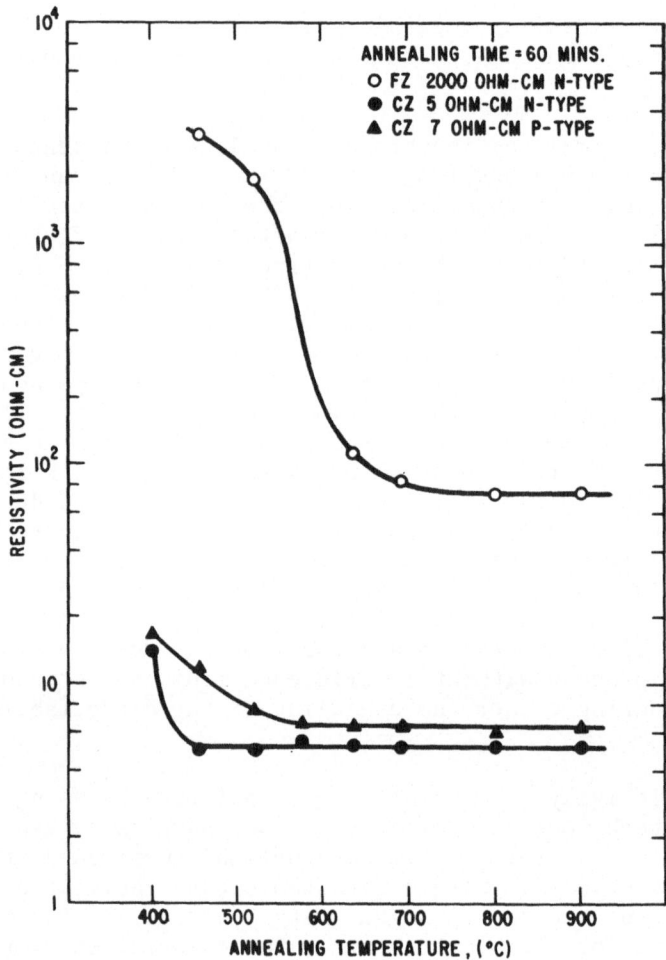

Fig. 1.   Isochronal annealing of resistivity in doped and undoped
          NTD-Si wafers.

Lifetime data obtained from the isochronal annealing experiments
is shown in Fig. 2 for a temperature range of 400°C to 900°C. It
can be seen that the lifetime is very low after annealing at 400°C
and monotonically increases to about 40 μsec when the annealing
temperature approaches 700°C. Beyond 700°C, the lifetime does not
increase any further indicating the annealing of the neutron induced
deep levels to concentrations below those introduced during diode
fabrication. When compared with the resistivity, the higher annealing temperature for recovery of the lifetime is to be expected because even small concentrations of the deep levels will cause

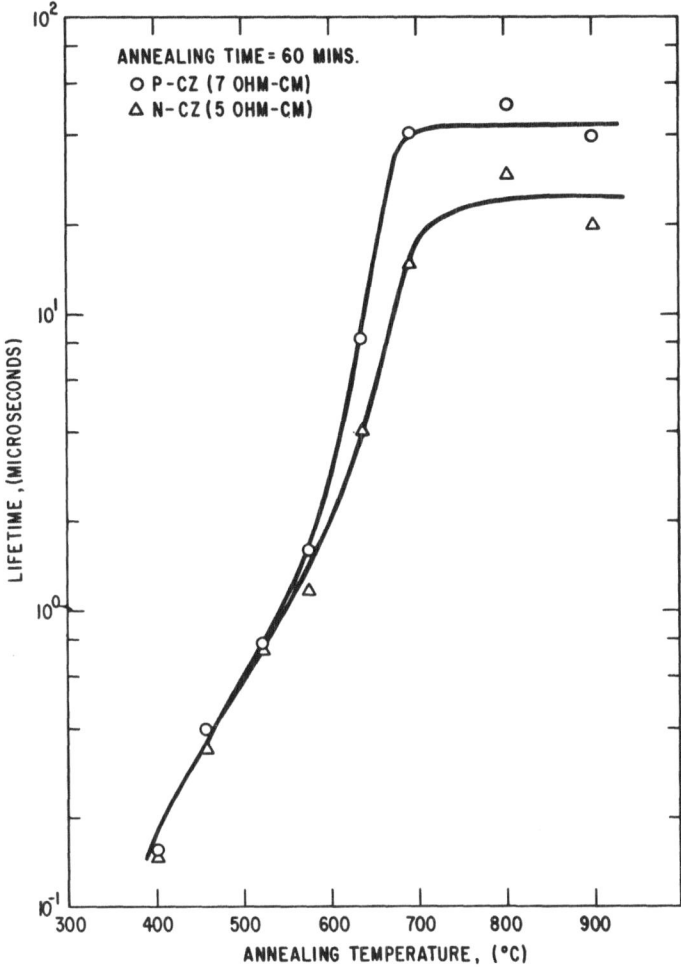

Fig. 2. Isochronal annealing of minority carrier lifetime.

significant lifetime reduction but will not cause any significant carrier removal.

The deep levels that control the behavior of both resistivity and lifetime during the annealing were identified by using deep level transient spectroscopy. Figure 3 shows the DLTS majority carrier trap spectrum for the n-type samples. Four deep levels have been observed. Emission rate measurements as a function of temperature have identified their location in the silicon energy gap as: $E_c - 0.211$ eV, $E_c - 0.218$ eV, $E_c - 0.30$ eV, and $E_c - 0.424$ eV.

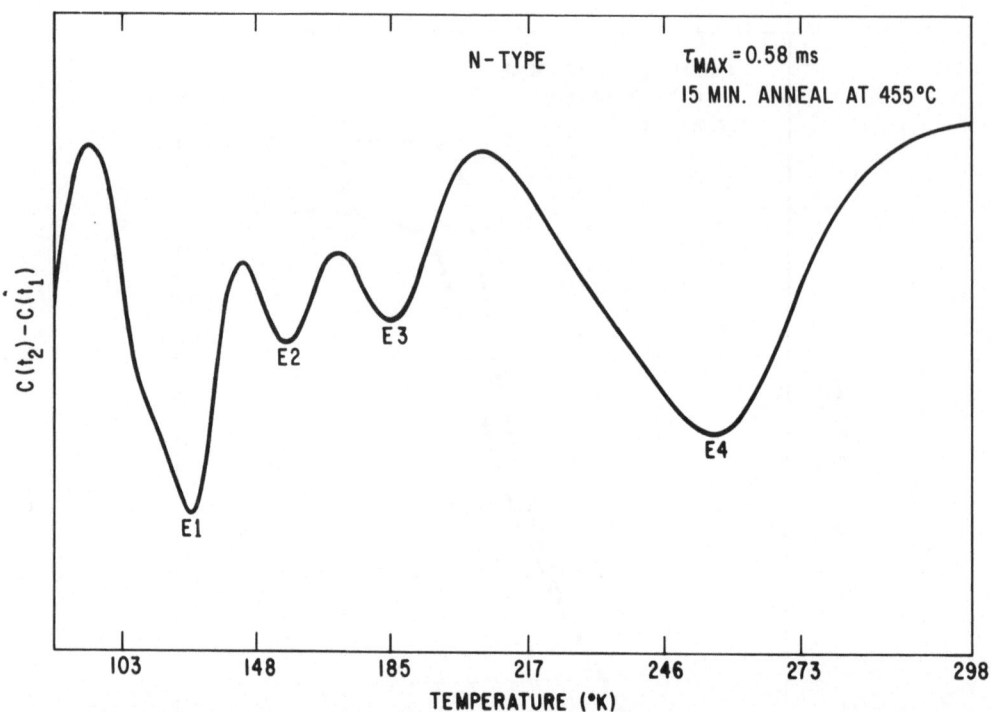

Fig. 3.　DLTS majority trap spectrum in n-type NTD samples.

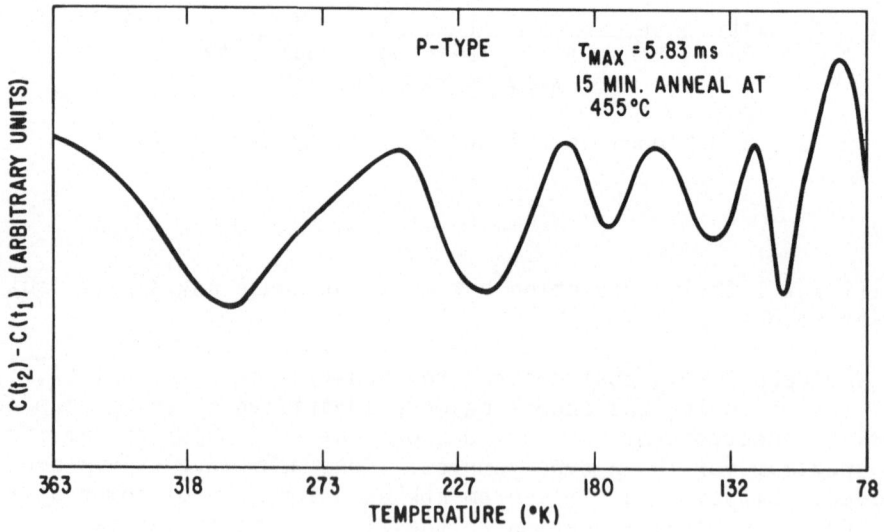

Fig. 4.　DLTS majority carrier trap spectrum in p-type NTD samples.

Figure 4 shows the DLTS majority carrier trap spectrum for the p-type samples. In this case five deep levels have been observed. Emission rate measurements as a function of temperature have identified the location of four of these levels in the silicon energy gap as: E_V + 0.20 eV (two levels), E_V + 0.32 eV and E_V + 0.385 eV. It is found that all these levels anneal at comparable rates and that a good correlation between the concentration of these levels and the lifetime cannot be made.

In order to determine the levels which control lifetime in the p- and n-type samples, lifetime measurements were, therefore, con-

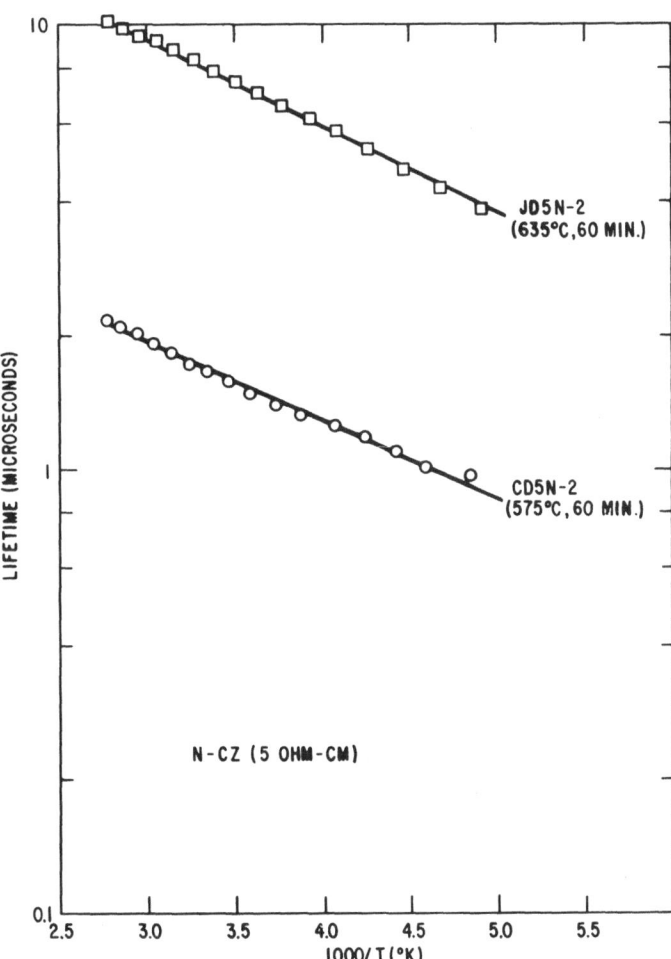

Fig. 5. Temperature dependence of minority carrier lifetime in n-type NTD-Si annealed at 575°C and 635°C for 60 min.

ducted as a function of ambient temperature. Figure 5 shows the
variation of the lifetime over a temperature range of -70°C to +90°C
for n-type samples annealed at 575°C and 635°C for 60 minutes. In
both cases, the data exhibits the same activation energy indicating
that the same recombination center controls lifetime under both
annealing conditions. The minority carrier lifetime in n-type
silicon is given by the expression

$$\tau = \tau_{po} \left[1 + \exp\left(\frac{E_r - E_F}{kT} \right) \right] + \tau_{no} \exp\left(\frac{2E_i - E_F - E_r}{kT} \right) \qquad (1)$$

where τ_{po}, τ_{no} are the minority carrier lifetimes in heavily doped
p- and n-type silicon, E_r is the recombination center location in
the energy gap, E_F is the Fermi level location in the energy gap,
E_i is the intrinsic level location, k is Boltzmann's constant and
T is the absolute temperature. Since the data shows an increase in
lifetime with increasing temperature, it can be concluded that the
first term in Eq. (1) can be neglected. From the data and the second
term in Eq. (1), the location of the recombination center can then
be determined as $(E_r - E_v) = 0.30$ eV.

Similar measurements have been performed for the p-type samples.
Figure 6 shows the variation in lifetime over a temperature range
of -70°C to +90°C for p-type samples annealed at 575°C and 635°C for
60 minutes. Again both samples show equal activation energies for the
the lifetime variation with temperature indicating that the same recombination center controls the lifetime in both cases. In the case
of p-type silicon, the minority carrier lifetime is given by the
expression

$$\tau = \tau_{po} \exp\left(\frac{E_r + E_F - 2E_i}{kT} \right) + \tau_{no} \left[1 + \exp\left(\frac{E_F - E_r}{kT} \right) \right]. \qquad (2)$$

Since the data shows an increase in lifetime with increasing temperature, it can be concluded that the second term in Eq. (2) can be
neglected. From the data and the first term in Eq. (2), the recombination center location can be determined as $(E_c - E_r) = 0.30$ eV.

4. DISCUSSION AND CONCLUSIONS

A comparison between the deep level data obtained by earlier
studies and our own measurements is provided in Fig. 7. The data
obtained by Tokuda and Usami is for the majority carrier traps due
to fast neutron damage as obtained by using the variation in FET
transconductance with frequency and temperature.[17-18] For n-type
silicon the four levels observed in this study correspond to the
levels observed at $E_c - 0.40$ eV, $E_c - 0.32$ eV, $E_c - 0.20$ eV and

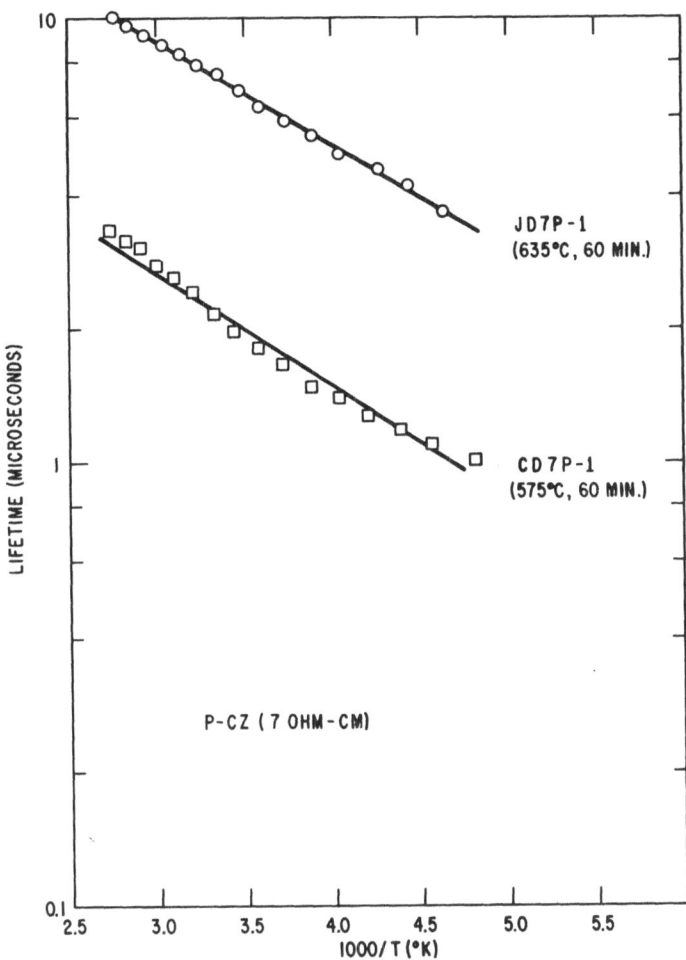

Fig. 6. Temperature dependence of minority carrier lifetime in p-
         type NTD-Si annealed at 575°C and 635°C for 60 min.

E_C - 0.19 eV by Guldberg. The level observed at E_C - 0.46 eV by
Guldberg is probably a surface effect as indicated by the author.
Guldberg has reported two shallow levels at E_C - 0.17 eV and at
E_C - 0.15 eV; the spectrum shown in Fig. 3 exhibits a broad shoulder
next to E1 (E_C - 0.211 eV) which we were unable to resolve. This
unresolved defect state may correspond to the level at E_C - 0.17 eV.
In the case of the p-type silicon, the four defect levels we characterized are in good agreement with the levels at E_V + 0.19 eV,
E_V + 0.21 eV, E_V + 0.30 eV, and E_V + 0.35 eV reported by Tokuda and
Usami. The fifth level observed by us would correspond to that
observed at E_V + 0.40 eV by Tokuda and Usami.

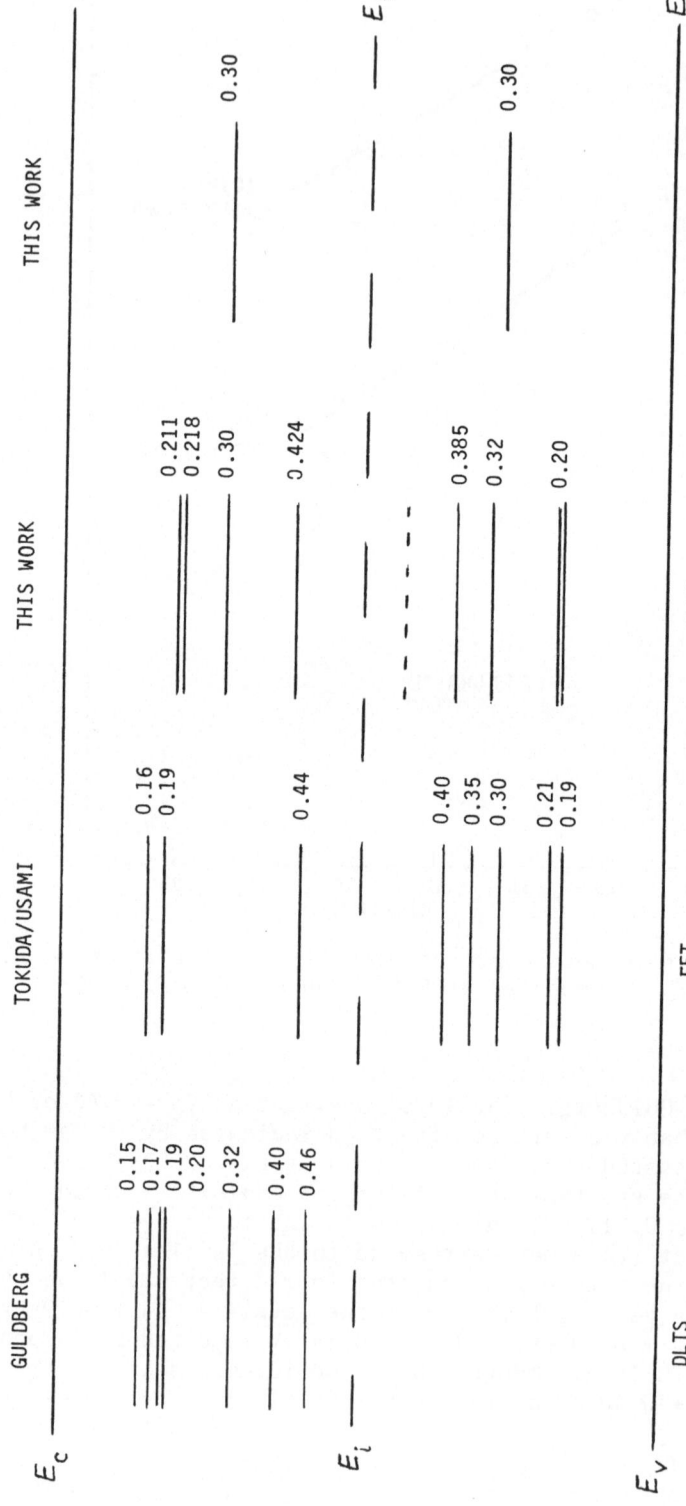

Fig. 7. Comparison chart of deep levels in neutron irradiated silicon obtained by different experimental methods.

The lifetime data indicates that the recombination center for n-type silicon lies at E_V + 0.30 eV during the annealing. This level correlates with the level at E_V + 0.32 eV observed by us in p-type silicon using DLTS. Similarly, the lifetime data for p-type silicon indicates that the recombination center lies at E_C - 0.30 eV during the annealing. This level correlates with the level at E_C - 0.30 eV observed by us in n-type silicon using DLTS. However, such correlations cannot be made with certainty because levels observed in the p- and n-type silicon may differ due to the interaction of defects with the dopant atoms. It is, therefore, necessary to obtain the minority carrier DLTS trap spectra for both p- and n-type silicon. This has been done by using the diode samples prepared in this study. Preliminary measurements have detected four minority carrier trap levels in n-type silicon samples and three minority carrier trap levels in p-type silicon samples. The location of these levels in the silicon energy gap has not yet been measured and will be reported in the future. It should be noted in closing that although these levels have been obtained by using Czochralski crystals, Curtis[19] has reported that there is no dependence of the lifetime upon oxygen content indicating that our results would also be applicable to float zone silicon.

REFERENCES

1) M. Tanenbaum and A. D. Mills, J. Electrochem. Soc. 108, 171 (1961).
2) H. M. Janus and O. Malmros, IEEE, ED-23, 797 (1976).
3) H. Herzer, Semiconductor Silicon 1977, ed. by H. R. Huff and E. Sirtl, p. 106 (The Electrochem. Soc., Princeton, New Jersey, 1977).
4) H. M. Liaw and C. J. Varker, Semiconductor Silicon 1977, ed. by H. R. Huff and E. Sirtl, p. 116 (The Electrochem. Soc., Princeton, New Jersey, 1977).
5) E. W. Hass and M. S. Schnoller, IEEE, ED-23, 803 (1976).
6) K. Platzoder and K. Loch, IEEE, ED-23, 805 (1976).
7) V. N. Mordkovich, S. P. Solov'ev, E. M. Temper, and V. A. Kharchenko, Sov. Phys. Semicon. 8, 139 (1974).
8) H. J. Stein, Phys. Rev. 163, 801 (1967).
9) H. J. Stein, J. Appl. Phys. 37, 3382 (1966).
10) K. Nakashima and Y. Inuishi, J. Phys. Soc. Japan 27, 397 (1969).
11) A. Senes, G. Sifre, and M. Breant, Semiconductor Silicon 1977, ed. by H. R. Huff and E. Sirtl, p. 135 (The Electrochem. Soc., Princeton, New Jersey, 1977).
12) D. V. Lang, J. Appl. Phys. 45, 3023 (1974).
13) J. Guldberg, Appl. Phys. Lett. 31, 578 (1977).
14) L. B. Valdes, Proc. IRE 42, 420 (1954).
15) L. J. Van der Pauw, Phillips Res. Repts. 13, 1 (1958).
16) D. C. Lewis, Solid St. Electronics 18, 87 (1975).

17)   Y. Tokuda and A. Usami, J. Appl. Phys. $\underline{47}$, 4952 (1976).
18)   Y. Tokuda and A. Usami, J. Appl. Phys. $\underline{49}$, 181 (1978).
19)   D. L. Curtis, IEEE, $\underline{NS-13}$, 33 (1966).

RESIDUAL RADIOACTIVITY MEASUREMENTS FOR HIGH PURITY SILICON IRRADIATED BY PILE NEUTRONS

Atsushi Yusa, Kazuhiro Hasebe and Yoshibumi Yatsurugi

Komatsu Electronic Metals Co., Ltd., Kanagawa, Japan

Hidio Higuchi

Japan Chemical Analysis Center, Tokyo, Japan

ABSTRACT

This paper describes residual radioactivity measurements for high purity silicon (purer than ten-nines) irradiated by pile neutrons to produce n-type semiconductor silicon. The silicon samples were irradiated in four different reactors with thermal neutron flux densities of 0.12, 1.45, 1.8, and 5.5 x 10^{13} n/cm^2/s. After irradiation, samples were dissolved in HF-HNO$_3$ mixed solution, and the solutions dried. The residual radioactivity was measured with a 2 π gas-flow low background G. M. counter. These results are listed in Table 1. The only radioactive species found was ^{32}P, which was produced by the consecutive reactions, ^{30}Si$(n,\gamma,\beta^-)^{31}$P followed by ^{31}P$(n,\gamma)^{32}$P. The observed production rate for ^{32}P was in good agreement with the calculated one, as shown in Fig. 1. This technique for residual radioactivity in irradiated silicon can be applied to estimate the amounts of not only doped phosphorus, but other trace impurities such as gold in irradiated silicon. Some samples, obtained from the tang end of the float zone purified silicon rod, contained about 0.01 atomic ppb of Au, but those from the middle of the rod contained no Au. The presented technique is easily applicable to check and characterize the rods for the production of neutron doped silicon.

Table 1. Results of residual radioactivity measurement of neutron irradiated silicon.

Sample No	Resistivity (300°K) undoped ($10^3 \Omega$cm)	Resistivity (300°K) doped (Ωcm)	Thermal neutron flux (10^{13}n/cm²·sec)	Neutron dose (10^{17}n/cm²)	Radioactivity at the end of irradiation (pCi/g of Si) ^{32}P	Radioactivity at the end of irradiation (pCi/g of Si) ^{198}Au
1	1.7	35	0.12	8.5	58 ± 4	–
2	9.8	100		2.9	6.5 ± 2	–
3	8.9	50	1.48	5.9	24 ± 4	–
4	8.0	26		1.2	107 ± 5	–
5	5.1	124		2.4	4.7 ± 1	–
6	5.3	63		4.7	13 ± 2	–
7	4.9	50	1.80	5.9	23 ± 2	–
8	3.5	40		7.4	67 ± 2	–
9	4.3	35		8.4	42 ± 2	–
10	3.5	37		7.9	6.5 ± 0.5	–
11*	1.5	37	5.5	7.9	6.5 ± 0.5	29 ± 3
12*	2.0	37		7.9	6.5 ± 0.5	94 ± 4

* The sample was cut out from a tail part of float-zone Si rods.

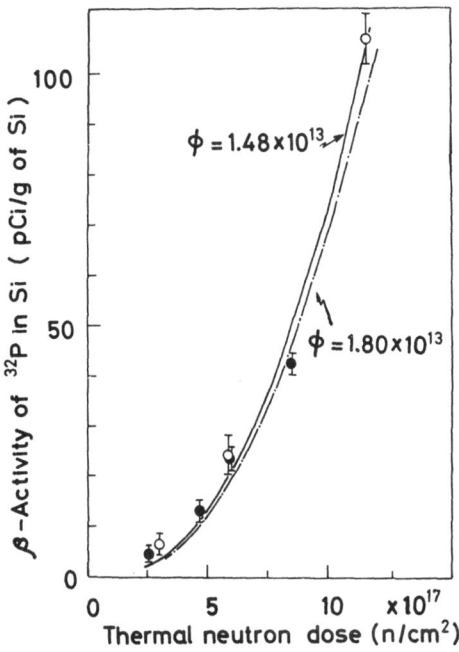

Fig. 1. NTD production of phosphorus as determined by β^- activity of ^{32}P vs. thermal neutron dose.

MAGNETO-OPTICAL STUDY OF SHALLOW DONORS IN TRANSMUTATION-DOPED GaAs

J.H.M. Stoelinga,[1] D. M. Larsen, W. Walukiewicz[2] and
R. L. Aggarwal[3]

Mass. Institute of Tech., Cambridge, MA 02173

C. O. Bozler

M.I.T. Lincoln Laboratory, Lexington, MA 02173

ABSTRACT

Lineshapes and peak positions of $1s \rightarrow 2p_{-1}$ donor transitions in epitaxial GaAs samples of relatively low compensation have been studied as functions of magnetic field by use of photoconductivity measurements. Some of these samples were produced by transmutation doping using thermal neutrons--a method which is useful for the controlled introduction of donor impurities in GaAs. Two new effects, tentatively attributed to van der Waals interactions between neutral donor atoms, are observed: (1) although both Se and Ge donors are introduced by thermal neutron transmutation, the Se line is much broader than the Ge line, and (2) deviations from isolated donor behavior occur in the magnetic field dependence of the chemical shift of the shallowest donor present. The separation of lines from two deeper donors, Ge and Si, verified the simple phenomenological theory of the magnetic field dependence of central cell corrections of isolated donors up to at least 10 T.

[1]On sabbatical leave from the K. University of Nijmegen, the Netherlands, with support of the Niels Stensen Stichting.
[2]Visitor from the Institute of Physics, Warsaw, under the interexchange program of the National Academy of Sciences and the Polish Academy of Sciences.
[3]Also Department of Physics, M.I.T.
This work was supported by the National Science Foundation and sponsored by the Department of the Air Force.

SHALLOW DEFECT LEVELS IN NEUTRON IRRADIATED EXTRINSIC P-TYPE SILICON

M. H. Young, O. J. Marsh, R. Baron

Hughes Research Laboratories

Malibu, CA 90265

ABSTRACT

Two shallow levels, at 0.027 eV and 0.039 eV from the valence band, measured by Hall effect vs. temperature, have been observed in neutron irradiated, float zone (FZ) grown Si:Ga. The neutron irradiation was for the purpose of counter-doping residual boron by producing phosphorus by neutron transmutation $^{30}Si(n,\gamma)^{31}Si \rightarrow {}^{31}P + \beta^-$ to allow the Si:Ga to be used as extrinsic Si detector material. These defect levels are observable after 575° to 625°C anneals. Annealing at 700°C - 850°C removes observable radiation defects. Analysis of the Hall effect vs. temperature data indicates that the two levels are acceptors with concentrations in the range 10^{14} - $10^{15}/cm^3$, in excess of the B concentration of $\lesssim 2 \times 10^{13}/cm^3$ measured before irradiation. A proportionality between defect concentration and Ga concentration is observed. The shallow levels appear also in photoconductivity spectral response measurements at 5K. Observations of shallow levels in FZ-grown Si:Al and Si:In will also be discussed.

1. INTRODUCTION

The emergence of facilities for phosphorus doping of silicon by neutron transmutation is stimulating active investigation of an increasing variety of new device applications. The extrinsic Si infrared photoconductive detector is one such device. P-type silicon, doped most often with Ga or In, is used in the fabrication of these detectors. For optimum device operation, it is necessary for the residual impurities to be as carefully controlled as the major dopant although residual impurities are typically three or four orders of magnitude lower in concentration than the major dopant. This control is necessary because at normal operating temperatures (25 - 65°K), these detectors are very sensitive to the proper balancing of the two most common residual impurities in silicon--boron and phosphorus. For proper operation of a p-type Si extrinsic photodetector, the concentration of phosphorus atoms must exceed that of boron. B must be over-compensated so that the conductivity of the material at the desired operating temperature is dominated by photoexcited carriers from the deeper major impurity level (i.e., Ga, In) and not by thermally excited carriers from the shallow B level. In addition, uniformity of detector response over a detector array, important for imaging applications, requires a uniform distribution of the net compensation (N_P - N_B for p-type Si). Since neutron transmutation allows the setting of a uniform P concentration precisely to a predetermined level, this doping method is ideal for counterdoping IR detector grade Si as long as radiation damage detrimental to device operation can be removed. Radiation defects characteristic of extrinsically doped p-type Si and associated with the transmutation doping process are the subject of this investigation. The principle result is the observation of very shallow electronic energy levels in Si:Ga. These levels, denoted A_1 and A_2, are located at 0.027 eV ± 0.001 eV and 0.039 ± 0.004 eV, respectively, measured from the valence band (E_V). Radiation induced levels are usually much deeper; these are particularly interesting since they are shallower than the most shallow of the column III acceptors (boron at 0.0457 eV).[1]

2.   EXPERIMENTAL PROCEDURES

In this work, float zone (FZ) grown Si:Ga, Si:In, and Si:Al samples doped in the 2×10^{16} - 2×10^{17}/cm^3 range were irradiated with a thermal neutron fluence of 2×10^{17}n/cm^2 to produce 5×10^{13} phosphorus/cm^3. The Cd ratio of the reactor facility[2] used is 22. The electrical properties of irradiated samples were investigated after anneals at temperatures between 500°C and 850°C. Hall measurements employed the van der Pauw technique[3] with a "cloverleaf" sample configuration to eliminate the effects of finite contact size. Ohmic contacts with negligible contact resistance were formed by boron ion implantations.[4] A computer controlled Hall measurement system was used in data acquisition.[5] The electric field in the samples did not exceed 2 V/cm and all measurements were made with the samples in thermal equilibrium. Samples of the unirradiated Si were fully characterized to determine the initial concentration of column III acceptors, compensating donors, and residual boron.

The dopant concentrations and acceptor activation energies (for those energy levels approached by the Fermi level at some temperature within the measured temperature range) were determined from the Hall effect data by a nonlinear least squares fit to the charge balance equation[6]

$$p + N_D = \left[\frac{N_{Ga}}{1 + \frac{pg_{Ga}}{N_V} \exp\left(\frac{E_{Ga}}{kT}\right)} \right] + \left[\frac{N_1}{1 + \frac{pg_1}{N_V} \exp\left(\frac{E_1}{kT}\right)} + \frac{N_2}{1 + \frac{pg_2}{N_V} \exp\left(\frac{E_2}{kT}\right)} \right] , \qquad (1)$$

where the last two terms are included to provide for the presence of A_1 and A_2. The effective density of states in the valance band, N_V, is calculated using the temperature dependent effective mass given by Barber.[7] N, g, and E are the density, degeneracy, and energy of the acceptors, and N_D is the compensating donor concentration. N_D includes initial donors, transmutation produced P, and compensators due to radiation damage (in partially annealed samples). The boron concentration (measured before irradiation as $\leq 2 \times 10^{13}/cm^3$ for all FZ samples used in this work) was negligible compared to the shallow acceptor defect densities N_1 and N_2, which we found to be in the $10^{14} - 10^{15}/cm^3$ range. In the analysis, the Hall scattering factor

$$r = \frac{\mu_{Hall}}{\mu_{Drift}}$$

(where μ = carrier mobility) was set equal to unity.[8] The degeneracies g_1 and g_2 were set equal to 4.[1] The RMS error of the fit of Eq. (1) to the data was 0.7 - 2.0% for the twelve Si:Ga samples in which the shallow electronic levels were investigated.

3. RESULTS AND DISCUSSION

Irradiated Si:Ga samples annealed in the 700°C - 850°C range show the same electrical characteristics as unirradiated samples except for the presence of the expected transmutation produced phosphorus. Since sufficient annealing removes all evidence of the shallow acceptors, these very interesting defects will not prevent the successful application of neutron transmutation doping to infrared detectors. For anneal temperatures below 700°C, the

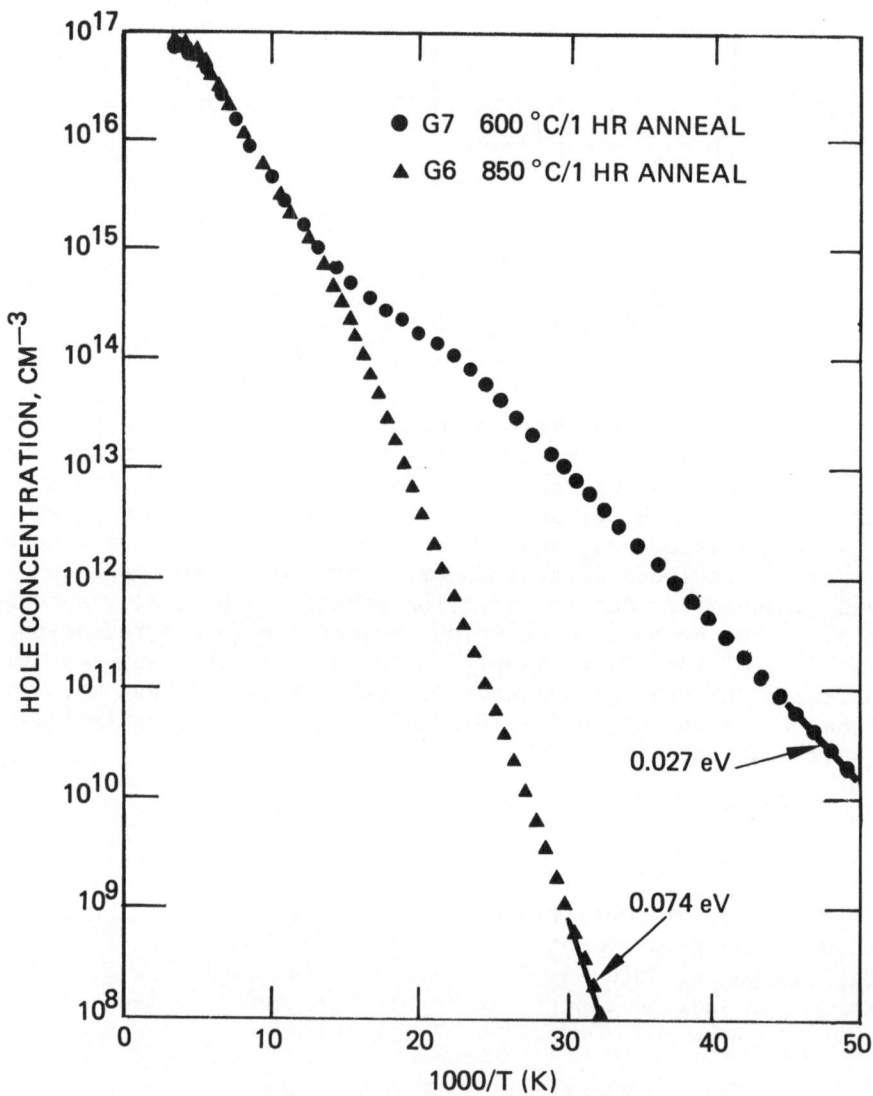

Fig. 1.   Free-carrier concentration from Hall effect measurements
          for two samples from neutron irradiated Si:Ga crystal
          Z2Ga for two anneal conditions.

net donor concentration measured exceeds the concentration of
initial donors plus transmutation produced phosphorus. However,
the concentration of these excess radiation-damage-produced donors
drops rapidly with increasing anneal temperature from 500°C to
700°C.

A_1 and A_2 were observed in twelve irradiated Si:Ga samples
annealed between 575°C and 625°C. Figure 1 compares carrier concentration measurements on a partially annealed sample (G7)
exhibiting the shallow acceptors and a fully annealed sample (G6)
displaying no evidence of shallow levels. The 0.027 eV slope of
the shallower defect level (A_1) appears explicitly at the lowest
temperatures for the 600°C anneal sample while the low temperature
slope of sample G6 (850°C anneal) is dominated by the 0.074 eV
level of Ga.[1] For most of the samples in which shallow levels
were studied, N_1 was found to exceed N_D and thus E_1 dominates the
slope at the lowest temperatures (as for G7). For two of the
twelve partially annealed samples studied, the condition $N_1 + N_2$
$> N_D > N_1$ was satisfied and E_2 was observed explicitly at the
lowest temperatures. Figure 1 also shows that at higher measurement temperatures (> 77K) where the half-slope and exhaustion
regions of Ga appear, the electrical properties are determined by
N_{Ga} and are unaffected by the presence of shallow levels since N_{Ga}
is much greater than N_1 and N_2.

The relative photoconductive spectral response beyond the cut-
off of Si:Ga ($\sim$ 17 μm) is shown in Fig. 2 for a neutron irradiated
sample at 5K. Analysis of Hall data on this sample clearly indicated the presence of both shallow electronic energy levels.
The spectrum of the response for $\lambda > 20$ μm is consistent with the
presence of two electronic energy levels near the valence band at
~ 0.027 eV and ~ 0.04 eV. This long wavelength response is not
seen in the "as grown" Si:Ga or fully annealed (850°C) neutron
irradiated Si:Ga.

A preliminary investigation of Si:Al and Si:In irradiated
and annealed under conditions identical to the Si:Ga samples also
indicates two new shallow defects for each dopant near the valence
band. Table 1 summarizes the initial results on Si:Al and Si:In
along with the Si:Ga results.

Results obtained so far indicate a proportionality between
shallow acceptor defect concentration produced and Ga concentration
in the range of 2.9 - 8.0 x 10^{16} Ga/cm^3. N_1 and N_2 vs. N_{Ga}
measured after one hour anneals at various temperatures is displayed
in Fig. 3. Unity slope lines are drawn through the points displayed
in this figure.

The apparent dependence of the shallow defect concentration
on N_{Ga} suggests a defect complex which contains the column III

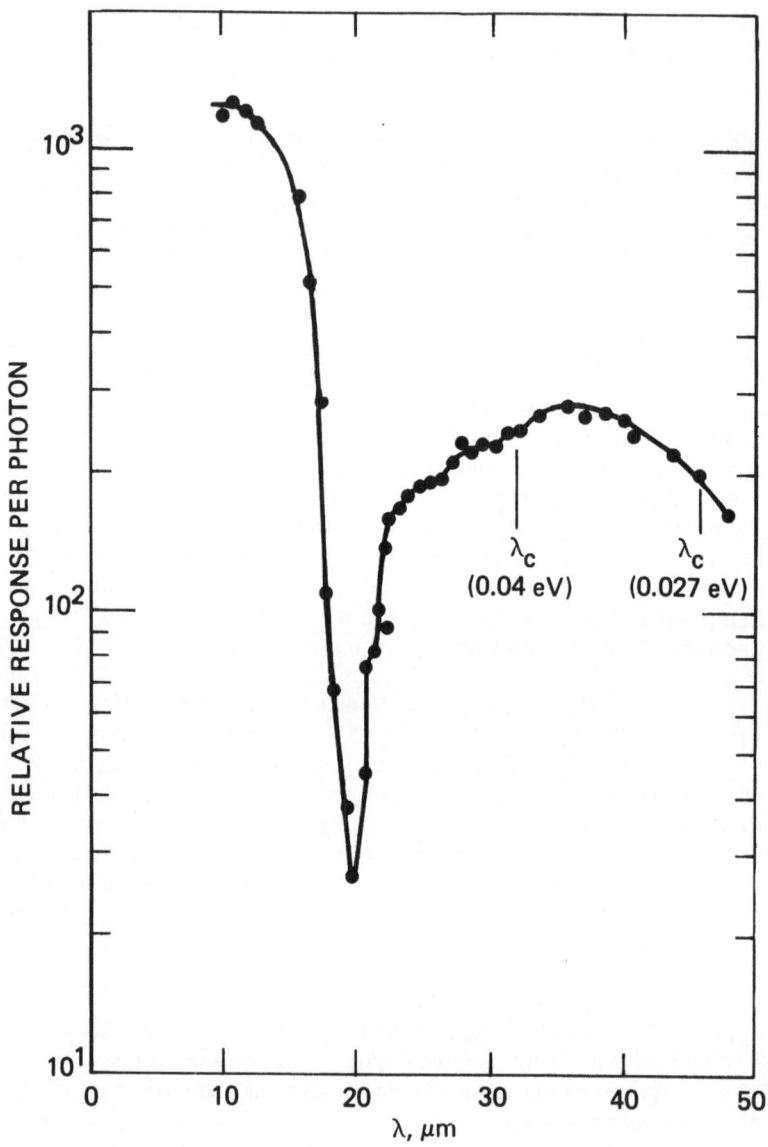

Fig. 2.   Photoconductivity spectrum at 5°K for neutron irradiated
          Si:Ga sample Z2Ga.G4 after a 600°C/1 hr anneal. The cut-
          off wavelengths   for 0.04 eV and 0.027 eV

$$\lambda_c (\mu m) = \frac{1.24}{E(ev)}$$

          levels are indicated.

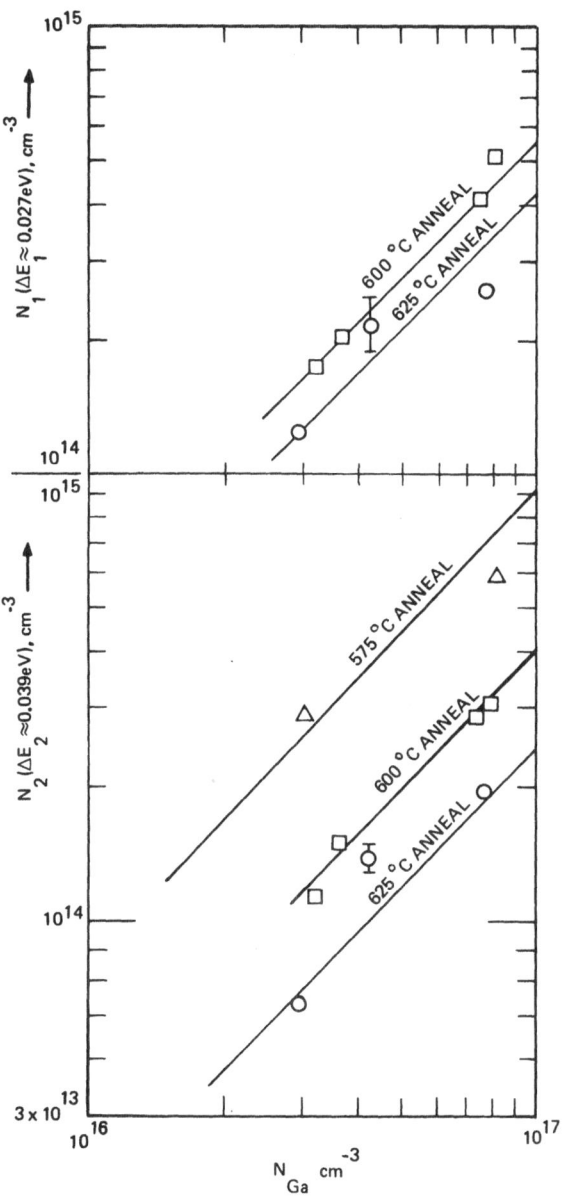

Fig. 3.   Shallow defect concentration N_1 and N_2 vs. Ga concentration for various anneal conditions.

Table 1. Observed activation energies of shallow radiation defects in Si:Al, In, and Ga. Al and In samples were annealed at 600°C/1 hr. Twelve Ga samples were annealed at temperatures from 575 - 625°C.

DOPANT	CONCENTRATION cm^{-3}	E_1 eV	E_2 eV
Al	2.5×10^{16}	0.030	0.041
In	1.3×10^{17}	0.031	0.045
Ga	$2.9 - 8.0 \times 10^{16}$[a]	0.027 ± 0.001	0.039 ± 0.004

[a.]TWELVE SAMPLES OVER INDICATED RANGE OF CONCENTRATIONS WERE MEASURED.

atom. Similar dependence on doping concentration occurs for the "E-center" in Si which involves vacancies and column V donor atoms (P, As, or Sb).[9] The electronic energy levels associated with the new acceptors we have found in irradiated p-type Si are shallower than any of the column III substitutional acceptor impurities in silicon; this particularly interesting result should stimulate theoretical discussions.

ACKNOWLEDGEMENTS

We gratefully acknowledge the many useful and stimulating discussions of this work with Professors T. C. McGill and O. M. Stafsudd. We thank D. J. O'Connor, A. F. Rabideau, R. Wong Quen, and J. J. Sheets for their skilled assistance in crystal growth and electrical and optical measurements, and Professor R. Hart for providing the neutron irradiations.

REFERENCES

1) A. Baldereschi and N. O. Lipari, Proceedings of the 13th International Conference on the Physics of Semiconductors, (edited by F. G. Fumi, Rome, 1976), p. 595.

2) Irradiations were performed under the direction of Professor R. Hart at Texas A&M University Nuclear Science Center.

3) L. J. van der Pauw, Phillips Res. Rep. 13, 1 (1958).

4) Boron implanted p$^+$ contact regions were formed during the 500°C - 850°C sample anneals. No separate contact anneal step was necessary.

5) R. Baron, M. H. Young, J. K. Neeland, and O. J. Marsh, Proceedings of the Third International Symposium on Silicon Materials Science and Technology, edited by H. R. Huff and E. Sirtl (The Electrochemical Society Inc., Princeton, N.J., 1977), p. 367.

6) G. S. Blakemore, Semiconductor Statistics (Pergamon, Oxford, 1962), Chap. 3.

7) H. D. Barber, Solid State Electron. 10, 1039 (1967).

8) Neglect of the temperature dependence of r above 100°K produces the largest error in the theoretical model used to fit the data. The degeneracy g_{Ga} was allowed to become an adjustable parameter in the least squares fit of Eq. (1) in order to compensate for r(T) above $\sim$ 100°K. Assumptions about g_{Ga} and r(T) have relatively little effect on the fit of the data over the temperature range dominated by the shallow defect levels (< 77°K).

9) See L. C. Kimerling, H. M. DeAngelis, and J. M. Diebold, Solid State Commun. 16, 171 (1975); A. W. Evwaraye, J. Appl. Phys. 48, 1840 (1977); A. O. Evwaraye, J. Appl. Phys. 48, 734 (1977) and references therein.

MEASUREMENTS OF ^{31}P CONCENTRATIONS PRODUCED BY NEUTRON TRANSMUTATION DOPING OF SILICON

R. R. Hart, L. D. Albert, and N. G. Skinner

Texas A&M University, College Station, Texas 77843

M. H. Young, R. Baron, and O. J. Marsh

Hughes Research Laboratories, Malibu, CA 90265

ABSTRACT

The absolute concentrations of ^{31}P, produced by the irradiations of float zone silicon samples in the Texas A&M University Research Reactor, have been measured to an accuracy of ± 10%. The neutron fluence was varied from 10^{16} to 10^{18}n/cm^2, which corresponded to ^{31}P concentrations in the 10^{12} to 10^{14} atoms/cm^3 range. The results are based on measurements of the absolute activities of ^{31}Si by detection of 1.266 MeV gamma rays. Secondary standards of Fe were required for the larger neutron fluences. The ^{31}P concentrations were compared to the concentrations of electrically active P following 850°C anneals for 1 hour. These concentrations were determined in the same samples from temperature-dependent Hall effect measurements at Hughes Research Laboratories. The two results agree to within experimental error, thus confirming that transmuted P in the 10^{12} to 10^{14} atoms/cm^3 range is completely electrically active following 850°C, 1 hour anneals of float zone silicon. In addition, a corrected value for the gamma abundance of ^{31}Si was established to be 5.6 x 10^{-4} ± 10%.

1. INTRODUCTION

Neutron transmutation doping has been demonstrated to be an effective method of introducing uniform and controllable concentrations of phosphorous into silicon.[1-3] The process is based on the ^{30}Si$(n,\gamma)^{31}$Si reaction with subsequent β decay of ^{31}Si to ^{31}P

with a half life of 2.62 hours. In approximately 0.07% of the
decays 1.266 MeV gamma rays are also emitted.[4]

The radiation damage produced during neutron bombardment and
during the decay of [31]Si may be removed by isothermal anneal at
temperatures greater than approximately 700°C.[1,5] This annealing
procedure also results in electrically active P. Although very
controllable resistivities are produced, it is not clear that
indeed all the neutron produced P is electrically active.

The object of this work is to measure the fraction of neutron
produced P that is electrically active following 850°C anneals
for 1 hour of float zone Si samples irradiated in the Texas A&M
University research reactor. The absolute concentrations of P
produced by neutron transmutations are determined from measurements
of [31]Si and ^{59}Fe activities. These results are compared to
electrically active P concentrations determined from temperature
dependent Hall effect measurements.

2. THEORY

The concentration of ^{31}P that is ultimately produced, N_P, is
given by the total number of (n,γ) reactions with [30]Si, i.e.,

$$N_P = \int_0^{t_i} \int_0^{\infty} N_{30}\sigma_{30}(E)\ \phi\ (E,t)\ dE\ dt, \tag{1}$$

where
$\quad N_{30}\quad$ = concentration of [30]Si,

$\quad \sigma_{30}(E)$ = energy dependent (n,γ) cross section of [30]Si,

$\quad \phi(E,t)$ = energy and time dependent neutron flux,

$\quad t_i\quad$ = irradiation time.

Negligible burnup of [31]Si and ^{31}P during the irradiation is assumed.
Since $\sigma_{30}(E)$ is not well known, nor is $\phi(E,t)$ sufficiently well
known in the reactor used for this work, it is necessary to solve
Eq. (1) experimentally.

An effective silicon flux, ϕ_{Si}, is defined by

$$N_P = N_{30}\sigma_{30}\phi_{Si}t_i, \tag{2}$$

where σ_{30} = the 2200 m/sec (n,γ) cross section of [30]Si. For irradiation times much less than the 2.62 hour half life of ^{31}Si the saturated activity of ^{31}Si is given by

$$A_{31} = N_{30}\sigma_{30}\phi_{Si}. \tag{3}$$

It is necessary to restrict Eq. (3) to short irradiations to account for possible changes in the actual neutron flux with time. Longer irradiations require the use of a standard which produces a long half life nuclide. A suitable standard is iron which results in ^{59}Fe having a 44.6 day half life. Following the same procedure as above, an effective iron flux is defined, ϕ_{Fe}, which may be used to give the saturated activity of ^{59}Fe, A_{59}, by

$$A_{59} = N_{58}\sigma_{58}\phi_{Fe}, \tag{4}$$

where

N_{58} = concentration of ^{58}Fe,

σ_{58} = the 2200 m/sec (n,γ) cross section of ^{58}Fe.

Equation (4) is valid for irradiation times much less than the 44.6 day half life of ^{59}Fe, even though the actual neutron flux may change with time.

Considering a simultaneous short irradiation of both Si and Fe, Eqs. (3) and (4) both hold. For this case the ratio of the effective Si flux to the effective Fe flux is given by

$$K = \frac{\phi_{Si}}{\phi_{Fe}} = \frac{N_{58}\sigma_{58}A_{31}}{N_{30}\sigma_{30}A_{59}}. \tag{5}$$

Thus, from measurements of the activities of both Si and Fe following a simultaneous short irradiation, the parameter, K, may be calculated. Note that K is considered constant since the neutron energy spectrum is expected to remain essentially constant, independent of time, at a given location in a specific reactor.

Once K has been experimentally determined then for any irradiation time much less than 44.6 days the activity of an Fe standard will give N_P. From Eqs. (1) and (5),

$$N_P = K N_{30}\sigma_{30}\phi_{Fe}t_i , \tag{6}$$

where ϕ_{Fe} is calculated using Eq. (4).

3. EXPERIMENTAL

The high resistivity, float zone Si used in this work was grown at Hughes Research Laboratories. The samples were 1 cm^2 in area by 0.020 in. thick. Prior to irradiations individual samples were sealed in close fitting polyethylene vials as were also Fe standards. Irradiations were performed in the thermal rotisserie, which is located in the cooling water near one face of the Texas A&M University Research Reactor. An Al fixture was used to position the vials at constant radius and height inside the rotisserie. This positioning and the 1 rpm rotational motion of the rotisserie insured that all samples and Fe standards received the same neutron fluence for a given irradiation. A reactor power of 1 MW was used for all irradiations. The cadmium ratio, using Si, was measured to be 22 ± 10%.

Following irradiations and cool down of sample activities to about 1 μCi, the samples and Fe standards were transferred to unirradiated vials for gamma counting. Gamma spectra were obtained using a Ge(Li) detector located in a well-shielded enclosure. A computer based 4096 channel analyzer was used for data acquisition and peak analysis. A detector efficiency curve, shown in Fig. 1, was determined using a National Bureau of Standards point source. The standard deviation of each point is less than 3%. Also indicated by the arrows in Fig. 1 are the gamma energies used for the ^{31}Si and ^{59}Fe activity measurements, i.e., 1.266 MeV and 1.099 MeV, respectively. The observed counts, C, were corrected to saturated activities using the equation

$$A = \frac{\lambda C (t_r/t_c)}{\varepsilon \gamma (1 - e^{-\lambda t_i})(e^{-\lambda t_w})(1 - e^{-\lambda t_r})} \quad . \tag{7}$$

where

ε = detector efficiency,

γ = gamma abundance,

λ = decay constant,

t_c = live time,

t_r = counting time,

t_w = wait time.

This equation is exact providing the dead time varies linearly during the counting period. In fact, the dead time remained essentially constant during the counting intervals for all of the present measurements.

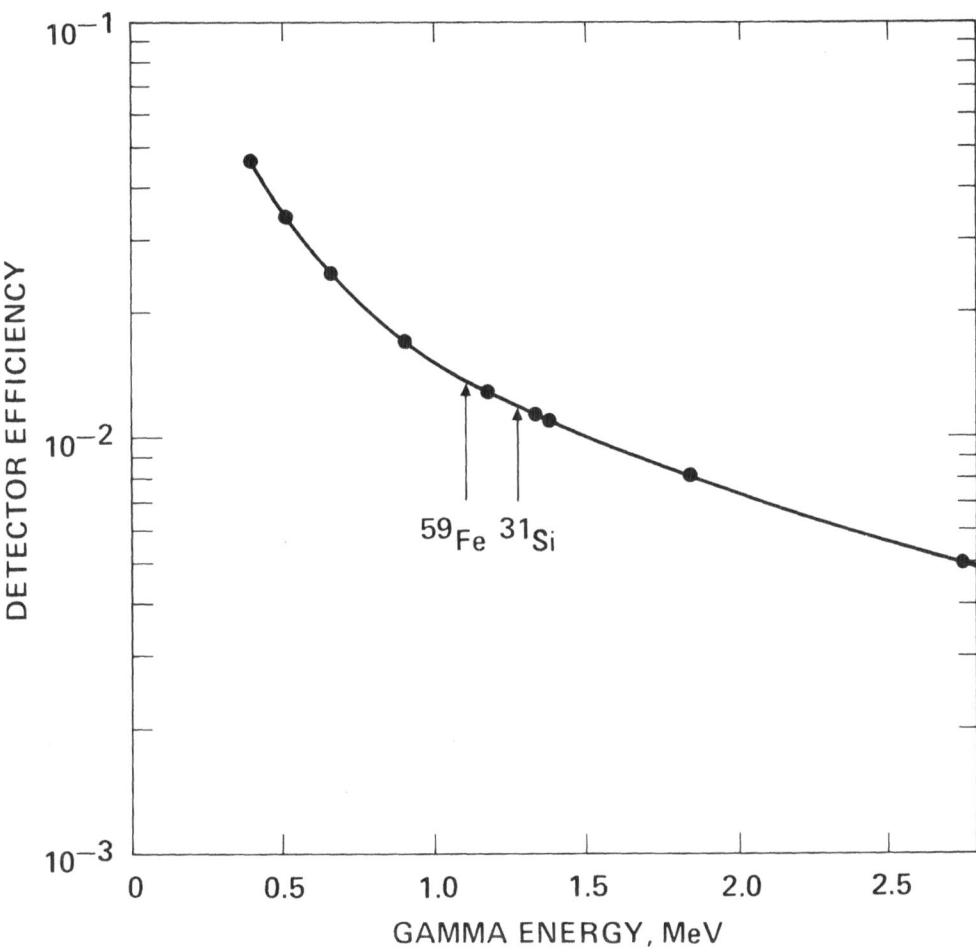

Fig. 1. Detector efficiency curve based on a National Bureau of Standards point source. The standard deviation of each point is less than 3%. The arrows indicate the gamma energies used for the ^{59}Fe and ^{31}Si activity measurements.

4. RESULTS

Three separate short irradiations of 30 min. each were used to evaluate the parameter K. As discussed previously, each case involved simultaneous irradiations of Si and Fe samples. Equation (7) was used to calculate the saturated activities which were then substituted into Eq. (5) to determine K. The average value of K was 0.80 ± 9%. Results of the three measurements were consistent to within counting statistics of ± 2%. The necessary nuclear parameters were obtained from references 6 and 7 and are known to within a few per cent except for the gamma abundance of ^{31}Si. This quantity has been measured to be 0.0007.[4] However, no experimental error was quoted in reference 4. Assuming one digit accuracy, the uncertainty in the gamma abundance of ^{31}Si was taken to be ± 7%. This estimated error dominates the other uncertainties in nuclear parameters and experimental measurements, a total of ± 5%, in arriving at the value of K = 0.80 ± 9%.

Five Si samples and associated Fe standards were irradiated from 3 to 28 hours. The measured activities of ^{59}Fe were corrected to saturated activities using Eq. (7). These were then substituted into Eq. (4) to give the effective iron fluxes. These values plus the value for K were substituted into Eq. (6) to determine the neutron produced concentrations of P. The results are plotted in Fig. 2 in terms of the Si fluence = $\phi_{Si}t_i = K\phi_{Fe}t_i$ and labeled, "Activation." The uncertainties in both the P concentration and the Si fluence are dominated by the uncertainty in K and are estimated to be ± 10%.

After implanting As contacts the samples were annealed at 850°C for 1 hour in an Ar atmosphere. Temperature dependent Hall effect measurements were then made to determine the electrically active P concentrations. These measurements were performed at Hughes Research Laboratories. Similar measurements are described in reference 5. The results were corrected for initial electrically active P concentrations (1 - 1.5 x 10^{13} atoms/cm^3). These were also determined by Hall effect measurements either on the same samples prior to irradiation or on adjacent samples from the same Si wafer. The resultant electrically active P concentrations are also plotted in Fig. 2 and labeled, "Hall Effect." The uncertainty in the results is estimated to be ± 25%. A straight line from the origin was drawn through the data by least squares fitting. As can be seen in Fig. 2 the resultant line gives electrically active P concentrations 30% greater than the results from activation measurements. This discrepancy is outside the estimated error limits.

A possible source of this discrepancy is larger error than estimated in the gamma abundance of ^{31}Si, which would lead to a larger error than estimated in the value of K. To check this

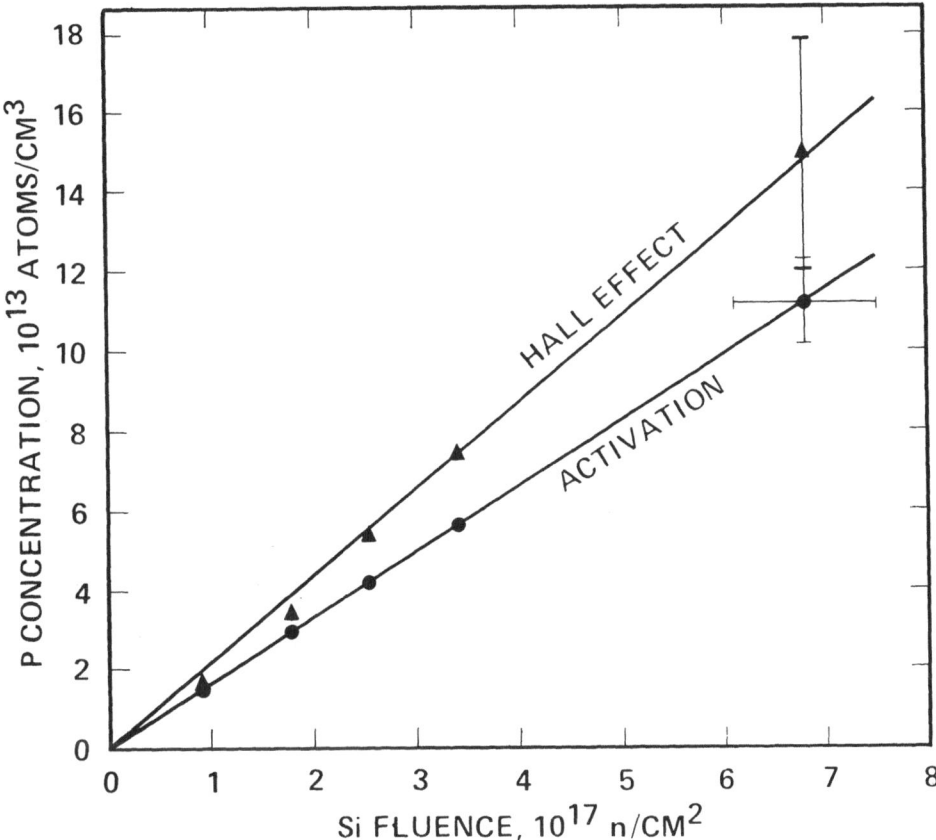

Fig. 2. Neutron produced P concentrations determined both by Hall
effect and activation measurements. The activation
measurements are based on a previously measured gamma
abundance of ^{31}Si of 0.0007.

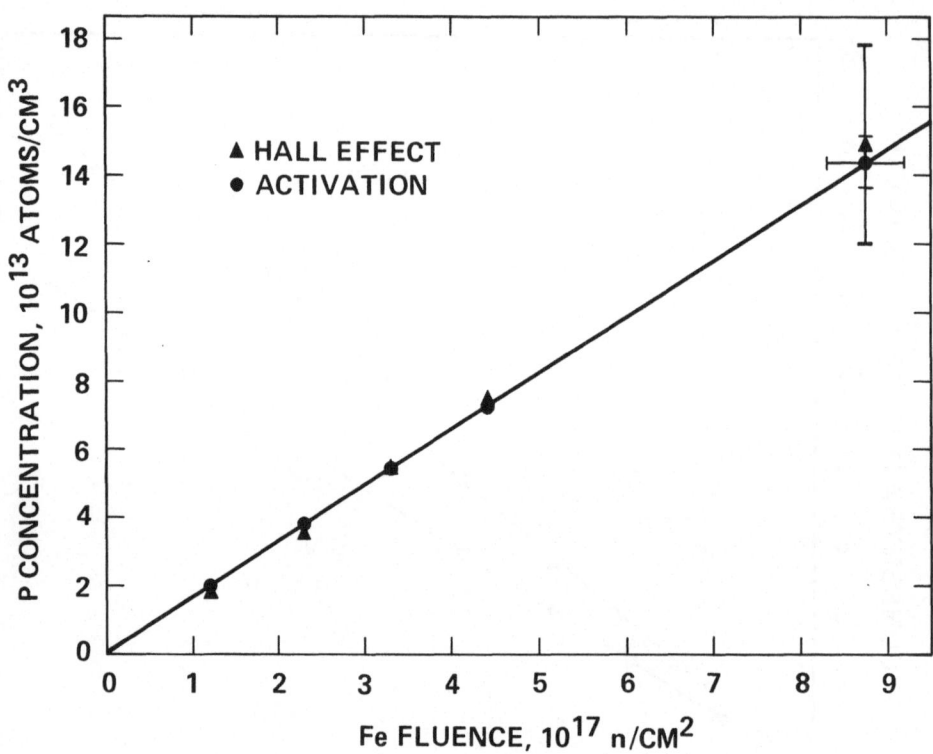

Fig. 3.   Neutron produced P concentrations determined both by Hall
          effect and activation measurements. The activation
          measurements are based on a corrected gamma abundance of
          ^{31}Si 5.6 x 10^{-4}.

possibility a theoretical expression for K was developed. Assuming 1/v cross sections below the Cd cut off of 0.5 eV and a 1/E slowing down spectrum,

$$K = \left[1 + \frac{I_{30}}{\sigma_{30}(R_{Cd} - 1)}\right] \Big/ \left[1 + \frac{I_{58}}{\sigma_{58}(R_{Cd} - 1)}\right], \qquad (8)$$

where

I_{30} = resonance integral of ^{30}Si,

I_{58} = resonance integral of ^{58}Fe,

R_{Cd} = cadmium ratio.

Based on the measured cadmium ratio of 22 and the tabulated resonance integrals,[6] K is evaluated to be 1.00. The error of this result is difficult to estimate since it is primarily due to limitations of the simple model used in the derivation of Eq. (8). We feel that an error of ± 10% is reasonable. Consequently, there is a significant difference between the calculated value of 1.00 ± 10% and the experimental value of 0.80 ± 8%. This discrepancy suggests that the previously measured value of 0.0007[4] for the gamma abundance of ^{31}Si is 25% too large. A more accurate value is 5.6 x 10^{-4} ± 10%.

Based on the corrected value for the gamma abundance of ^{31}Si, or equivalently, the theoretical value for K = 1.00 ± 10%, the effective Si flux equals the effective Fe flux to within ± 10%. Again using Eq. (6) the P concentrations produced in the Si samples were calculated and are plotted using solid circles in Fig. 3 as a function of Fe fluence - $\phi_{Fe} t_i$. The experimental error in Fe fluence of ± 5%, as well as the ± 5% effect this error has on the P concentration are indicated by the error bars. Of course the estimated ± 10% error in K also leads to an overall error in P concentration of about ± 10%.

Also plotted in Fig. 3 are the Hall effect data, given by solid triangles. The agreement between the activation and Hall effect results is well within the experimental errors. Therefore, it is concluded that transmuted P in the 10^{12} to 10^{14} atoms/cm^3 range, produced by reactor neutrons having a cadmium ratio of 22, is completely electrically active following 850°C anneals for 1 hour of float zone Si.

REFERENCES

1) H. Herzer, Proceedings of the Third International Symposium
    on Silicon Materials Science and Technology, edited by H. R.
    Huff and E. Sirtl (The Electrochemical Society, Inc.,
    Princeton, N. J., 1977), p. 106 and references therein.
2) Hans Mørk Janus and Olof Malmros, IEEE Transactions on
    Electron Devices, ED-23, 797 (1976).
3) Ernst W. Haas and Manfred S. Schnoller, IEEE Transactions on
    Electron Devices, ED-23, 803 (1976).
4) W. S. Lyon and J. J. Manning, Phys. Rev. 93, 501 (1954).
5) M. H. Young, O. J. Marsh, and R. Baron, J. Appl. Phys., to
    be published.
6) S. F. Mughabghab and D. I. Garber, BNL-325 (1973).
7) W. W. Bowman and K. W. MacMurdo, Atomic Data and Nuclear
    Data Tables 13, 89 (1974).

"PURSUIT OF THE ULTIMATE JUNCTION"

J. M. Meese

University of Missouri Research Reactor

Columbia, MO 65211

John Cleland, the father of transmutation doping in semi-conductors, very appropriately presented an excellent summary of the conference as a final talk at the meeting in April. John, of course, had little time to prepare his impressions into a talk and no time at all to write them down. Because I have had the advantage of rereading the manuscripts before me, I will try to fulfill this same function for the published proceedings. This summary, therefore, represents my own opinions and prejudices.

I would like again to thank all of you who participated in the conference. In rereading the manuscripts, it is apparent that a great deal of technical information has been exchanged which will be of interest to those who pursue further applications of the transmutation doping technique. We sincerely appreciate the excellent response of the authors of the papers presented in this volume.

Because NTD is in its infancy as a semiconductor device technique, it was not at all clear to any of us on the conference committee, prior to the conference, that it would be possible to prepare this volume which is, at present, unique to the semi-conductor literature. It is unfortunate that only one name appears on the cover of this volume because the real effort was made by those authors whose names are found in the contents. This book would not have been published without a very strong desire to do so by those scientists.

Perhaps Gerry Huth has best summarized the NTD process as a major step in "pursuit of the ultimate junction." This is a topic

Gerry has been studying long before NTD-Si made an appearance as a commercial product. The paper by Huth, Hikin, and Rodov presents directly observable experimental evidence that semiconductor junctions break down more uniformly when the junctions are fabricated on NTD-Si. It is clear, however, that other processes such as defect generation during diffusion still contribute to a degradation of that ultimate elusive device.

NTD-Si is a significant step in the right direction and the papers by Phillips, Chu and Johnson, and Janus, have discussed the device industry experience in using NTD-Si for high power devices. Judging by NTD sales, the experience is rewarding. Janus points out that the medium power device manufacturers also are finding NTD-Si a useful material and he predicts that NTD will make a future impact on the LSI integrated circuit industry. NTD compensation of detector material has also proven to be useful as reported by the Westinghouse and Hughes groups.

Because of the rapidly growing acceptance of this new semiconductor product, which is produced by a radiation technology, it is very important to be concerned about industrial self-regulation of radiation hazards. It is apparent from the papers in Chapters 2 and 4 and the paper by Hobgood et al., in Chapter 3 that considerable effort is being made by most to remain under exempt limits of radioactivity by a good safety margin prior to wafer or device processing. Since activity decays to still lower limits before NTD devices become available to consumers, the radioactivity levels are expected to be immeasurable at the final product stage. The devices, themselves, fortunately will also contain a built-in safety factor since generation of excess carriers by radioactivity would in many cases cause the devices to function improperly.

Chapter 4 contains descriptions of various facilities for transmutation doping. It is apparent that there are as many different approaches taken in processing NTD-Si as there are reactors doing the work. The device manufacturer might rightfully wonder how this vast variety of irradiation techniques in a variety of reactor neutron spectra will affect his device performances and yields. I believe that the current consensus of opinion of those who have been using NTD material is that they can find little difference in materials' parameters or device performance for NTD-Si irradiated at different facilities. In the papers by Malmros and also by Cleland et al., both have studied material irradiated at many different facilities and the conclusion seems to be that the similarities far outweigh the differences. Both papers suggest that the minority carrier lifetimes in silicon irradiated in a hard neutron spectrum (a neutron spectrum containing relatively more fast neutrons) seem to be slightly higher than similar material irradiated in a highly thermal spectrum. The differences, however,

are not spectactular and the reason for this effect is clearly an
area for additional research.

It is clear from the paper by Kirk et al. that the fast
neutron damage dominates for the harder neutron spectrum in terms
of the number of displacements produced. One would expect
differences in the annealing of material from different reactors
and differences in material irradiated to different doses in the
same reactor. The experimental evidence, however, suggests that
these differences are relatively minor in terms of the final end-
product. It is my personal opinion that contamination during
annealing, or perhaps structural changes induced in trace impurity
defect complexes involving oxygen and carbon, are dominating the
minority carrier lifetime in NTD material presently and that these
problems must be solved before the radiation damage effects can
be studied properly.

It is clear from the paper by Stein that many of the effects
of neutron radiation damage in silicon are well understood, however, a lack of knowledge exists on the types of defect structures
which remain after annealing at 500°C and above. There is also
room for further research on the defects generated by annealing
unirradiated material above these temperatures. The experimental
difficulties due to low defect concentrations have been discouraging
since ESR and optical studies are usually not possible. The DLTS
technique utilized by the G.E. group is, however, very sensitive
to small defect concentrations as is minority carrier lifetime and
resistivity measurements on high resistivity samples. Unfortunately, these techniques do not produce microscopic models of the
defects. A classification of different defects can still be made
using these techniques on the basis of annealing kinetics and
energy levels even if the defect structures themselves can not be
identified. NTD-Si should stimulate defect studies of this kind
in the future.

Many papers allude to the fact that defect annealing in NTD-Si
is very, very active in the range of 500°C. Dominate concentrations
of acceptor defects are found in this temperature range which have
not been studied extensively by the radiation damage community.
Additionally, oxygen can become electrically active and epitaxial
regrowth of amorphous regions is observed in ion implanted samples
which have been made amorphous by massive radiation damage. Stein's
paper in this conference has demonstrated, however, that amorphous
zone formation in neutron damage clusters is not likely to occur
in NTD-Si. We must look elsewhere for reasons for the great defect
activity in this temperature range.

Related to this defect annealing activity near 500°C is the
sixty-four dollar question, "Will NTD-Czochralski silicon become

a commercial product in the near future?" The Oak Ridge group
has discussed annealing problems associated with donor activation
in the 500°C range at this conference. It appears that maintaining the high doping accuracies associated with float zone is at
present very difficult because of this excess donor production.
My guess, however, is that NTD-CZ will become a commercial product
very soon. The radial uniformity and higher breakdown voltages
obtainable by NTD will outweight any target resistivity errors
which might result from this excess donor formation. If this
occurs, the capacities of many reactors now producing NTD-Si will
be strained considerably and new techniques will need to be
devised to handle the much larger volumes of material than we are
accustomed to for the float zone market.

What does NTD technology promise for the future? The answer
is very difficult to predict, however, the paper by the M.I.T.
group illustrates that neutron doping of GaAs is indeed possible
with little difficulty. The device applications for other NTD
materials need to be further explored. Clearly, NTD research will
move toward the compound semiconductors and here, dopant production rates, activity decay, annealing problems due to crystal
decomposition, doping uniformity and a host of new device
applications will be of interest.

Can the efficiencies of GaAs solar cells be improved by NTD
technology? Can LED efficiencies be increased? Can higher
efficiencies be obtained from microwave power devices? And what
about our ultimate junction? Will a combination of ion implantation and NTD technolgy produce even higher breakdown voltages?
Do NTD devices perform better if the quality of the starting
material is higher or will NTD cover up a multitude of materials
problems? Only future research can answer these questions. It
is my hope that this series of conferences will become a source
of stimulation for this research and that a multitude of applications for this technology will be found.

PARTICIPANTS

Alger, D. M.	University of Missouri Research Reactor, Columbia, Missouri 65211
Armbruster, K.	University of Missouri Research Reactor, Columbia, Missouri 65211
Austerman, S.	Rockwell International, 3370 Miralomar, Anaheim, California 92803
Badham, K. A.	Atomic Energy of Canada, Ltd., Post Office Box 6300, Ottawa, Ontario K2A3W3, Canada
Baker, J.	Dow Corning Corp., Geddes Road, Hemlock, Michigan 48626
Baliga, B. J.	General Electric Co., Building K-1, One River Road, Schenectady, New York 12301
Bartko, J.	Westinghouse Research & Development Center, Pittsburgh, Pennsylvania 15738
Baldwin, T. O.	Southern Illinois University, Department of Physics, Edwardsville, Illinois 62026
Bauman, B.	University of Missouri Research Reactor, Columbia, Missouri 65211
Berliner, R.	University of Missouri Research Reactor Columbia, Missouri 65211
Bickford, N.	National Bureau of Standards, Bldg. 235, Room A106, Washington, D.C. 20234
Blewitt, T. H.	Argonne National Laboratory, 9700 South Cass Street, Argonne, Illinois 60439
Bosack, D.	Union Carbide Corporation, Tarrytown, New York 10591

Bourdon, J. L.   Joint Research Center Euratom,
                                  21020 Ispra, Italy

Breant, P.   Commissariat a l'Energie Atomique,
                                  B. P. 510--29 rue de la Federation,
                                  75752 Paris Cedex 15, France

Brugger, R. M.   University of Missouri Research Reactor,
                                  Columbia, Missouri 65211

Bryson, E. L.   The Bendix Corporation, 2000 E. 95th St.
                                  Kansas City, Missouri 64141

Burd, J.   Monsanto Company, P. O. Box 8,
                                  St. Peters, Missouri 63376

Carter, R. S.   National Bureau of Standards, Bldg. 235,
                                  Room A106, Washington, D.C. 20234

Chandrasekhar, H. R.   University of Missouri-Columbia, Physics
                                  Department, Columbia, Missouri 65211

Chandrasekhar, M.   University of Missouri-Columbia, Physics
                                  Department, Columbia, Missouri 65211

Chang, M.   General Electric, 248 Blackberry Road,
                                  Liverpool, New York 13018

Charlson, E. J.   University of Missouri-Columbia, Electrical
                                  Engineering Dept., Columbia, MO 65211

Charlson, E.   University of Missouri-Columbia, Electrical
                                  Engineering Dept., Columbia, MO 65211

Chu, C. K.   Westinghouse, Semiconductor Division,
                                  Youngwood, Pennsylvania 15697

Cleland, J. W.   Oak Ridge National Laboratory, P.O. Box X,
                                  Oak Ridge, Tennessee 37830

Closs, D. C.   Dow Corning Corporation, Corporate Center
                                  M5-2318, Midland, Michigan 48640

Cook, G.   University of Michigan, Phoenix Memorial
                                  Laboratory, 2301 Bonisteel Blvd., Ann
                                  Arbor, Michigan 48109

Ede, E. J.   Monsanto Company, 990 Washington Street,
                                  Dedham, Massachusetts 02026

Emmons, A. H.	University of Missouri, 309 University Hall Columbia, Missouri 65211
Evwaraye, A.	General Electric Company, P.O. Box 8, Schenectady, New York 12301
Farmer, J.	University of Dayton, Physics Department, Dayton, Ohio 45469
Forbeck, J. L.	Monsanto Company, P. O. Box 8, St. Peters, Missouri 63376
Glairon, P.	University of Missouri Research Reactor, Columbia, Missouri 65211
Grabowski, Z. W.	Purdue University, Physics Department, West Lafayette, Indiana 47907
Green, W. W.	Atomic Energy of Canada, Ltd., Commercial Products, P. O. Box 6300, Station J, Ottawa, Ontario, Canada
Gros, D.	Centre d'Etudes Nucleaires, 85X-38 041 Grenoble, Cedex, France
Gunn, S. L.	University of Missouri Research Reactor, Columbia, Missouri 65211
Haas, E. W.	Kraftwerk Union AG., 852 Erlangen, R537 Federal Republic of Germany
Hale, E. B.	University of Missouri-Rolla, Physics Department, Rolla, Missouri 65401
Hart, R. R.	Texas A&M University, Department of Nuclear Engineering, College Station, Texas 77843
Hasbe, K.	Komatsu Electronic Metals Co., 2612 Shinomiga Hiratsuka-shi, Kanagawa, Japan
Henry, K. J.	Harwell Laboratory, UKAEA, Harwell, Oxfordshire, England
Hensley, G.	University of Missouri-Columbia, Physics Department, Columbia, Missouri 65211
High, D. A.	1007 Edgeworth Avenue, Kirkwood, Missouri 63172

Hill, D. E.   Monsanto Company, 800 N. Lindbergh Blvd.
                                     St. Louis, Missouri 63166

Hobgood, H. M.   Westinghouse Research & Development Center,
                                     1310 Beulah Road, Pittsburgh, PA 15235

Huth, G.   University of Southern California, Medical
                                     Imaging Science Group, 4676 Admiralty Way,
                                     Marina Del Ray, California 90291

Hyder, A.   Ortec, Inc., EG & G, 100 Midland Road
                                     Oak Ridge, Tennessee 37830

Hysell, R. E.   General Electric Company, 520 Pusey Avenue,
                                     Collingdale, Pennsylvania 19023

Iredale, P.   Harwell, UKAEA, Harwell,
                                     Oxfordshire, England

Jagannath, C.   Purdue University, Department of Physics,
                                     West Lafayette, Indiana 47907

Janus, H. M.   Topsil A/S, Linderupvej, DK-3600,
                                     Frederikssund, Denmark

Kaltenborn, N.   Institutt for Atomenergi, P.O. Box 40,
                                     2007 Kjeller, Norway

Katz, L. E.   University of Missouri-Rolla, Physics
                                     Department, Rolla, Missouri 65401

Kawamoto, S.   Komatsu Electronic Metals Company, 75-52
                                     Kessel Street, Forest Hills, New York 11375

Keig, G. A.   Union Carbide Corporation, 270 Park Avenue,
                                     New York, New York 10017

Kennedy, R. W.   General Electric Corporation
                                     Auburn, New York

King, J. S.   University of Michigan, Nuclear Engineering
                                     Department, Ann Arbon, Michigan 48109

Kirk, M.   Argonne National Laboratory, 9700 South
                                     Cass Street, Argonne, Illinois 60439

Lamb, R. D.   Monsanto Company, 3400 Hillview Avenue,
                                     Palo Alto, California 94304

Larrabee, R. D.	National Bureau of Standards, Bldg. 225, Room A327, Washington, D.C. 20234
Larsen, D. M.	Massachusetts Institute of Technology, National Magnet Laboratory, 170 Albany Street, Cambridge, Massachusetts 02139
Lenie, C.	International Rectifier Corporation, 233 Kansas Street, El Segundo, CA 90245
Lerch, R.	Monsanto Company, 3400 Hillview Avenue, Palo Alto, California 94304
Lindley, R.	University of Missouri Research Reactor, Columbia, Missouri 65211
McCarver, M.	Monsanto Company, P.O. Box 8, St. Peters, Missouri 63376
Magee, T. J.	Advanced Research & Applications Corp., 1223 East Arques Avenue, Sunnyvale, California 94086
Malmros, O.	Topsil A/S, P.O. Box 93, DK-3600, Frederikssund, Denmark
Marsh, O. J.	Hughes Research Laboratories, 3011 Malibu Canyon Road, Malibu, California 90265
Matlock, J.	Monsanto Company, P.O. Box 8, St. Peters, Missouri 63376
Mayfield, R. E.	Monsanto Company, P. O. Box 8, St. Peters, Missouri 63376
Meese, J. M.	University of Missouri Research Reactor, Columbia, Missouri 65211
Messier, J.	Centre d'Etudes Nucleaires de Saclay, SES/Lera, BP N2, 91190 Gif-sur-Yvette, France
Meyer, A. J.	University of Missouri Research Reactor Columbia, Missouri 65211
Mildner, D.	University of Missouri Research Reactor Columbia, Missouri 65211

Mirat, P.   Commissariat a l'Energie Atomique, BP 510-
                                29 rue de la Federation,
                                Paris Cedex 15, France

Morrissey, J. E.   General Electric Company, P.O. Box 460,
                                Pleasanton, California 94566

Moyson, J.   FMC Corporation, Semiconductor Products
                                Division, 800 Hoyt Street,
                                Broomfield, Colorado 80020

'Nicholson, G.   Atomic Energy of Canada, Ltd., Chalk River
                                Nuclear Laboratories, Ontario K03130,
                                Canada

Ohmer, M. C.   Air Force Materials Laboratory, Wright
                                Patterson Air Force Base, Ohio 45433

Phillips, R. W.   National Electronics, P.O. Box 269,
                                Geneva, Illinois 60134

Reed, B.   Monsanto Company, P. O. Box 8,
                                St. Peters, Missouri 63376

Restelli, G.   Joint Research Center Euratom,
                                Ispra, Italy

Riemann, R.   Monsanto Company, P. O. Box 8,
                                St. Peters, Missouri 63376

Robinson, F. L.   Bendix Corporation, P. O. Box 1159,
                                Kansas City, Missouri 64141

Saltvedt, K.   AB Atomenergi, Fack, Studsvik, S-611 01
                                Nykoping, Sweden

Smith, S. R.   University of Dayton Research Institute,
                                300 College Park Drive, Dayton, Ohio 45469

Smith, T. G. G.   Harwell Laboratory, UKAEA, Harwell,
                                Oxfordshire, England

Snyderman, N.   Electronic News, 7 East 12th Street,
                                New York, New York 10003

Spry, R. S.   Air Force Materials Laboratory, AFML/LPO
                                Wright Patterson Air Force Base, Ohio 45433

Stein, H.	Sandia Laboratories, Division 5112, P.O. Box 5800, Albuquerque, NM 87185
Stone, B.	Monsanto Company, P.O. Box 8, St. Peters, Missouri 63376
Sweep, H. C.	Wacker Chemical Corporation, 11307 Hindry Avenue, Los Angeles, California 90045
Takahashi, H.	Dow Corning Corporation Midland, Michigan 48640
Teszner, J. L.	French Department of Defense, 26 B D Victor Paris, Armees 75996, France
Thoele, H.	Monsanto Company, 3318 Russwood Lane, Garland, Texas 75042
Thomas, H. C.	Texas Tech University, P.O. Box 4180 Lubbock, Texas 79409
Thomas, R. N.	Westinghouse Research & Development Center, Pittsburgh, Pennsylvania 15235
Vesaghi, M. A.	University of Chicago, 5640 S. Ellis, Chicago, Illinois 60634
Virtue, R.	Monsanto Company, P. O. Box 8, St. Peters, Missouri 63376
Watkins, G. D.	Lehigh University, Physics Department, Sherman Fairchild Laboratory #161, Bethlehem, Pennsylvania 18015
Werner, S.	University of Missouri-Columbia, Physics Department, Columbia, Missouri 65211
Whydrew, J.	System Planning Corporation, 1500 Wilson Blvd., Suite 1500, Arlington, VA 22209
Wiles, A.	Topsil A/S, 730 Distel Drive, Suite C-1, Los Altos, California 94022
Wolfram, T.	University of Missouri-Columbia, Physics Department, Columbia, Missouri 65211
Yaggy, L.	Hughes Aircraft Company, 6155 El Camino Real, Carlsbad, California 92008

Yatsurugi, T.   Irradiation Development Association, c/o
                           Jaeri Tokai, Ibaraki, Japan

Yelon, W. B.   University of Missouri Research Reactor,
                           Columbia, Missouri 65211

Young, M. H.   Hughes Research Laboratories, 3011 Malibu
                           Canyon Road, Malibu, California 90265

Yusa, A.   Komatsu Electronic Metals Company, 2612
                           Shinomiya Kirastuka 254, Japan

SUBJECT INDEX